CLYMER®
MANUALS

SKI-DOO
SNOWMOBILE SHOP MANUAL
1985-1989

WHAT'S IN YOUR TOOLBOX?

More information available at haynes.com

Phone: 805-498-6703

Haynes Group Limited
Haynes North America, Inc.

ISBN-10: 0-89287-521-6
ISBN-13: 978-0-89287-521-4
Library of Congress: 90-55393

Technical Photography: Ron Wright
Technical Illustrations: Steve Amos, Diana Kirkland and Carl Rohkar

S829, 23-400

Common spark plug conditions

NORMAL

Symptoms: Brown to grayish-tan color and slight electrode wear. Correct heat range for engine and operating conditions.
Recommendation: When new spark plugs are installed, replace with plugs of the same heat range.

WORN

Symptoms: Rounded electrodes with a small amount of deposits on the firing end. Normal color. Causes hard starting in damp or cold weather and poor fuel economy.
Recommendation: Plugs have been left in the engine too long. Replace with new plugs of the same heat range. Follow the recommended maintenance schedule.

CARBON DEPOSITS

Symptoms: Dry sooty deposits indicate a rich mixture or weak ignition. Causes misfiring, hard starting and hesitation.
Recommendation: Make sure the plug has the correct heat range. Check for a clogged air filter or problem in the fuel system or engine management system. Also check for ignition system problems.

ASH DEPOSITS

Symptoms: Light brown deposits encrusted on the side or center electrodes or both. Derived from oil and/or fuel additives. Excessive amounts may mask the spark, causing misfiring and hesitation during acceleration.
Recommendation: If excessive deposits accumulate over a short time or low mileage, install new valve guide seals to prevent seepage of oil into the combustion chambers. Also try changing gasoline brands.

OIL DEPOSITS

Symptoms: Oily coating caused by poor oil control. Oil is leaking past worn valve guides or piston rings into the combustion chamber. Causes hard starting, misfiring and hesitation.
Recommendation: Correct the mechanical condition with necessary repairs and install new plugs.

GAP BRIDGING

Symptoms: Combustion deposits lodge between the electrodes. Heavy deposits accumulate and bridge the electrode gap. The plug ceases to fire, resulting in a dead cylinder.
Recommendation: Locate the faulty plug and remove the deposits from between the electrodes.

TOO HOT

Symptoms: Blistered, white insulator, eroded electrode and absence of deposits. Results in shortened plug life.
Recommendation: Check for the correct plug heat range, over-advanced ignition timing, lean fuel mixture, intake manifold vacuum leaks, sticking valves and insufficient engine cooling.

PREIGNITION

Symptoms: Melted electrodes. Insulators are white, but may be dirty due to misfiring or flying debris in the combustion chamber. Can lead to engine damage.
Recommendation: Check for the correct plug heat range, over-advanced ignition timing, lean fuel mixture, insufficient engine cooling and lack of lubrication.

HIGH SPEED GLAZING

Symptoms: Insulator has yellowish, glazed appearance. Indicates that combustion chamber temperatures have risen suddenly during hard acceleration. Normal deposits melt to form a conductive coating. Causes misfiring at high speeds.
Recommendation: Install new plugs. Consider using a colder plug if driving habits warrant.

DETONATION

Symptoms: Insulators may be cracked or chipped. Improper gap setting techniques can also result in a fractured insulator tip. Can lead to piston damage.
Recommendation: Make sure the fuel anti-knock values meet engine requirements. Use care when setting the gaps on new plugs. Avoid lugging the engine.

MECHANICAL DAMAGE

Symptoms: May be caused by a foreign object in the combustion chamber or the piston striking an incorrect reach (too long) plug. Causes a dead cylinder and could result in piston damage.
Recommendation: Repair the mechanical damage. Remove the foreign object from the engine and/or install the correct reach plug.

Contents

Quick Reference Data

GENERAL ENGINE SPECIFICATIONS

Bore	
MX, MX (H/A) and MX LT	69.5 mm (2.736 in.)
Plus and Plus LT	72.0 mm (2.835 in.)
Mach I	76.0 mm (2.992 in.)
Stroke	
MX, MX (H/A) and MX LT	61.0 mm (2.402 in.)
Plus and Plus LT	64.0 mm (2.520 in.)
Mach I	64.0 mm (2.520 in.)
Displacement	
MX, MX (H/A) and MX LT	462.8 cc (28.2 cu. in.)
Plus and Plus LT	521.2 cc (31.8 cu. in.)
Mach I	580.8 cc (35.4 cu. in.)
Compression ratio	
MX, MX (H/A) and MX LT	7.5:1
Plus	
1985-1988	6.5:1
Plus and Plus LT	
1989	6.1:1
Mach I	5.9:1

SPARK PLUGS

Model	Plug type	Gap mm (in.)
1985	NGK BR9ES	0.40 (0.016)
1986		
Formula MX	NGK BR10ES	0.40 (0.016)
Formula MX (H/A)	NGK BR10ES	0.40 (0.016)
Formula Plus	NGK BR9ES	0.40 (0.016)
1987-1988		
All models	NGK BR9ES	0.40 (0.016)
1989		
Formula Mach I	NGK BR9ES	0.45 (0.018)
All other models	NGK BR9ES	0.40 (0.016)

FUEL RECOMMENDATIONS

1985-1987	Regular—leaded or unleaded
1988	
MX and MX LT	Regular—leaded or unleaded
Plus	Super
1989	Regular unleaded

APPROXIMATE REFILL CAPACITY

Rotary valve reservoir	
1985-1988	455 cc (16 oz.)
1989	N.A.
Chaincase	
1985-1988	256 cc (9 oz.)
1989	200 cc (7 oz.)
Oil injection reservoir	
All models	2.9 L (98 oz.)
Cooling system	
All models	4.2 L (142 oz.)
Fuel tank	
All models	28.6 L (7.6 gal.)

RECOMMENDED LUBRICANTS

Item	Lubricant type
Countershaft bearing, hub bearings, bogie wheels, ski legs, idler bearings, leaf spring cushion pads, etc.	A
Oil seal interior lips	A
Engine injection oil	B
Chaincase	C
Rotary valve lubricant (1985-1988)	D

Lubricant legend:
A. Bombardier bearing grease or equivalent multi-purpose lithium base grease for use through a temperature range of −40° to 95° C (−40° to 200° F). This grease will be referred to as a 'low temperature grease' throughout this manual.
B. Bombardier injection oil or equivalent. Injection oil must flow at −40° C (−40° F).
C. Bombardier chaincase oil or equivalent. Make sure equivalent oil provides lubrication at low temperatures.
D. Bombardier injection oil or equivalent.
* WD-40 can be used as a general lubricant.

REPLACEMENT BULBS

	Watt
Headlight	
MX, MX (H/A) and MX LT	60/60
Plus and Plus LT	60/55*
Mach I	60/55*
Taillight	5/21
Tachometer and speedometer	5
Fuel and temperature gauge	2

* Halogen bulb.

CARBURETOR PILOT AIR SCREW ADJUSTMENT

	Turns out*
1985	
Formula Plus	2.0
All other models	1 1/2
1986	
Formula Plus	1.0
All other models	1 1/2
1987-1988	
Formula Plus	1.0
All other models	1 1/2
1989	
All models	1 1/2
* (± 1/8 turn)	

IGNITION COIL TEST SPECIFICATIONS

Primary coil resistance	
All models	0.23-0.43 ohms
Secondary coil resistance	
1985-1986	2.45-4.55 K ohms
1987-on	3.85-7.15 K ohms

CARBURETOR IDLE SPEED

All models	1,800-2,000 rpm

MAGNETO COIL TESTING

Low speed charge coil	120-180 ohms
High speed charge coil	2.8-4.2 ohms
Lighting coil	0.21-0.31 ohms

MAINTENANCE TIGHTENING TORQUES

	N·m	ft.-lb.
Cylinder head		
1985-1988	20	15
1989	22	16
Engine mounts		
1985-1987	11	8
1988-on	38	28
Rear idler wheel bolt	48	35

FLYWHEEL NUT TIGHTENING TORQUE

	N·m	ft.-lb.
1985-1988	100	74
1989	105	77

DRIVE SYSTEM SPECIFICATIONS

Engagement rpm
  1985-1986
    MX and MX (H/A)   3,100-3,400 rpm
    Plus   3,600-3,900 rpm
  1987-1988
    MX and MX LT   3,500-3,700 rpm
    Plus   3,700-3,900 rpm
  1989
    MX and MX LT   3,500-3,700 rpm
    Plus andd Plus LT   3,400-3,600 rpm
    Mach I   3,000-3,200 rpm
Drive belt width
  1985-1987   34.92 mm (1 3/8 in.)
  1988-on   34.5 mm (1 23/64 in.)
Drive belt deflection
  1985   30.2-38.1 mm (1 3/16-1 1/2 in.)
  1986-1987   25.4-32.0 mm (1-1 1/4 in.)
  1988-on   32.0 mm (1 1/4 in.)

CLYMER®

SKI-DOO

SNOWMOBILE SHOP MANUAL
1985-1989

Chapter One

General Information

This Clymer shop manual covers the 1985-1989 Ski-Doo Formula MX, Formula MX (H/A), Formula MX LT, Formula Plus, Formula Plus LT and Formula Mach I models.

Troubleshooting, tune-up, maintenance and repair are not difficult, if you know what tools and equipment to use and what to do. Step-by-step instructions guide you through jobs ranging from simple maintenance to complete engine and suspension overhaul.

This manual can be used by anyone from a first time do-it-yourselfer to a professional mechanic. Detailed drawings and clear photographs give you all the information you need to do the work right.

Some of the procedures in this manual require the use of special tools. The resourceful mechanic can, in many cases, think of acceptable substitutes for special tools. However, using a substitute for a special tool is not recommended as it may damage the part and can be dangerous. If you find that a tool can be designed and safely made, but will require some type of machine work, you may want to search out a local community college or high school that has a machine shop curriculum. Shop teachers sometimes welcome outside work that can be used as practical shop applications for advanced students.

Table 1 lists model number coverage.

General specifications are listed in **Table 2**.

Table 3 lists vehicle weight.

Metric and U.S. standards are used throughout this manual. U.S. to metric conversions are given in **Table 4**.

Critical torque specifications are found in tables at the end of each chapter (as required). The general torque specifications listed in **Table 5** can be used when a torque specification is not listed for a specific component or assembly.

A list of general technical abbreviations are given in **Table 6**.

Metric tap drill sizes can be found in **Table 7**.

Table 8 lists wind chill factors.

Tables 1-8 are found at the end of the chapter.

MANUAL ORGANIZATION

This chapter provides general information useful to snowmobile owners and mechanics. In addition, information in this chapter discusses the tools and techniques for preventive maintenance, troubleshooting and repair.

Chapter Two provides methods and suggestions for quick and accurate diagnosis and repair of problems. Troubleshooting procedures discuss typical symptoms and logical methods to pinpoint the trouble.

Chapter Three explains all periodic lubrication and routine maintenance necessary to keep your snowmobile operating well. Chapter Three also includes recommended tune-up procedures, eliminating the need to constantly consult other chapters on the various assemblies.

Subsequent chapters describe specific systems, providing disassembly, repair, assembly and adjustment procedures in simple step-by-step form. If a repair is impractical for a home mechanic, it is so indicated. It is usually faster and less expensive to take such repairs to a dealer or competent repair shop. Specifications concerning a specific system are included at the end of the appropriate chapter.

NOTES, CAUTIONS AND WARNINGS

The terms *NOTE, CAUTION* and *WARNING* have specific meanings in this manual. A *NOTE* provides additional information to make a step or procedure easier or clearer. Disregarding a *NOTE* could cause inconvenience, but would not cause damage or personal injury.

A *CAUTION* emphasizes areas where equipment damage could occur. Disregarding a *CAUTION* could cause permanent mechanical damage; however, personal injury is unlikely.

A *WARNING* emphasizes areas where personal injury or even death could result from negligence. Mechanical damage may also occur. *WARNINGS* are to be taken *seriously*. In some cases, serious injury and death have resulted from disregarding similar warnings.

SAFETY FIRST

Professional mechanics can work for years and never sustain a serious injury. If you observe a few rules of common sense and safety, you can enjoy many safe hours servicing your own machine. If you ignore these rules, you can hurt yourself or damage the equipment.

1. Never use gasoline as a cleaning solvent.
2. Never smoke or use a torch in the vicinity of flammable liquids, such as cleaning solvent, in open containers.
3. If welding or brazing is required on the machine, remove the fuel tank to a safe distance, at least 50 feet away.
4. Use the proper sized wrenches to avoid damage to fasteners and injury to yourself.
5. When loosening a tight or stuck nut, be guided by what would happen if the wrench should slip. Be careful; protect yourself accordingly.
6. When replacing a fastener, make sure to use one with the same measurements and strength

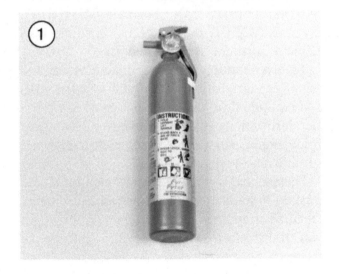

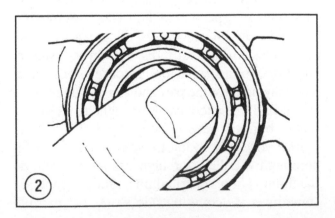

as the old one. Incorrect or mismatched fasteners can result in damage to the snowmobile and possible personal injury. Beware of fastener kits that are filled with cheap and poorly made nuts, bolts, washers and cotter pins. Refer to *Fasteners* in this chapter for additional information.

7. Keep all hand and power tools in good condition. Wipe greasy and oily tools after using them. They are difficult to hold and can cause injury. Replace or repair worn or damaged tools.

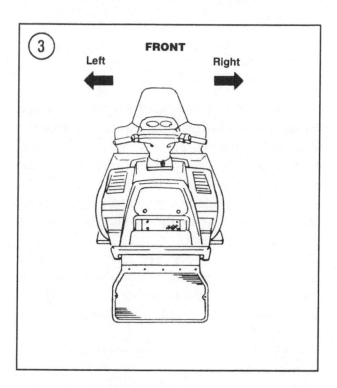

FRONT
Left   Right

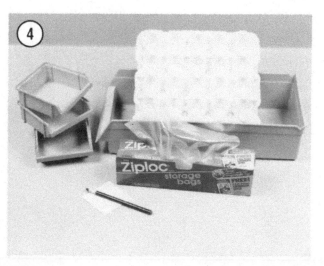

8. Keep your work area clean and uncluttered.

9. Wear safety goggles during all operations involving drilling, grinding, the use of a cold chisel or *any* time you feel unsure about the safety of your eyes. Safety goggles should also be worn any time solvent and compressed air is used to clean parts.

10. Keep an approved fire extinguisher (**Figure 1**) nearby. Be sure it is rated for gasoline (Class B) and electrical (Class C) fires.

11. When drying bearings or other rotating parts with compressed air, never allow the air jet to rotate the bearing or part. The air jet is capable of rotating them at speeds far in excess of those for which they were designed. The bearing or rotating part is very likely to disintegrate and cause serious injury and damage. To prevent bearing damage when using compressed air, hold the inner bearing race (**Figure 2**) by hand.

SERVICE HINTS

Most of the service procedures covered are straightforward and can be performed by anyone reasonably handy with tools. It is suggested, however, that you consider your own capabilities carefully before attempting any operation involving major disassembly.

1. "Front," as used in this manual, refers to the front of the snowmobile; the front of any component is the end closest to the front of the snowmobile. The "left-" and "right-hand" sides refer to the position of the parts as viewed by a rider sitting and facing forward. For example, the throttle control is on the right-hand side. These rules are simple, but confusion can cause a major inconvenience during service. See **Figure 3**.

2. When disassembling any engine or drive component, mark the parts for location and mark all parts which mate together. Small parts, such as bolts, can be identified by placing them in plastic sandwich bags (**Figure 4**). Seal the bags and label them with masking tape and a marking pen. When reassembly will take place

immediately, an accepted practice is to place nuts and bolts in a cupcake tin or egg carton in the order of disassembly.

3. Finished surfaces should be protected from physical damage or corrosion. Keep gasoline off painted surfaces.

4. Use penetrating oil on frozen or tight bolts, then strike the bolt head a few times with a hammer and punch (use a screwdriver on screws). Avoid the use of heat where possible, as it can warp, melt or affect the temper of parts. Heat also ruins finishes, especially paint and plastics.

5. No parts removed or installed (other than bushings and bearings) in the procedures given in this manual should require unusual force during disassembly or assembly. If a part is difficult to remove or install, find out why before proceeding.

6. Cover all openings after removing parts or components to prevent dirt, small tools, etc. from falling in.

7. Read each procedure *completely* while looking at the actual parts before starting a job. Make sure you *thoroughly* understand what is to be done and then carefully follow the procedure, step-by-step.

8. Recommendations are occasionally made to refer service or maintenance to a snowmobile dealer or a specialist in a particular field. In these cases, the work will be done more quickly and economically than if you performed the job yourself.

9. In procedural steps, the term "replace" means to discard a defective part and replace it with a new or exchange unit. "Overhaul" means to remove, disassemble, inspect, measure, repair or replace defective parts, reassemble and install major systems or parts.

10. Some operations require the use of a hydraulic press. It would be wiser to have these operations performed by a shop equipped for such work, rather than to try to do the job yourself with makeshift equipment that may damage your machine.

11. Repairs go much faster and easier if your machine is clean before you begin work. There are many special cleaners on the market, like Bel-Ray Degreaser, for washing the engine and related parts. Follow the manufacturer's directions on the container for the best results. Clean all oily or greasy parts with cleaning solvent as you remove them.

> *WARNING*
> *Never use gasoline as a cleaning agent. It presents an extreme fire hazard. Be sure to work in a well-ventilated area when using cleaning solvent. Keep a fire extinguisher, rated for gasoline fires, handy in any case.*

12. Much of the labor charges for repairs made by dealers are for the time involved during the removal, disassembly, assembly, and reinstallation of other parts in order to reach the defective part. It is frequently possible to perform the preliminary operations yourself and then take the defective unit to the dealer for repair at considerable savings.

13. If special tools are required, make arrangements to get them before you start. It is frustrating and time-consuming to get partly into a job and then be unable to complete it.

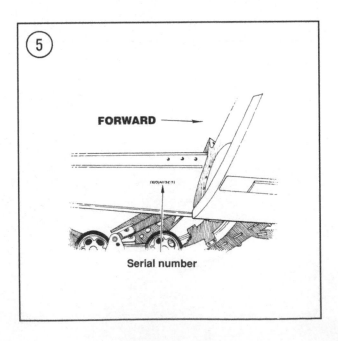

14. Make diagrams (or take a Polaroid picture) wherever similar-appearing parts are found. For instance, crankcase bolts are often not the same length. You may think you can remember where everything came from—but mistakes are costly. There is also the possibility that you may be sidetracked and not return to work for days or even weeks—in which time carefully laid out parts may have become disturbed.

15. When assembling parts, be sure all shims and washers are replaced exactly as they came out.

16. Whenever a rotating part butts against a stationary part, look for a shim or washer. Use new gaskets if there is any doubt about the condition of the old ones. A thin coat of silicone sealant on non-pressure type gaskets may help them seal more effectively.

17. If it is necessary to make a cover gasket and you do not have a suitable old gasket to use as a guide, you can use the outline of the cover and

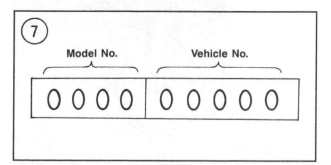

Model No.   Vehicle No.

0 0 0 0   0 0 0 0 0

gasket material to make a new gasket. Apply engine oil to the cover gasket surface. Then place the cover on the new gasket material and apply pressure with your hands. The oil will leave a very accurate outline on the gasket material that can be cut around.

CAUTION
When purchasing gasket material to make a gasket, measure the thickness of the old gasket (at an uncompressed point) and purchase gasket material with the same approximate thickness.

18. Heavy grease can be used to hold small parts in place if they tend to fall out during assembly. However, keep grease and oil away from electrical components.

19. A carburetor is best cleaned by disassembling it and cleaning the parts in hot soapy water. Never soak gaskets and rubber parts in commercial carburetor cleaners. Never use wire to clean out jets and air passages. They are easily damaged. Use compressed air to blow out the carburetor only if the float has been removed first.

20. Take your time and do the job right. Do not forget that a newly rebuilt engine must be broken in just like a new one.

ENGINE AND CHASSIS SERIAL NUMBERS

Ski-Doo snowmobiles are identified by frame and engine identification numbers. The frame number is stamped on the right-hand side of the tunnel just below the front of the seat (**Figure 5**). The engine number is stamped on the left-hand side of the crankcase as shown in **Figure 6**. **Figure 7** shows the breakdown of the vehicle serial number found on Ski-Doo snowmobiles covered in this manual. The first 4 digits represent the vehicle's model number. The model numbers are listed in **Table 1**. The last 5 digits list the vehicle number.

Factory installed tracks have a serial number stamped on the outside of the track.

Write down all serial and model numbers applicable to your machine and carry the numbers with you when you order parts from a dealer. Always order by year and engine and machine numbers. If possible, compare the old parts with the new ones before purchasing them. If the parts are not alike, have the parts manager explain the reason for the difference and insist on assurance that the new parts will fit and are correct.

ENGINE OPERATION

The following is a general discussion of a typical 2-stroke piston-ported engine. The same principles apply to rotary valve engines, except that during the intake cycle, the fuel/air mixture passes through a rotary valve assembly into the crankcase. During this discussion, assume that the crankshaft is rotating counterclockwise in **Figure 8**. As the piston travels downward, a transfer port (A) between the crankcase and the cylinder is uncovered. The exhaust gases leave the cylinder through the exhaust port (B), which is also opened by the downward movement of the piston. A fresh fuel/air charge, which has previously been compressed slightly, travels from the crankcase (C) to the cylinder through the transfer port (A) as the port opens. Since the incoming charge is under pressure, it rushes into the cylinder quickly and helps to expel the exhaust gases from the previous cycle.

Figure 9 illustrates the next phase of the cycle. As the crankshaft continues to rotate, the piston moves upward, closing the exhaust and transfer ports. As the piston continues upward, the air/fuel mixture in the cylinder is compressed. Notice also that a vacuum is created in the crankcase at the same time. Further upward movement of the piston uncovers the intake port (D). A fresh fuel/air charge is then drawn into

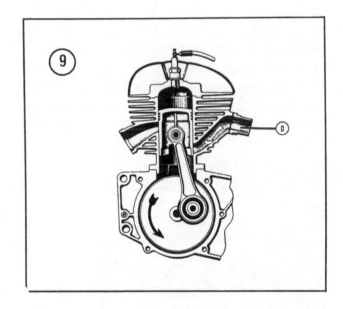

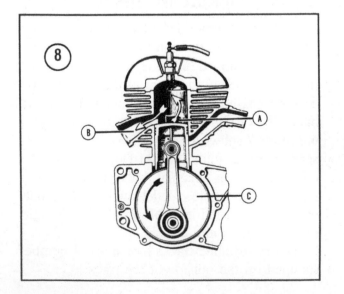

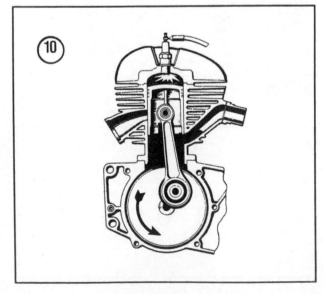

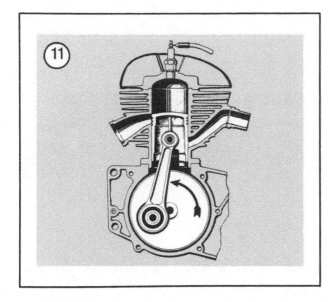

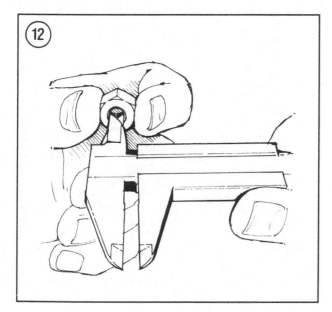

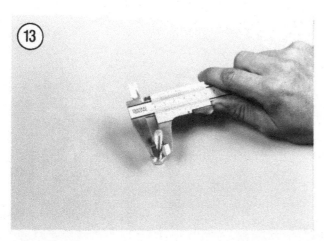

the crankcase through the intake port because of the vacuum created by the upward piston movement.

The third phase is shown in **Figure 10**. As the piston approaches top dead center, the spark plug fires, igniting the compressed mixture. The piston is then driven downward by the expanding gases.

When the top of the piston uncovers the exhaust port, the fourth phase begins, as shown in **Figure 11**. The exhaust gases leave the cylinder through the exhaust port. As the piston continues downward, the intake port is closed and the mixture in the crankcase is compressed in preparation for the next cycle. It can be seen from this discussion that every downward stroke of the piston is a power stroke.

TORQUE SPECIFICATIONS

Torque specifications throughout this manual are given in Newton-meters (N·m) and foot-pounds (ft.-lb.).

Table 5 lists general torque specifications for nuts and bolts that are not listed in the respective chapters. To use the table, first determine the size of the nut or bolt by measuring it with a vernier caliper. **Figure 12** and **Figure 13** show how to do this.

FASTENERS

The materials and designs of the various fasteners used on your snowmobile are not arrived at by chance or accident. Fastener design determines the type of tool required to work the fastener. Fastener material is carefully selected to decrease the possibility of physical failure (**Figure 14**).

Nuts, bolts and screws are manufactured in a wide range of thread patterns. To join a nut and bolt, the diameter of the bolt and the diameter of the hole in the nut must be the same and that the threads on both parts are the same.

The best way to tell if the threads on 2 fasteners are matched is to turn the nut on the bolt (or the bolt into the threaded hole in a piece of equipment) with fingers only. Be sure both pieces are clean. If much force is required, check the thread condition on each fastener. If the thread condition is good but the fasteners jam, the threads are not compatible. A thread pitch gauge (**Figure 15**) can also be used to determine pitch. Ski-Doo snowmobiles are manufactured with ISO (International Organization for Standardization) metric fasteners. The threads are cut differently than that of American fasteners (**Figure 16**). Most threads are cut so that the fastener must be turned clockwise to tighten it. These are called right-hand threads. Some fasteners have left-hand threads; they must be turned counterclockwise to be tightened. Left-hand threads are used in locations where normal rotation of the equipment would tend to loosen a right-hand threaded fastener.

ISO Metric Screw Threads

ISO (International Organization for Standardization) metric threads come in 3 standard thread sizes: coarse, fine and constant pitch. The ISO coarse pitch is used for most all common fastener applications. The fine pitch thread is used on certain precision tools and instruments. The constant pitch thread is used mainly on machine parts and not for fasteners. The constant pitch thread, however, is used on all metric thread spark plugs.

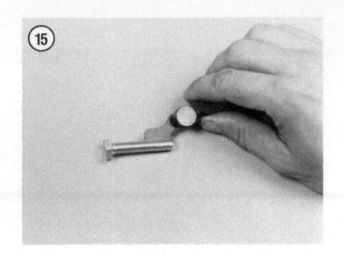

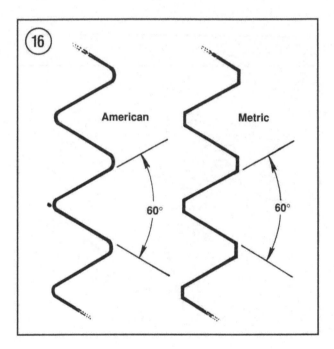

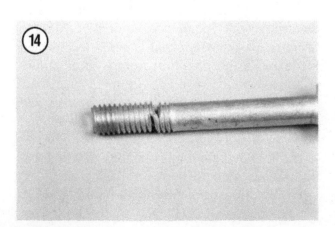

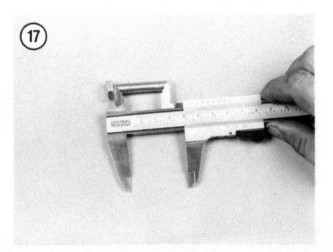

ISO metric threads are specified by the capital letter M followed by the diameter in millimeters and the pitch (or the distance between each thread) in millimeters separated by the sign "×". For example a M8 × 1.25 bolt is one that has a diameter of 8 millimeters with a distance of 1.25 millimeters between each thread. The measurement across 2 flats on the head of the bolt indicates the proper wrench size to be used. **Figure 13** shows how to determine bolt diameter.

NOTE
When purchasing a bolt from a dealer or parts store, it is important to know how to specify bolt length. The correct way to measure bolt length is by measuring the length starting from underneath the bolt head to the end of the bolt (*Figure 17*). *Always measure bolt length in this manner to prevent from purchasing bolts that are too long.*

Machine Screws

There are many different types of machine screws. **Figure 18** shows a number of screw heads requiring different types of turning tools. Heads are also designed to protrude above the metal (round) or to be slightly recessed in the metal (flat). See **Figure 19**.

Bolts

Commonly called bolts, the technical name for these fasteners is cap screw. Metric bolts are described by the diameter and pitch (or the distance between each thread).

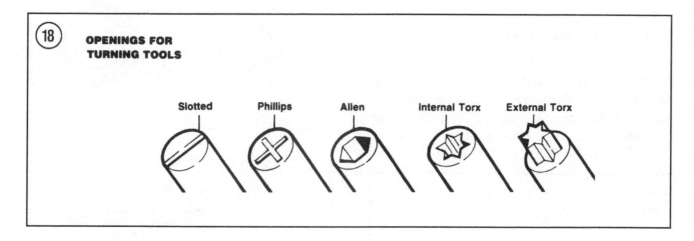

(18) **OPENINGS FOR TURNING TOOLS**

Slotted   Phillips   Allen   Internal Torx   External Torx

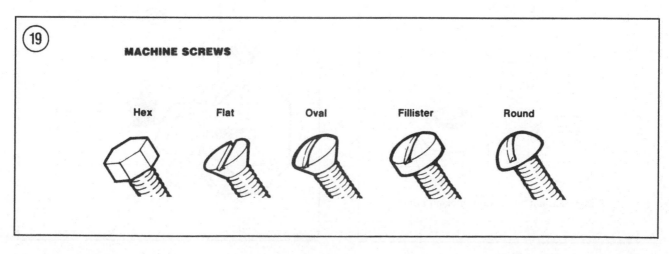

(19) **MACHINE SCREWS**

Hex   Flat   Oval   Fillister   Round

Nuts

Nuts are manufactured in a variety of types and sizes. Most are hexagonal (6-sided) and fit on bolts, screws and studs with the same diameter and pitch.

Figure 20 shows several types of nuts. The common nut is generally used with a lockwasher. Self-locking nuts have a nylon insert which prevents the nut from loosening; no lockwasher is required. Wing nuts are designed for fast removal by hand. Wing nuts are used for convenience in non-critical locations.

To indicate the size of a metric nut, manufacturers specify the diameter of the opening and the thread pitch. This is similar to bolt specifications, but without the length dimension. The measurement across 2 flats on the nut indicates the proper wrench size to be used.

Prevailing Torque Fasteners

Several types of bolts, screws and nuts incorporate a system that develops an interference between the bolt, screw, nut or tapped hole threads. Interference is achieved in various ways: by distorting threads, coating threads with dry adhesive or nylon, distorting the top of an all-metal nut, using a nylon insert in the center or at the top of a nut, etc.

Prevailing torque fasteners offer greater holding strength and better vibration resistance. Some prevailing torque fasteners can be reused if in good condition. Others, like the nylon insert nut, form an initial locking condition when the nut is first installed; the nylon forms closely to the bolt thread pattern, thus reducing any

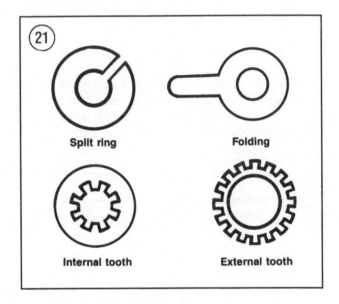

Split ring

Folding

Internal tooth

External tooth

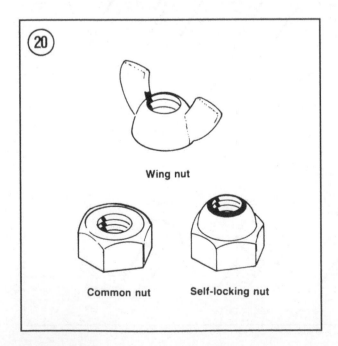

Wing nut

Common nut

Self-locking nut

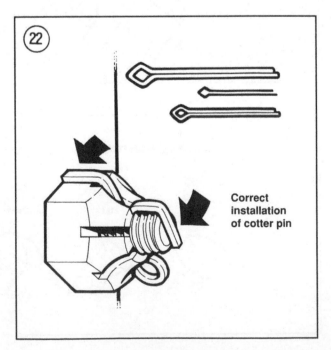

Correct installation of cotter pin

tendency for the nut to loosen. When the nut is removed, the locking efficiency is greatly reduced. For greatest safety, it is recommended that you install new prevailing torque fasteners whenever they are removed.

Washers

There are 2 basic types of washers: flat washers and lockwashers. Flat washers are simple discs with a hole to fit a screw or bolt. Lockwashers are designed to prevent a fastener from working loose due to vibration, expansion and contraction. **Figure 21** shows several types of washers. Washers are also used in the following functions:

a. As spacers.
b. To prevent galling or damage of the equipment by the fastener.

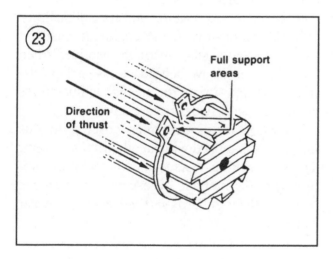

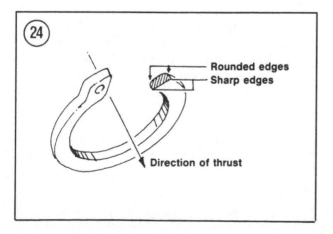

c. To help distribute fastener load during torquing.
d. As seals.

Note that flat washers are often used between a lockwasher and a fastener to provide a smooth bearing surface. This allows the fastener to be turned easily with a tool.

Cotter Pins

Cotter pins (**Figure 22**) are used to secure special kinds of fasteners. The threaded stud must have a hole in it; the nut or nut lock piece has castellations around which the cotter pin ends wrap. Cotter pins should not be reused after removal.

Circlips

Circlips can be internal or external design. They are used to retain items on shafts (external type) or within tubes (internal type). In some applications, circlips of varying thicknesses are used to control the end play of parts assemblies. These are often called selective circlips. Circlips should be replaced during installation, as removal weakens and deforms them.

Two basic styles of circlips are available: machined and stamped circlips. Machined circlips (**Figure 23**) can be installed in either direction (shaft or housing) because both faces are machined, thus creating two sharp edges. Stamped circlips (**Figure 24**) are manufactured with one sharp edge and one rounded edge. When installing stamped circlips in a thrust situation, the sharp edge must face away from the part producing the thrust. When installing circlips, observe the following:

a. Circlips should be removed and installed with circlip pliers. See *Circlip Pliers* in this chapter.
b. Compress or expand circlips only enough to install them.
c. After the circlip is installed, make sure it is completely seated in its groove.

LUBRICANTS

Periodic lubrication assures long life for any type of equipment. The *type* of lubricant used is just as important as the lubrication service itself. The following paragraphs describe the types of lubricants most often used on snowmobiles. Be sure to follow the manufacturer's recommendations for lubricant types.

Generally, all liquid lubricants are called "oil." They may be mineral-based (including petroleum bases), natural-based (vegetable and animal bases), synthetic-based or emulsions (mixtures). "Grease" is an oil to which a thickening base has been added so that the end product is semi-solid. Grease is often classified by the type of thickener added; lithium soap is commonly used.

Engine Oil

Four-stroke

Four-stroke oil for automotive engines is graded by the American Petroleum Institute (API) and the Society of Automotive Engineers (SAE) in several categories. Oil containers display these ratings on the top or label.

API oil grade is indicated by letters; oils for gasoline engines are identified by an "S".

Viscosity is an indication of the oil's thickness. The SAE uses numbers to indicate viscosity; thin oils have low numbers while thick oils have high numbers. A "W" after the number indicates that the viscosity testing was done at a low temperature to simulate cold-weather operation. Engine oils fall into the 5W-30 and 20W-50 range.

Multi-grade oils (for example 10W-40) are less viscous (thinner) at low temperatures and more viscous (thicker) at high temperatures. This allows the oil to perform efficiently across a wide range of engine operating conditions. The lower the number, the better the engine will start in cold climates. Higher numbers are usually recommended for engines running in hot weather conditions.

> *CAUTION*
> *Four-stroke oils are only discussed to provide a comparison. Ski-Doo snowmobile engines are two-stroke engines, thus only two-stroke oil should be used.*

Two-stroke

Lubrication for a two-stroke engine is provided by oil mixed with the incoming fuel/air mixture. Some of the oil mist settles out in the crankcase, lubricating the crankshaft and lower end of the connecting rods. The rest of the oil enters the combustion chamber to lubricate the piston rings and cylinder walls. This oil is burned during the combustion process.

Engine oil must have several special qualities to work well in a two-stroke engine. It must mix easily and stay in suspension in gasoline. When burned, it can't leave behind excessive deposits. It must be appropriate for the high temperatures associated with two-stroke engines.

All Ski-Doo snowmobiles covered in this manual are equipped with an oil injection system. The oil injection system does not require pre-mixing of the fuel and oil except during engine break-in. Refer to *Break-in Procedure* in Chapter Three.

> *NOTE*
> *The injection oil used in Ski-Doo snowmobile engines must be able to flow at temperatures of -40° C (-40° F). See Chapter Three under* **Lubrication** *for additional information.*

Grease

Greases are graded by the National Lubricating Grease Institute (NLGI). Greases are graded by number according to the consistency of the grease; these range from No. 000 to No. 6, with No. 6 being the most solid. A typical multipurpose grease is NLGI No. 2. For specific applications, equipment manufacturers may require grease with an additive such as molybdenum disulfide (MOS2).

NOTE
A low temperature grease should be used wherever grease is required on the snowmobile. Chapter Three lists the low temperature greases recommended by Ski-Doo.

RTV GASKET SEALANT

Room temperature vulcanizing (RTV) sealant is used on some pre-formed gaskets and to seal some components. RTV is a silicone gel supplied in tubes and can be purchased in a number of different colors. For most snowmobile use, the clear color is more preferable.

Moisture in the air causes RTV to cure. Always place the cap on the tube as soon as possible when using RTV. RTV has a shelf life of one year and will not cure properly when the shelf life has expired. Check the expiration date on RTV tubes before using and keep partially used tubes tightly sealed.

Applying RTV Sealant

Clean all gasket residue from mating surfaces. Surfaces should be clean and free of oil and dirt. Remove all RTV gasket material from blind

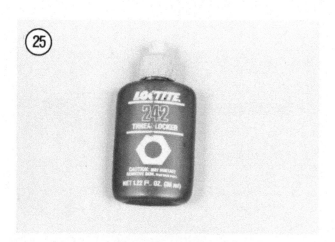

attaching holes, as it can cause a "hydraulic" effect and affect bolt torque.

Apply RTV sealant in a continuous bead 2-3 mm (0.08-0.12 in.) thick. Circle all mounting holes unless otherwise specified. Torque mating parts within 10 minutes after application.

THREADLOCK

Because of the snowmobile's operating conditions, a threadlock (**Figure 25**) is required to help secure many of the fasteners. A threadlock will lock fasteners against vibration loosening and seal against leaks. Loctite 242 (blue) and 271 (red) are recommended for many threadlock requirements described in this manual.

Loctite 242 (blue) is a medium strength threadlock for general purpose use. Component disassembly can be performed with normal hand tools. Loctite 271 (red) is a high strength threadlock that is normally used on studs or critical fasteners. Heat or special tools, such as a press or puller, may be required for component disassembly.

Applying Threadlock

Surfaces should be clean and free of oil and dirt. If a threadlock was previously applied to the component, this residue should also be removed.

Shake the Loctite container thoroughly and apply to both parts. Assemble parts and/or tighten fasteners.

GASKET REMOVER

Stubborn gaskets can present a problem during engine service as they can take a long time to remove. Consequently, there is the added problem of secondary damage occurring to the gasket mating surfaces from the incorrect or accidental use of a gasket scraping tool. To quickly and safely remove stubborn gaskets, use a spray gasket remover. Spray gasket remover can be purchased through Ski-Doo dealers and automotive parts houses. Follow the manufacturer's directions for use.

BASIC HAND TOOLS

Many of the procedures in this manual can be carried out with simple hand tools and test equipment familiar to the mechanic. Keep your tools clean and in a tool box. Keep them organized with the sockets and related drives together, the open-end combination wrenches together, etc. After using a tool, wipe off dirt and grease with a clean cloth and return the tool to its correct place.

Top quality tools are essential; they are also more economical in the long run. If you are now starting to build your tool collection, stay away from the "advertised specials" featured at some parts houses, discount stores and chain drug stores. These are usually a poor grade tool that can be sold cheaply and that is exactly what they are—*cheap*. They are usually made of inferior material, and are thick, heavy and clumsy. Their rough finish makes them difficult to clean and they usually don't last very long. If it is ever your misfortune to use such tools, you will probably find out that the wrenches do not fit the heads of bolts and nuts correctly, thus damaging the fastener.

Quality tools are made of alloy steel and are heat treated for greater strength. They are lighter and better balanced than cheap ones. Their surface is smooth, making them a pleasure to work with and easy to clean. The initial cost of good quality tools may be more but they are cheaper in the long run. Don't try to buy everything in all sizes in the beginning; do it a little at a time until you have the necessary tools.

The following tools are required to perform virtually any repair job. Each tool is described and the recommended size given for starting a tool collection. Additional tools and some duplicates may be added as you become familiar with the vehicle. Ski-Doo snowmobiles are built

with metric standard fasteners—so if you are starting your collection now, buy metric sizes.

Screwdrivers

The screwdriver is a very basic tool, but if used improperly it will do more damage than good. The slot on a screw has a definite dimension and shape. A screwdriver must be selected to conform with that shape. Use a small screwdriver for small screws and a large one for large screws or the screw head will be damaged.

Two basic types of screwdriver are required: common (flat-blade) screwdrivers (**Figure 26**) and Phillips screwdrivers (**Figure 27**).

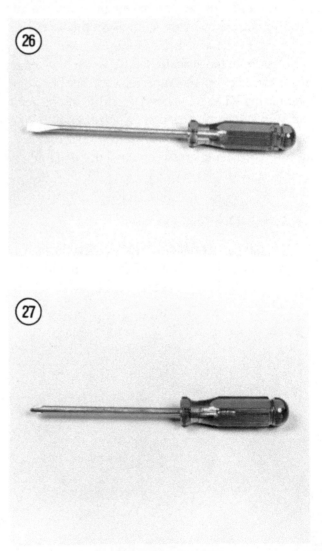

Screwdrivers are available in sets which often include an assortment of common and Phillips blades. If you buy them individually, buy at least the following:

   a. Common screwdriver—5/16 × 6 in. blade.
   b. Common screwdriver—3/8 × 12 in. blade.
   c. Phillips screwdriver—size 2 tip, 6 in. blade.
   d. Phillips screwdriver—size 3 tip, 6 in. blade.

Use screwdrivers only for driving screws. Never use a screwdriver for prying or chiseling metal. Do not try to remove a Phillips or Allen head screw with a common screwdriver (unless the screw has a combination head that will accept either type); you can damage the head so that the proper tool will be unable to remove it.

Keep screwdrivers in the proper condition and they will last longer and perform better. Always keep the tip of a common screwdriver in good condition. **Figure 28** shows how to grind the tip to the proper shape if it becomes damaged. Note the symmetrical sides of the tip.

Pliers

Pliers come in a wide range of types and sizes. Pliers are useful for cutting, bending and crimping. They should never be used to cut hardened objects or to turn bolts or nuts. **Figure 29** shows several pliers useful in snowmobile repair.

Each type of pliers has a specialized function. Slip-joint pliers are used mainly for holding things and for bending. Needlenose pliers are used to hold or bend small objects. Water pump pliers (commonly referred to as channel lock pliers) can be adjusted to hold various sizes of objects; the jaws remain parallel to grip around

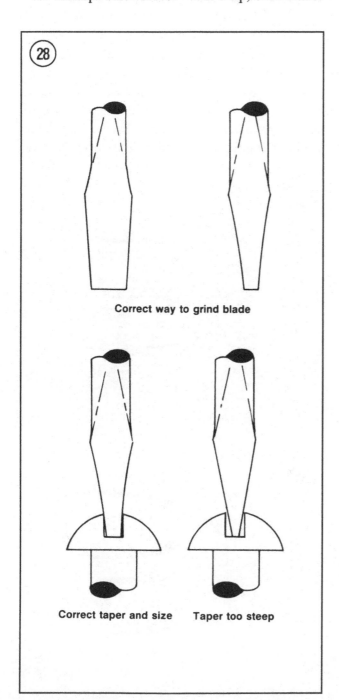

Correct way to grind blade

Correct taper and size   Taper too steep

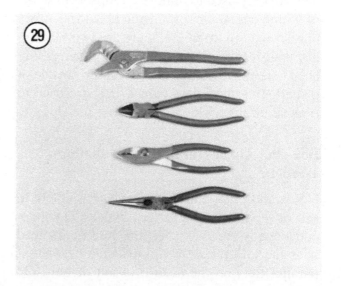

objects such as pipe or tubing. There are many more types of pliers.

CAUTION
Pliers should not be used for loosening or tightening nuts or bolts. The pliers sharp teeth will grind off the nut or bolt corners and damage the fastener.

CAUTION
If slip-joint pliers are going to be used to hold an object with a finished surface that can be easily damaged, wrap the object with tape or cardboard for protection.

Vise Grips

Vise Grips (**Figure 30**) are used to hold objects very tightly while another task is performed on the object. While Vise Grips work well, caution should be followed with their use. Because Vise Grips exert more force than regular pliers, their sharp jaws will permanently scar the object. In addition, when Vise Grips are locked in position, they can crush or deform thin wall material.

Vise Grips are available in many types for more specific tasks.

Circlip Pliers

Circlip pliers (**Figure 31**) are special in that they are only used to remove circlips from shafts or within engine or suspension housings. When purchasing circlip pliers, there are two kinds to distinguish from. External pliers (spreading) are used to remove circlips that fit on the outside of a shaft. Internal pliers (squeezing) are used to remove circlips which fit inside a housing.

Box-end, Open-end and Combination Wrenches

Box and open-end wrenches are available in sets or separately in a variety of sizes. On open-end and box-end wrenches, the number stamped near the end refers to the distance between 2 parallel flats on the head of a nut or bolt. On combination wrenches, the number is stamped near the center.

Open-end wrenches are speedy and work best in areas with limited overhead access. Their wide jaws make them unsuitable for situations where the bolt or nut is sunken in a well or close to the edge of a casting. These wrenches only grip on two flats of a fastener so that if either the fastener head or wrench jaws are worn, the wrench may slip off.

Box-end wrenches require clear overhead access to the fastener, but can work well in situations where the fastener head is close to another part. They grip on all six edges of a fastener for a very secure grip. They are available in either 6-point or 12-point. The 6-point gives

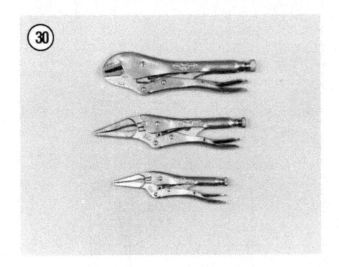

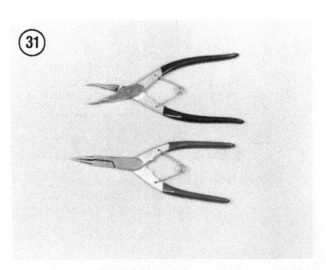

superior holding power and durability, but requires a greater swinging radius. The 12-point works better in situations with limited swinging radius.

Combination wrenches (**Figure 32**), which are open on one side and boxed on the other, are also available.

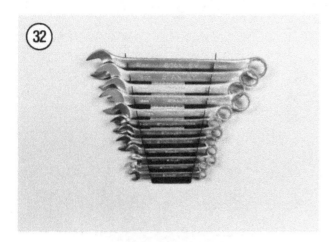

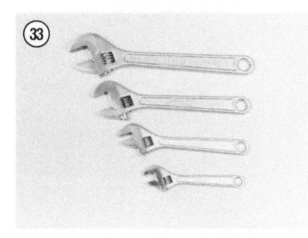

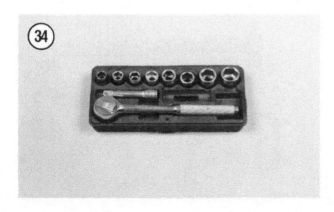

No matter what style of wrench you choose, proper use is important to prevent personal injury. When using a wrench, get in the habit of pulling the wrench toward you. This reduces the risk of injuring your hand if the wrench should slip. If you have to push the wrench away from you to loosen or tighten a fastener, open and push with the palm of your hand. This technique gets your fingers and knuckles out of the way should the wrench slip. Before using a wrench, always think ahead as to what could happen if the wrench should slip or if the bolt strips or breaks.

Adjustable Wrenches

An adjustable wrench (sometimes called a Crescent wrench) can be adjusted to fit nearly any nut or bolt head which has clear access around its entire perimeter. Adjustable wrenches are best used as a backup wrench to keep a large nut or bolt from turning while the other end is being loosened or tightened with a proper wrench. See **Figure 33**.

Adjustable wrenches have only two gripping surfaces which makes them more subject to slipping off the fastener and damaging the part and possibly your hand. See *Box-end, Open-end and Combination Wrenches* in this chapter for proper wrench usage.

These wrenches are directional; the solid jaw must be the one transmitting the force. If you use the adjustable jaw to transmit the force, it will loosen and possibly slip off.

Adjustable wrenches come in all sizes but something in the 6 to 8 inch range is recommended as an all-purpose wrench.

Socket Wrenches

This type is undoubtedly the fastest, safest and most convenient to use. Sockets which attach to a ratchet handle are available with 6-point or 12-point openings and 1/4, 3/8, 1/2 and 3/4 in. drives (**Figure 34**). The drive size indicates the size of the square hole which mates with the ratchet handle.

Torque Wrench

A torque wrench (**Figure 35**) is used with a socket to measure how tightly a nut or bolt is installed. They come in a wide price range and with either 3/8 or 1/2 in. square drive. The drive size indicates the size of the square drive which mates with the socket.

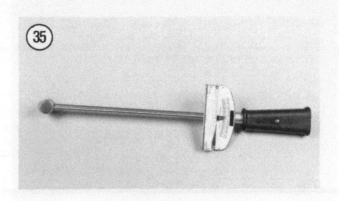

Impact Driver

This tool makes removal of tight fasteners easy and eliminates damage to bolts and screw slots. Impact drivers and interchangeable bits (**Figure 36**) are available at most large hardware, snowmobile and motorcycle dealers. Sockets can also be used with a hand impact driver. However, make sure the socket is designed for impact use. Do not use regular hand type sockets, as they may shatter (**Figure 37**).

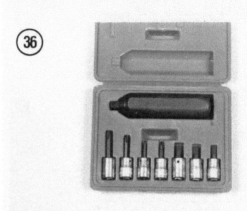

Hammers

The correct hammer (**Figure 38**) is necessary for repairs. Use only a hammer with a face (or head) of rubber or plastic or the soft-faced type that is filled with buckshot. These are sometimes necessary in engine teardowns. *Never* use a metal-faced hammer on engine or suspension parts, as severe damage will result in most cases. You can always produce the same amount of force with a soft-faced hammer. A metal-faced hammer, however, will be required when using a hand impact driver.

PRECISION MEASURING TOOLS

Measurement is an important part of snowmobile service. When performing many of the service procedures in this manual, you will be required to make a number of measurements. These include basic checks such as engine compression and spark plug gap. As you get deeper into engine disassembly and service, measurements will be required to determine the condition of the piston and cylinder bore, crankshaft runout and so on. When making these

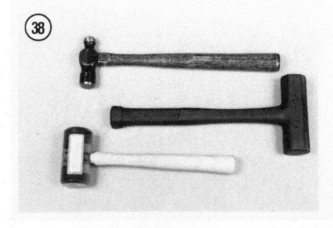

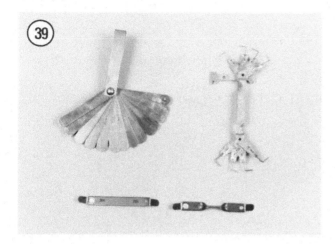

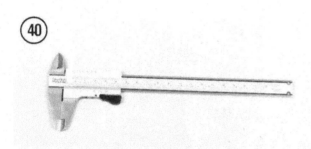

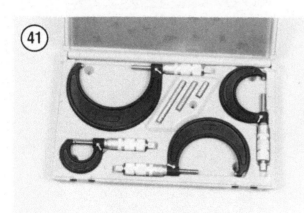

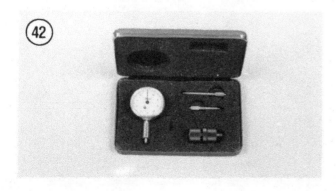

measurements, the degree of accuracy will dictate which tool is required. Precision measuring tools are expensive. If this is your first experience at engine service, it may be more worthwhile to have the checks made at a dealer. However, as your skills and enthusiasm increase for doing your own service work, you may want to begin purchasing some of these specialized tools. The following is a description of the measuring tools required in order to perform engine service described in this manual.

Feeler Gauge

The feeler gauge (**Figure 39**) is made of either a piece of a flat or round hardened steel of a specified thickness. Wire gauges are used to measure spark plug gap. Flat gauges are used for all other measurements.

Vernier Caliper

This tool is invaluable when reading inside, outside and depth measurements to within close precision. See **Figure 40**.

Outside Micrometers

One of the most reliable tools used for precision measurement is the outside micrometer. Outside micrometers will be required to measure piston diameter. Outside micrometers are also used with other tools to measure cylinder bore. Micrometers can be purchased individually or as a set (**Figure 41**).

Dial Indicator

Dial indicators (**Figure 42**) are precision tools used to check ignition timing and runout limits. For snowmobile repair, select a dial indicator with a continuous dial (**Figure 43**). This type of dial is required to accurately measure ignition timing and can be used for runout and height measurement checks.

Degree Wheel

A degree wheel (**Figure 44**) is a specific tool used to measure parts of a circle and angles. For Ski-Doo snowmobiles, a degree wheel will be required to mark the rotary valve timing position.

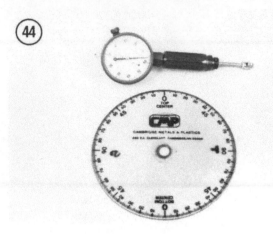

Cylinder Bore Gauge

The cylinder bore gauge is a very specialized precision tool. The gauge set shown in **Figure 45** is comprised of a dial indicator, handle and a number of length adapters to adapt the gauge to different bore sizes. The bore gauge can be used to make cylinder bore measurements such as bore size, taper and out-of-round. An outside micrometer must be used together with the bore gauge to determine bore dimensions.

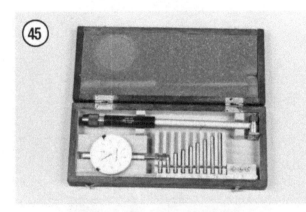

Small Hole Gauges

A set of small hole gauges (**Figure 46**) allows you to measure a hole, groove or slot ranging

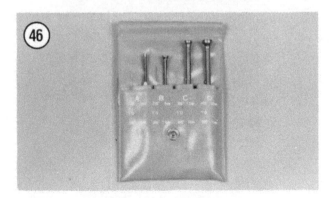

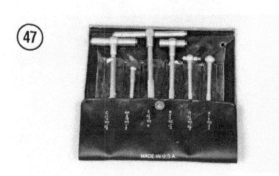

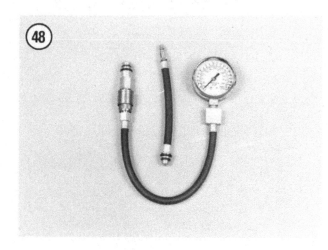

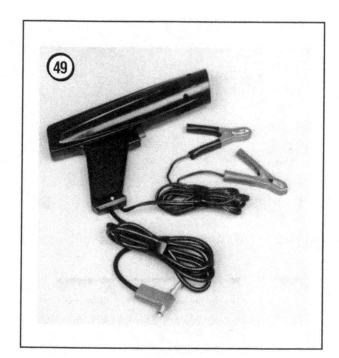

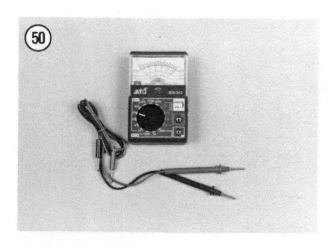

in size up to 13 mm (0.500 in.). An outside micrometer must be used together with the small hole gauge to determine bore dimensions.

Telescoping Gauges

Telescoping gauges (**Figure 47**) can be used to measure hole diameters from approximately 8 mm (5/16 in.) to 150 mm (6 in.). Like the small hole gauge, the telescoping gauge does not have a scale gauge for direct readings. An outside micrometer must be used together with the telescoping gauge to determine bore dimensions.

Compression Gauge

An engine with low compression cannot be properly tuned and will not develop full power. A compression gauge (**Figure 48**) measures engine compression. The one shown has a flexible stem with an extension that can allow you to hold it while starting the engine. Open the throttle all the way when checking engine compression. See Chapter Three.

Two-stroke Pressure Tester

Refer to *Two-stroke Pressure Testing* in Chapter Two.

Strobe Timing Light

This instrument is useful for checking ignition timing. By flashing a light at the precise instant the spark plug fires, the position of the timing mark can be seen. The flashing light makes a moving mark appear to stand still opposite a stationary mark.

Suitable lights range from inexpensive neon bulb types to powerful xenon strobe lights. See **Figure 49**. A light with an inductive pickup is recommended to eliminate any possible damage to ignition wiring.

Multimeter or VOM

This instrument (**Figure 50**) is invaluable for electrical system troubleshooting.

Screw Pitch Gauge

A screw pitch gauge (**Figure 51**) determines the thread pitch of bolts, screws, studs, etc. The gauge is made up of a number of thin plates. Each plate has a thread shape cut on one edge to match one thread pitch. When using a screw pitch gauge to determine a thread pitch size, try to fit different blade sizes onto the bolt thread until both threads match.

Magnetic Stand

A magnetic stand (**Figure 52**) is used to securely hold a dial indicator when checking the runout of a round object or when checking the end play of a shaft.

V-blocks

V-blocks (**Figure 53**) are precision ground blocks used to hold a round object when checking its runout or condition.

Surface Plate

A surface plate is used to check the flatness of parts. While industrial quality surface plates are quite expensive, the home mechanic can improvise. A piece of thick metal can be put to use as a surface plate. The metal surface plate in **Figure 54** shows a piece of sandpaper glued to its surface that is used for cleaning and smoothing cylinder head and crankcase mating surfaces.

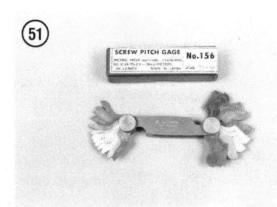

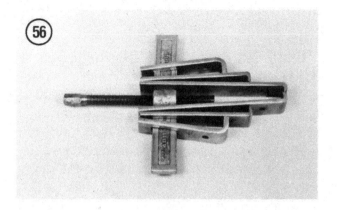

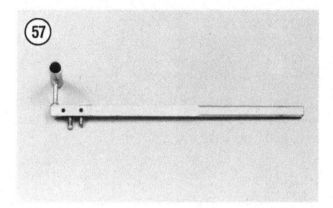

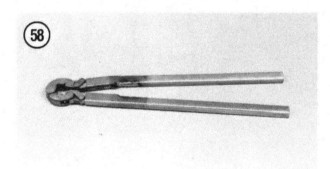

NOTE
Check with a local machine shop, fabricating shop or school offering a machine shop course for the availability of a metal plate that can be resurfaced and used as a surface plate.

SPECIAL TOOLS

This section describes special tools unique to snowmobile service and repair.

Flywheel Puller

A flywheel puller (**Figure 55**) will be required whenever it is necessary to remove the flywheel and service the stator plate assembly or when adjusting the ignition timing. In addition, when disassembling the engine, the flywheel must be removed before the crankcases can be split. There is no satisfactory substitute for this tool. Because the flywheel is a taper fit on the crankshaft, makeshift removal often results in crankshaft and flywheel damage. Don't think about removing the flywheel without this tool.

Flywheel Holder

The flywheel holder is used to hold the flywheel during removal.

Wheel Bearing Pullers

A puller set with long arms (**Figure 56**) will be required to remove suspension wheel bearings.

Track Clip Remover

This tool is be used to remove track cleats (**Figure 57**).

Track Clip Installer

A track clip installer will be required to install track clips. See **Figure 58** and **Figure 59**.

Spring Scale

A spring scale (**Figure 60**) will be required to check track tension.

Clutch Tools

A number of special tools will be required for clutch service. These are described in Chapter Eleven.

Expendable Supplies

Certain expendable supplies are also required. These include grease, oil, gasket cement, shop rags and cleaning solvent. Ask your dealer for the special locking compounds, silicone lubricants and lube products which make vehicle maintenance simpler and easier. Cleaning solvent is available at some service stations.

> *WARNING*
> *Having a stack of clean shop rags on hand is important when performing engine work. However, to prevent the possibility of fire damage from spontaneous combustion from a pile of solvent soaked rags, store them in a lid-sealed metal container until they can be washed or properly discarded.*

> *NOTE*
> *To prevent absorbing solvent and other chemicals into your skin while cleaning parts, wear a pair of petroleum-resistant rubber gloves. These can be purchased through industrial supply houses or well-equipped hardware stores.*

MECHANIC'S TIPS

Removing Frozen Nuts and Screws

When a fastener rusts and cannot be removed, several methods may be used to loosen it. First, apply penetrating oil such as Liquid Wrench or WD-40 (available at hardware or auto supply stores). Apply it liberally and let it penetrate for 10-15 minutes. Rap the fastener several times with a small hammer; do not hit it hard enough to

cause damage. Reapply the penetrating oil if necessary.

For frozen screws, apply penetrating oil as described, then insert a screwdriver in the slot and rap the top of the screwdriver with a hammer. This loosens the rust so the screw can be removed in the normal way. If the screw head is too chewed up to use this method, grip the head with Vise Grips pliers and twist the screw out.

Avoid applying heat unless specifically instructed, as it may melt, warp or remove the temper from parts.

Removing Broken Screws or Bolts

When the head breaks off a screw or bolt, several methods are available for removing the remaining portion.

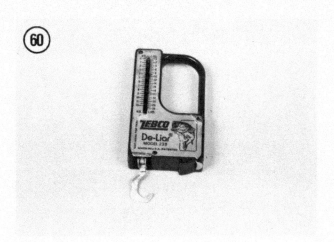

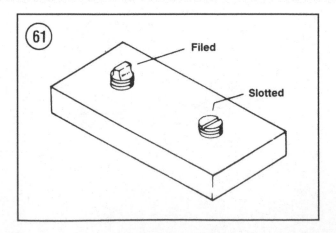

If a large portion of the remainder projects out, try gripping it with Vise Grips. If the projecting portion is too small, file it to fit a wrench or cut a slot in it to fit a screwdriver. See **Figure 61**.

If the head breaks off flush, use a screw extractor. To do this, centerpunch the exact center of the remaining portion of the screw or bolt. Drill a small hole in the screw and tap the extractor into the hole. Back the screw out with a wrench on the extractor. See **Figure 62**.

Remedying Stripped Threads

Occasionally, threads are stripped through carelessness or impact damage. Often the threads can be cleaned up by running a tap (for internal threads on nuts) or die (for external threads on bolts) through the threads. See **Figure 63**. To clean or repair spark plug threads, a spark plug tap can be used.

NOTE
Tap and dies can be purchased individually or in a set as shown in Figure 64.

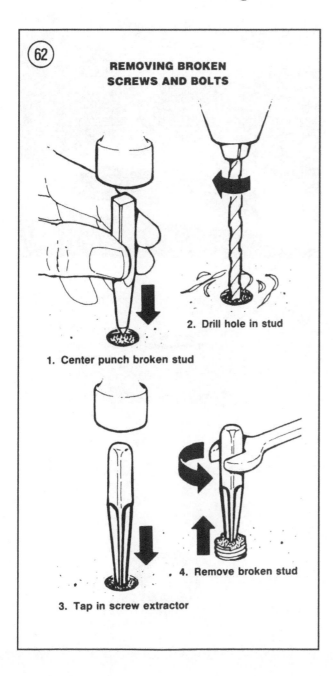

REMOVING BROKEN SCREWS AND BOLTS

1. Center punch broken stud

2. Drill hole in stud

3. Tap in screw extractor

4. Remove broken stud

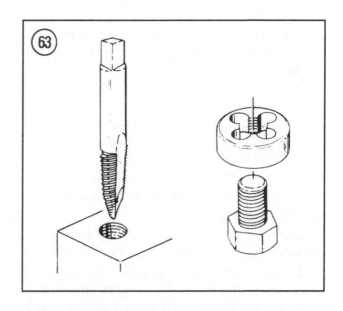

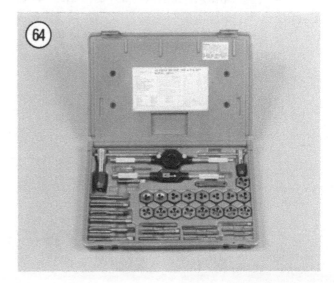

If an internal thread is damaged, it may be necessary to install a Helicoil (**Figure 65**) or some other type of thread insert. Follow the manufacturer's instructions when installing their insert.

If it is necessary to drill and tap a hole, refer to **Table 7** for metric tap drill sizes.

Removing Broken or Damaged Studs

If a stud is broken (**Figure 66**) or the threads severely damaged, perform the following. A tube of Loctite 271 (red), 2 nuts, 2 wrenches and a new stud will be required during this procedure (**Figure 67**).

1. Thread two nuts onto the damaged stud. Then tighten the 2 nuts against each other so that they are locked.

> *NOTE*
> *If the threads on the damaged stud do not allow installation of the 2 nuts, you will have to remove the stud with a pair of Vise Grips.*

2. Turn the bottom nut counterclockwise and unscrew the stud.
3. Clean the threads with solvent or electrical contact cleaner and allow to dry thoroughly.
4. Install 2 nuts on the top half of the new stud as in Step 1. Make sure they are locked securely.
5. Coat the bottom half of a new stud with Loctite 271 (red).
6. Turn the top nut clockwise and thread the new stud securely.

7. Remove the nuts and repeat for each stud as required.
8. Follow Loctite's directions on cure time before assembling the component.

BALL BEARING REPLACEMENT

Ball bearings (**Figure 68**) are used throughout the snowmobile engine and chassis to reduce

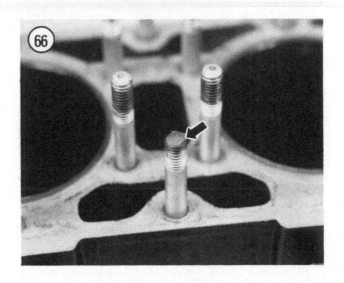

power loss, heat and noise resulting from friction. Because ball bearings are precision made parts, they must be maintained by proper lubrication and maintenance. When a bearing is found to be damaged, it should be replaced immediately. However, when installing a new bearing, care should be taken to prevent damage to the new bearing. While bearing replacement is described in the individual chapters where applicable, the following should be used as a guideline.

NOTE
Unless otherwise specified, install bearings with the manufacturer's mark or number facing outward.

Bearing Removal

While bearings are normally removed only when damaged, there may be times when it is necessary to remove a bearing that is in good condition. However, improper bearing removal will damage the bearing and maybe the shaft or case half. Note the following when removing bearings.

1. When using a puller to remove a bearing on a shaft, care must be taken so that shaft damage does not occur. Always place a piece of metal between the end of the shaft and the puller screw. In addition, place the puller arms next to the inner bearing race. See **Figure 69**.

2. When using a hammer to remove a bearing on a shaft, do not strike the hammer directly

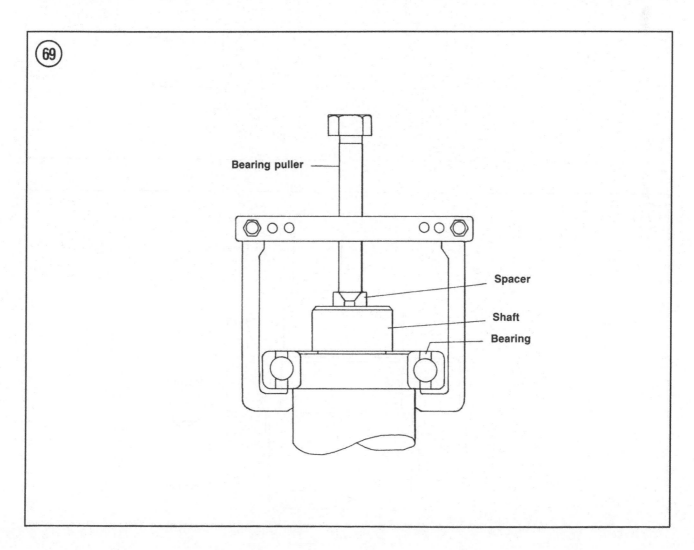

Bearing puller

Spacer

Shaft

Bearing

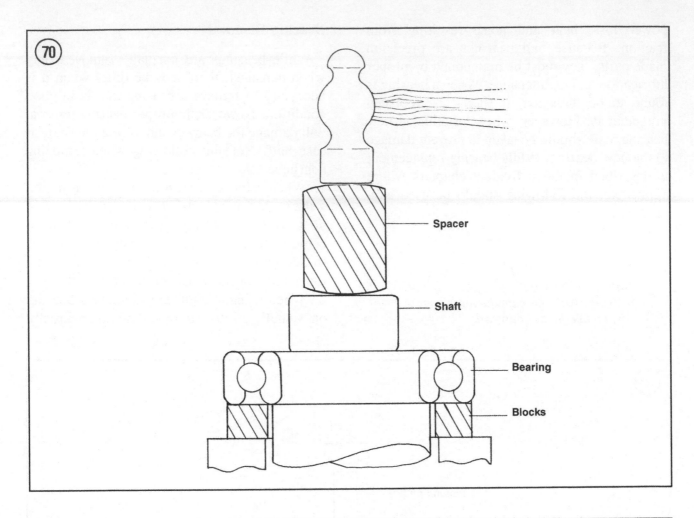

(70)

Spacer

Shaft

Bearing

Blocks

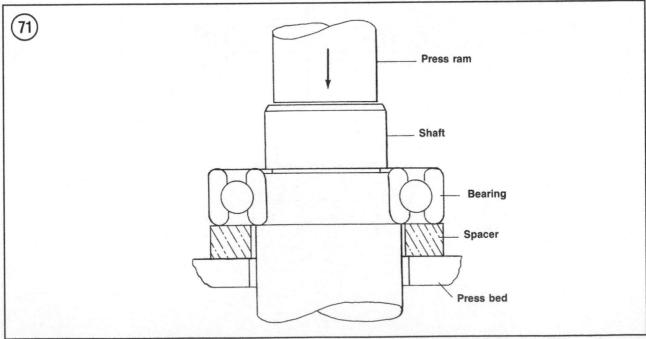

(71)

Press ram

Shaft

Bearing

Spacer

Press bed

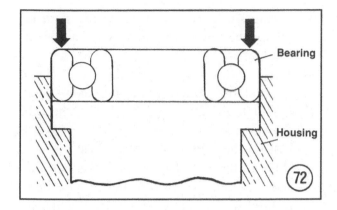

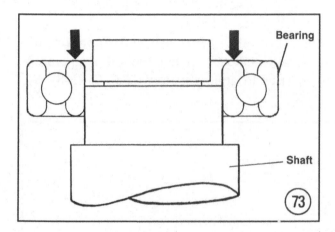

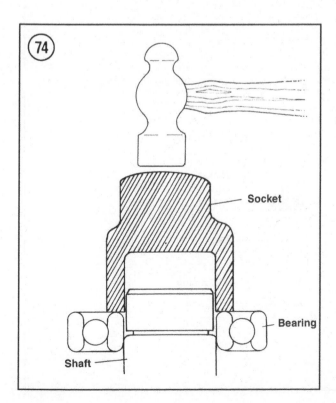

against the shaft. Instead, use a brass or aluminum rod between the hammer and shaft (**Figure 70**). In addition, make sure to support both bearing races with wood blocks as shown in **Figure 70**.

3. The most ideal method of bearing removal is with a hydraulic hand press. However, certain procedures must be followed or damage may occur to the bearing, shaft or case half. Note the following when using a press:

   a. Always support the inner and outer bearing races with a suitable size wood or aluminum ring (**Figure 71**). If only the outer race is supported, the balls and/or the inner race will be damaged.

   b. Always make sure the press ram (**Figure 71**) aligns with the center of the shaft. If the ram is not centered, it may damage the bearing and/or shaft.

   c. The moment the shaft is free of the bearing, it will drop to the floor. Secure or hold the shaft to prevent it from falling.

Bearing Installation

1. When installing a bearing in a housing, pressure must be applied to the *outer* bearing race (**Figure 72**). When installing a bearing on a shaft, pressure must be applied to the *inner* bearing race (**Figure 73**).

2. When installing a bearing as described in Step 1, some type of driver will be required. Never strike the bearing directly with a hammer or the bearing will be damaged. When installing a bearing, a piece of pipe or a socket with an outer diameter that matches the bearing race will be required. **Figure 74** shows the correct way to use a socket and hammer when installing a bearing.

3. Step 1 describes how to install a bearing in a case half and over a shaft. However, when installing over a shaft and into a housing at the same time, a snug fit will be required for both outer and inner bearing races. In this situation, a spacer must be installed underneath the driver tool so that pressure is applied evenly across *both*

races. See **Figure 75**. If the outer race is not supported as shown in **Figure 75**, the balls will push against the outer bearing track and damage it.

Shrink Fit

1. *Installing a bearing over a shaft:* When a tight fit is required, the bearing inside diameter will be smaller than the shaft. In this case, driving the bearing on the shaft using normal methods may cause bearing damage. Instead, the bearing should be heated before installation. Note the following:

  a. Secure the shaft so that it can be ready for bearing installation.

  b. Clean the bearing surface on the shaft of all residue. Remove burrs with a file or sandpaper.

  c. Fill a suitable pot or beaker with clean mineral oil. Place a thermometer (rated higher than 120° C [248° F]) in the oil. Support the thermometer so that it does not rest on the bottom or side of the pot.

  d. Remove the bearing from its wrapper and secure it with a piece of heavy wire bent to hold it in the pot. Hang the bearing in the pot so that it does not touch the bottom or sides of the pot.

  e. Turn the heat on and monitor the thermometer. When the oil temperature rises to approximately 120° C (248° F),

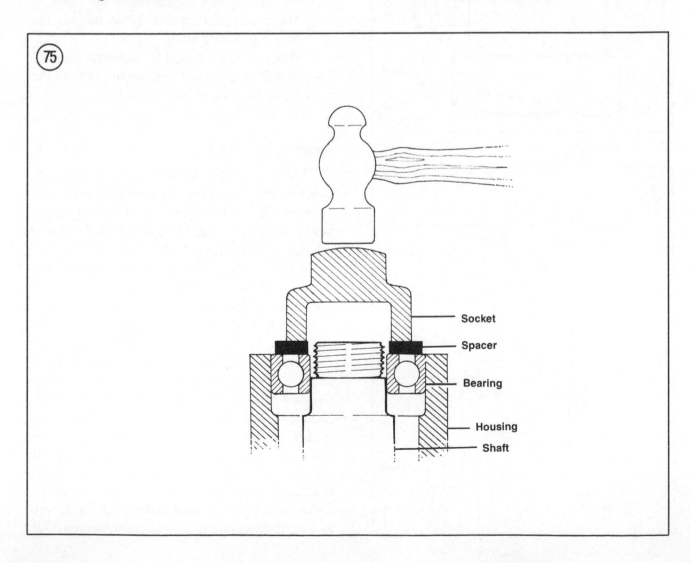

(75)

Socket

Spacer

Bearing

Housing

Shaft

remove the bearing from the pot and quickly install it. If necessary, place a socket on the inner bearing race and tap the bearing into place. As the bearing chills, it will tighten on the shaft so you must work quickly when installing it. Make sure the bearing is installed all the way.

2. *Installing a bearing in a housing:* Bearings are generally installed in a housing with a slight interference fit. Driving the bearing into the housing using normal methods may damage the housing or cause bearing damage. Instead, the housing should be heated before the bearing is installed. Note the following:

CAUTION
Before heating the crankcases in this procedure to remove the bearings, wash the cases thoroughly with detergent and water. Rinse and rewash the cases as required to remove all traces of oil and other chemical deposits.

a. The housing must be heated to a temperature of about 100° C (212° F) in an oven or on a hot plate. An easy way to check to see that it is at the proper temperature is to drop tiny drops of water on the case; if they sizzle and evaporate immediately, the temperature is correct. Heat only one housing at a time.

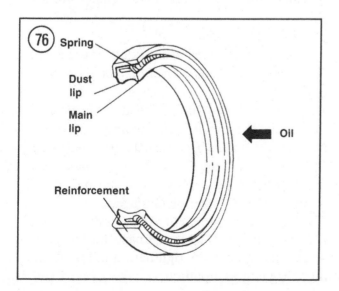

76 Spring
Dust lip
Main lip
Oil
Reinforcement

CAUTION
Do not heat the housing with a torch (propane or acetylene)—never bring a flame into contact with the bearing or housing. The direct heat will destroy the case hardening of the bearing and will likely warp the housing.

b. Remove the housing from the oven or hot plate and hold onto the housing with a kitchen pot holder, heavy gloves, or heavy shop cloths—*it is hot.*

NOTE
A suitable size socket and extension works well for removing and installing bearings.

c. Hold the housing with the bearing side down and tap the bearing out. Repeat for all bearings in the housing.

d. While heating up the housing halves, place the new bearings in a freezer if possible. Chilling them will slightly reduce their overall diameter while the hot housing assembly is slightly larger due to heat expansion. This will make installation much easier.

NOTE
Always install bearings with the manufacturer's mark or number facing outward.

e. While the housing is still hot, install the new bearing(s) into the housing. Install the bearings by hand, if possible. If necessary, lightly tap the bearing(s) into the housing with a socket placed on the outer bearing race. *Do not* install new bearings by driving on the inner bearing race. Install the bearing(s) until it seats completely.

OIL SEALS

Oil seals (**Figure 76**) are used to contain oil, grease or combustion gases in a housing or shaft. Improper removal of a seal can damage the housing or shaft. Improper installation of the seal can damage the seal. Note the following:

a. Prying is generally the easiest and most effective method of removing a seal from a housing. However, always place a rag underneath the pry tool to prevent damage to the housing.

b. A low temperature grease should be packed in the seal lips before the seal is installed.

c. Oil seals should always be installed so that the manufacturer's numbers or marks face out.

d. Oil seals should be installed with a socket placed on the outside of the seal as shown in **Figure 77**. Make sure the seal is driven squarely into the housing. Never install a seal by hitting against the top of the seal with a hammer.

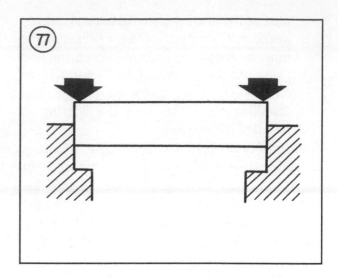

SNOWMOBILE OPERATION

Snowmobiles are ideal machines for getting around during winter months. However, because snowmobiles are often operated in extreme weather conditions and over rough terrain, they should be checked before each ride and maintained on a periodic basis.

> *WARNING*
> *Never lean into a snowmobile's engine compartment while wearing a scarf or other loose clothing when the engine is running or when the driver is attempting to start the engine. If the scarf or clothing should catch in the drive belt or clutch, severe injury or death could result.*

Pre-start Inspection

A pre-start inspection should always be performed before heading out on your snowmobile. While the following list may look exhaustive, it can be performed rather quickly after a few times.

1. Familiarize yourself with your snowmobile.

2. Clean the windshield with a clean, damp cloth. Do not use gasoline, solvents or abrasive cleaners.

3. Check track tension (Chapter Three) and adjust if necessary.

4. Check the tether switch and the emergency cut-out switch for proper operation. If your machine is new or if you are using a friend's machine, practice using the tether or stop switch a few times so that its use will be automatic during an emergency.

5. Check the brake operation. Ensure that the brake system is correctly adjusted.

6. Check the fuel level and top it up if necessary.

7. Check the injection oil tank. Make sure it is full.

8. Check the coolant level.

9. Check fan belt tension and adjust if necessary.

10. Operate the throttle lever. It should open and close smoothly.

11. Open the belt guard and visually inspect the drive belt. If the belt seems worn or damaged, replace it. Chapter Twelve lists drive belt wear limit specifications. Close the belt guard after inspecting the belt. Make sure the belt guard mounts are not loose or damaged.

12. While the engine shroud is open, visually inspect all hoses, fittings and parts for looseness or damage. Check the tightness of all bolts and nuts. Tighten as required.

13. Check the handlebar and steering components for looseness or damage. Do not ride the vehicle if any steering component is damaged. Tighten loose fasteners as required.

14. After closing the shroud, make sure the shroud latches are fastened securely.

15. Check the skis for proper alignment (Chapter Three). Check the ski pivot bolt for tightness or damage.

> *WARNING*
> *When starting the engine, be sure that no bystanders are in front or behind the snowmobile. A sudden lurch of the machine could cause serious injury.*

16. Make sure that all lights are working.

> *NOTE*
> *If abnormal noises are detected after starting the engine, locate and repair the problem before starting out.*

> *NOTE*
> *Refer to the appropriate chapter for tightening torques and service procedures.*

Tools and Spare Parts

Before leaving on a trip, make sure that you carry tools and spare parts in case of emergency. A tool kit should include the following:
  a. Flashlight.
  b. Rope.
  c. Tools.
  d. Tape.
A spare parts kit should include the following:
  a. Drive belt.
  b. Emergency starter rope.
  c. Light bulbs.
  d. Spark plugs.
  e. Main jets.
  f. Throttle cable.
  g. Brake cable.
  h. A good book...just in case.

If you are going out on a long trip, extra oil and fuel should be carried.

Emergency Starting

If your recoil starter rope should break (and the electric starter does not work, if so equipped), the engine can be started with an emergency rope that is wrapped around the primary sheave cap.

1. Open the shroud.

> *WARNING*
> *The drive belt guard must be removed when starting the engine with the emergency starter rope. **Never** lean into a snowmobile's engine compartment while wearing a scarf or other loose clothing while the engine is running or when attempting to start the engine. If the scarf or clothing should catch in the drive belt or clutch, severe injury or death could result.*

2. Remove the drive belt guard.

3. Remove the spare starter rope from your tool kit.

4. Wind the rope around the primary sheave cap and start the engine.

5. Reinstall the drive belt guard after starting the engine.

6. Close and secure the shroud.

7. Store the spare starter rope in your tool kit.

Clearing the Track

If the snowmobile has been operated in deep or slushy snow, it is necessary to clear the track after stopping to prevent the track from freezing. This condition would make starting and running difficult.

> *WARNING*
> *Make sure no one is behind the machine when clearing the track. Ice and rocks thrown from the track can cause injury.*

Tip the snowmobile on its side until the track clears the ground *completely*. Run the track at a moderate speed until all the ice and snow is thrown clear.

CAUTION
If the track does freeze, it must be broken loose manually with the engine turned off. Attempting to force a frozen track with the engine will burn and damage the drive belt.

SNOWMOBILE SAFETY

Proper Clothing

Warm and comfortable clothing are a must to provide protection from frostbite. Even mild temperatures can be very uncomfortable and dangerous when combined with a strong wind or when traveling at high speeds. See **Table 8** for wind chill factors. Always dress according to what the wind chill factor is, not the temperature. Check with an authorized dealer for suggested types of snowmobile clothing.

WARNING
To provide additional warmth as well as protection against head injury, always wear an approved helmet when snowmobiling.

Emergency Survival Techniques

1. Do not panic in the event of an emergency. Relax, think the situation over, then decide on a course of action. You may be within a short distance of help. If possible, repair your snowmobile so you can drive to safety. Conserve your energy and stay warm.
2. Keep hands and feet active to promote circulation and avoid frostbite while servicing your machine.
3. Mentally retrace your route. Where was the last point where help could be located? Do not attempt to walk long distances in deep snow. Make yourself comfortable until help arrives.
4. If you are properly equipped for your trip, you can turn any undesirable area into a suitable campsite.

5. If necessary, build a small shelter with tree branches or evergreen boughs. Look for a sheltered area against a hill or cliff. Even burrowing in the snow offers protection from the cold and wind.
6. Prepare a signal fire using evergreen boughs and snowmobile oil. If you cannot build a fire, make an S-O-S in the snow.
7. Use a policeman's whistle or beat cooking utensils to attract attention.
8. When your camp is established, climb the nearest hill and determine your whereabouts. Observe landmarks on the way, so you can find your way back to your campsite. Do not rely on your footprints. They may be covered by blowing snow.

SNOWMOBILE CODE OF ETHICS

1. I will be a good sportsman and conservationist. I recognize that people judge all snowmobilers by my actions. I will use my influence with other snowmobile owners and operators to promote sportsmanlike conduct.
2. I will not litter any trails or areas, nor will I pollute streams or lakes. I will carry out what I carry in.
3. I will not damage living trees, shrubs or other natural features.
4. I will respect other people's properties and rights.
5. I will lend a helping hand when I see someone in need.
6. I will make myself and my vehicle available to assist in search and rescue operations.
7. I will not interfere with the activities of other winter sportsmen. I will respect their right to enjoy their recreational activity.
8. I will know and obey all federal, state or provincial and local rules regulating the operation of snowmobiles in areas where I use my vehicle.
9. I will not harass wildlife.
10. I will not snowmobile where prohibited.

Table 1 SKI-DOO MODEL NUMBER

Year	Model	Model number
1985	Formula MX	3720
	Formula Plus	3721
1986	Formula MX	3725
	Formula Plus	3726
	Formula MX (H/A*)	3727
1987	Formula MX	3728
	Formula Plus	3729
	Formula MX LT	3730
1988	Formula MX	3732
	Formula Plus	3733
	Formula MX LT	3734
1989	Formula MX	3735
	Formula MX LT	3736
	Formula Plus	3737
	Formula Plus LT	3738
	Formula Mach I	3739

*High altitude

Table 2 GENERAL DIMENSIONS

	cm	in.
Overall length		
1985-1987	271.8	107
1988-on	276.5	109
Overall width	104.1	41
Overall height		
1985-1986	91.4	36
1987-on	99	39

Table 3 VEHICLE WEIGHT

	kg	lb.
1985		
Formula MX	198.7	438
Formula Plus	203.2	448
1986		
Formula MX	198.7	438
Formula MX (H/A)	207.7	458
Formula Plus	203.2	448
1987		
Formula MX	204.1	450
Formula MX LT	208.6	460
Formula Plus	212.2	468

(continued)

Table 3 VEHICLE WEIGHT (continued)

	kg	lb.
1988		
Formula MX	222.3	489
Formula MX LT	250	528
Formula Plus	226.8	499
1989		
Formula MX	222.3	489
Formula MX LT	233	513
Formula Plus	229.1	505
Formula Plus LT	235	518
Formula Mach I	234.3	517

Table 4 DECIMAL AND METRIC EQUIVALENTS

Fractions	Decimal in.	Metric mm	Fractions	Decimal in.	Metric mm
1/64	0.015625	0.39688	33/64	0.515625	13.09687
1/32	0.03125	0.79375	17/32	0.53125	13.49375
3/64	0.046875	1.19062	35/64	0.546875	13.89062
1/16	0.0625	1.58750	9/16	0.5625	14.28750
5/64	0.078125	1.98437	37/64	0.578125	14.68437
3/32	0.09375	2.38125	19/32	0.59375	15.08125
7/64	0.109375	2.77812	39/64	0.609375	15.47812
1/8	0.125	3.1750	5/8	0.625	15.87500
9/64	0.140625	3.57187	41/64	0.640625	16.27187
5/32	0.15625	3.96875	21/32	0.65625	16.66875
11/64	0.171875	4.36562	43/64	0.671875	17.06562
3/16	0.1875	4.76250	11/16	0.6875	17.46250
13/64	0.203125	5.15937	45/64	0.703125	17.85937
7/32	0.21875	5.55625	23/32	0.71875	18.25625
15/64	0.234375	5.95312	47/64	0.734375	18.65312
1/4	0.250	6.35000	3/4	0.750	19.05000
17/64	0.265625	6.74687	49/64	0.765625	19.44687
9/32	0.28125	7.14375	25/32	0.78125	19.84375
19/64	0.296875	7.54062	51/64	0.796875	20.24062
5/16	0.3125	7.93750	13/16	0.8125	20.63750
21/64	0.328125	8.33437	53/64	0.828125	21.03437
11/32	0.34375	8.73125	27/32	0.84375	21.43125
23/64	0.359375	9.121812	55/64	0.859375	21.82812
3/8	0.375	9.52500	7/8	0.875	22.22500
25/64	0.390625	9.92187	57/64	0.890625	22.62187
13/32	0.40625	10.31875	29/32	0.90625	23.01875
27/64	0.421875	10.71562	59/64	0.921875	23.41562
7/16	0.4375	11.11250	15/16	0.9375	23.81250
29/64	0.453125	11.50937	61/64	0.953125	24.20937
15/32	0.46875	11.90625	31/32	0.96875	24.60625
31/64	0.484375	12.30312	63/64	0.984375	25.00312
1/2	0.500	12.70000	1	1.00	25.40000

Table 5 GENERAL TORQUE SPECIFICATIONS

Item	N·m	ft.-lb.
Bolt		
6 mm	6	4.3
8 mm	15	11
10 mm	30	22
12 mm	55	40
14 mm	85	61
16 mm	130	94
Nut		
6 mm	6	4.3
8 mm	15	11
10 mm	30	22
12 mm	55	40
14 mm	85	61
16 mm	130	94

Table 6 TECHNICAL ABBREVIATIONS

ABDC	After bottom dead center
ATDC	After top dead center
BBDC	Before bottom dead center
BDC	Bottom dead center
BTDC	Before top dead center
C	Celsius (Centigrade)
cc	Cubic centimeters
CDI	Capacitor discharge ignition
cu. in.	Cubic inches
F	Fahrenheit
ft.-lb.	Foot-pounds
gal.	Gallons
H/A	High altitude
hp	Horsepower
in.	Inches
kg	Kilogram
kg/cm²	Kilograms per square centimeter
kgm	Kilogram meters
km	Kilometer
l	Liter
m	Meter
MAG	Magneto
mm	Millimeter
N·m	Newton-meters
oz.	Ounce
psi	Pounds per square inch
PTO	Power take off
pts.	Pints
qt.	Quarts
rpm	Revolutions per minute

Table 7 METRIC TAP DRILL SIZES

Metric (mm)	Drill size	Decimal equivalent	Nearest fraction
3×0.50	No. 39	0.0995	3/32
3×0.60	3/32	0.0937	3/32
4×0.70	No. 30	0.1285	1/8
4×0.75	1/8	0.125	1/8
5×0.80	No. 19	0.166	11/64
5×0.90	No. 20	0.161	5/32
6×1.00	No. 9	0.196	13/64
7×1.00	16/64	0.234	15/64
8×1.00	J	0.277	9/32
8×1.25	17/64	0.265	17/64
9×1.00	5/16	0.3125	5/16
9×1.25	5/16	0.3125	5/16
10×1.25	11/32	0.3437	11/32
10×1.50	R	0.339	11/32
11×1.50	3/8	0.375	3/8
12×1.50	13/32	0.406	13/32
12×1.75	13/32	0.406	13/32

Table 8 WIND CHILL FACTORS

Estimated Wind Speed in MPH	Actual Thermometer Reading (°F)											
	50	40	30	20	10	0	−10	−20	−30	−40	−50	−60
	Equivalent Temperature (°F)											
Calm	50	40	30	20	10	0	−10	−20	−30	−40	−50	−60
5	48	37	27	16	6	−5	−15	−26	−36	−47	−57	−68
10	40	28	16	4	−9	−21	−33	−46	−58	−70	−83	−95
15	36	22	9	−5	−18	−36	−45	−58	−72	−85	−99	−112
20	32	18	4	−10	−25	−39	−53	−67	−82	−96	−110	−124
25	30	16	0	−15	−29	−44	−59	−74	−88	−104	−118	−133
30	28	13	−2	−18	−33	−48	−63	−79	−94	−109	−125	−140
35	27	11	−4	−20	−35	−49	−67	−82	−98	−113	−129	−145
40	26	10	−6	−21	−37	−53	−69	−85	−100	−116	−132	−148

*

Little Danger (for properly clothed person)   **Increasing Danger**   **Great Danger**

*Danger from freezing of exposed flesh.

*Wind speeds greater than 40 mph have little additional effect.

Chapter Two

Troubleshooting

Diagnosing mechanical problems is relatively simple if you use orderly procedures and keep a few basic principles in mind. The first step in any troubleshooting procedure is to define the symptoms as closely as possible and then localize the problem. Subsequent steps involve testing and analyzing those areas which could cause the symptoms. A haphazard approach may eventually solve the problem, but it can be very costly in terms of wasted time and unnecessary parts replacement.

Proper lubrication, maintenance and periodic tune-ups as described in Chapter Three will reduce the necessity for troubleshooting. Even with the best of care, however, all snowmobiles are prone to problems which will require troubleshooting.

Never assume anything. Do not overlook the obvious. If the engine won't start, check the position of the emergency cut-out switch and the tether switch. Is the engine flooded with fuel from using the primer too much?

If the engine suddenly quits, check the easiest, most accessible problem first. Is there gasoline in the tank? Has a spark plug wire broken or fallen off?

If nothing obvious turns up in a quick check, look a little further. Learning to recognize and describe symptoms will make repairs easier for you or a mechanic at the shop. Describe problems accurately and fully.

Gather as many symptoms as possible to aid in diagnosis. Note whether the engine lost power gradually or all at once, what color smoke came from the exhaust and so on. Remember that the more complicated a machine is, the easier it is to troubleshoot because symptoms point to specific problems.

After the symptoms are defined, areas which could cause problems are tested and analyzed. Guessing at the cause of a problem may provide the solution, but it can easily lead to frustration, wasted time and a series of expensive, unnecessary parts replacements.

You do not need fancy equipment or complicated test gear to determine whether repairs can be attempted at home. A few simple checks could save a large repair bill and lost time while your snowmobile sits in a dealer's service department. On the other hand, be realistic and do not attempt repairs beyond your abilities. Service departments tend to charge heavily for putting together a disassembled engine that may have been abused. Some won't even take on such a job—so use common sense, don't get in over your head.

Electrical specifications are listed in **Table 1** and **Table 2** at the end of this chapter.

OPERATING REQUIREMENTS

An engine needs 3 basics to run properly: correct fuel/air mixture, compression and a spark at the right time (**Figure 1**). If one basic requirement is missing, the engine will not run. Two-stroke engine operating principles are described in Chapter One under *Engine Operation*. The ignition system is the weakest link of the 3 basics. More problems result from ignition breakdowns than from any other source. Keep that in mind before you begin tampering with carburetor adjustments and the like.

If the snowmobile has been sitting for any length of time and refuses to start, check and clean the spark plugs. Then check the condition of the battery (if so equipped) to make sure it has an adequate charge. If these are okay, then look to the gasoline delivery system. This includes the tank, fuel shutoff valve, fuel pump and fuel line to the carburetor. Gasoline deposits may have gummed up carburetor jets and air passages. Gasoline tends to lose its potency after standing for long periods. Condensation may contaminate it with water. Drain the old gas and try starting with a fresh tankful.

TROUBLESHOOTING INSTRUMENTS

Chapter One lists the instruments needed and detailed instruction on their use.

TESTING ELECTRICAL COMPONENTS

Most dealers and parts houses will not accept returns on electrical parts purchased through their business. When testing electrical components, make sure that you perform the test procedures as described in this chapter and that your test equipment is working properly. If a test

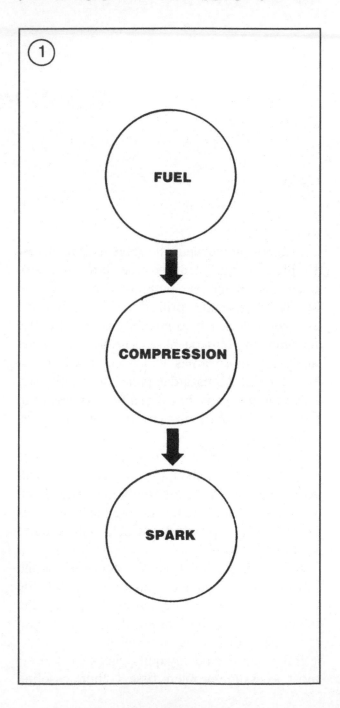

result shows that the component is defective, but the reading is close to the service limit, have the component tested by a Ski-Doo dealer to verify the test result before purchasing a new component.

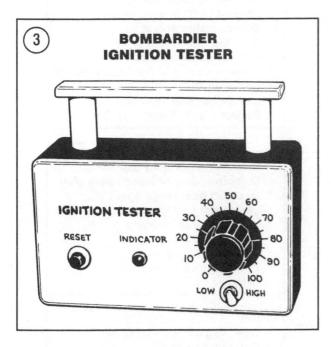

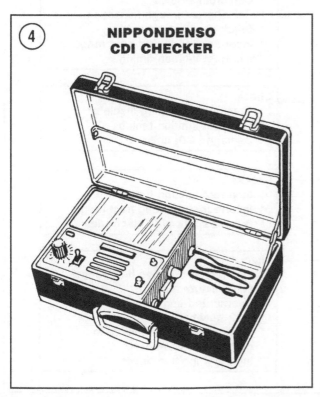

ENGINE ELECTRICAL SYSTEM TROUBLESHOOTING

All models are equipped with a capacitor discharge ignition system. This section describes complete ignition and charging system troubleshooting.

This solid-state system uses no contact breaker point or other moving parts. Because of the solid-state design, problems with the capacitor discharge system are relatively few. However, when problems arise, they stem from one of the following:

  a. Weak spark.

  b. No spark.

It is possible to check CDI systems that:

  a. Do not spark.

  b. Have broken or damaged wires.

  c. Have a weak spark.

It is difficult to check CDI systems that malfunction due to:

  a. Vibration problems.

  b. Components that malfunction only when the engine is hot or under a load.

General troubleshooting procedures are provided in **Figure 2**.

Test Equipment

Complete testing of the engine electrical system will require the use of the Bombardier ignition tester (part No. 419 0033 00) (**Figure 3**) and the Nippondenso CDI checker (part No. 419 0084 00) (**Figure 4**). *Basic* testing of the electrical system can be performed with an accurate ohmmeter.

If you do not have access to the special tools shown in **Figure 3** and **Figure 4**, you can use visual inspection and an ohmmeter to pinpoint electrical problems caused by dirty or damaged connectors, faulty or damaged wiring or electrical components that may have cracked or broken. If basic checks fail to locate the problem, take your snowmobile to a Ski-Doo dealer and have them troubleshoot the electrical system.

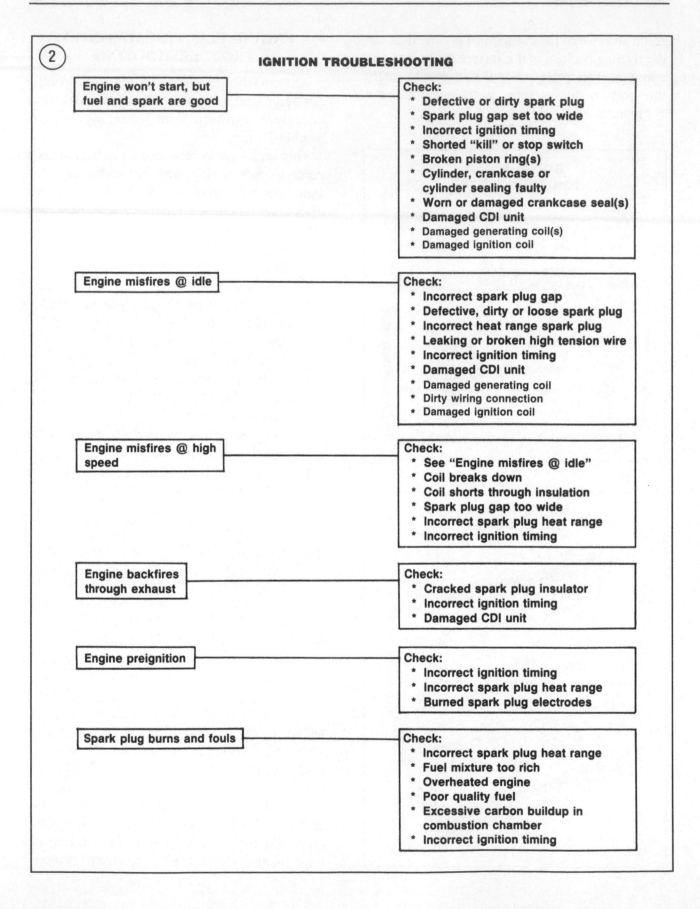

② IGNITION TROUBLESHOOTING

Engine won't start, but fuel and spark are good

Check:
* Defective or dirty spark plug
* Spark plug gap set too wide
* Incorrect ignition timing
* Shorted "kill" or stop switch
* Broken piston ring(s)
* Cylinder, crankcase or cylinder sealing faulty
* Worn or damaged crankcase seal(s)
* Damaged CDI unit
* Damaged generating coil(s)
* Damaged ignition coil

Engine misfires @ idle

Check:
* Incorrect spark plug gap
* Defective, dirty or loose spark plug
* Incorrect heat range spark plug
* Leaking or broken high tension wire
* Incorrect ignition timing
* Damaged CDI unit
* Damaged generating coil
* Dirty wiring connection
* Damaged ignition coil

Engine misfires @ high speed

Check:
* See "Engine misfires @ idle"
* Coil breaks down
* Coil shorts through insulation
* Spark plug gap too wide
* Incorrect spark plug heat range
* Incorrect ignition timing

Engine backfires through exhaust

Check:
* Cracked spark plug insulator
* Incorrect ignition timing
* Damaged CDI unit

Engine preignition

Check:
* Incorrect ignition timing
* Incorrect spark plug heat range
* Burned spark plug electrodes

Spark plug burns and fouls

Check:
* Incorrect spark plug heat range
* Fuel mixture too rich
* Overheated engine
* Poor quality fuel
* Excessive carbon buildup in combustion chamber
* Incorrect ignition timing

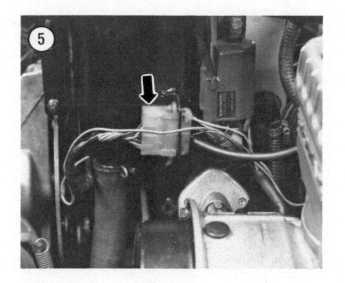

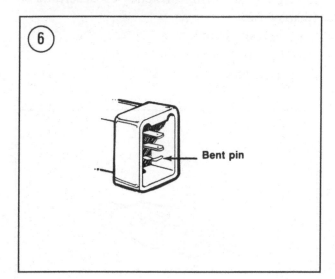

Bent pin

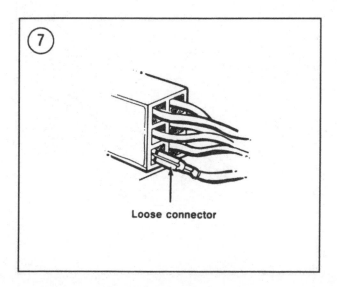

Loose connector

Precautions

Certain measures must be taken to protect the capacitor discharge system. Instantaneous damage to the semiconductors in the system will occur if the following is not observed.

1. Do not crank the engine if the CDI unit is not grounded to the engine.

2. Do not touch or disconnect any ignition components when the engine is running.

3. Keep all connections between the various units clean and tight. Be sure that the wiring connectors are pushed together firmly.

Troubleshooting Preparation

NOTE
To test the wiring harness for poor connections in Step 1, bend the molded rubber connector while checking each wire for resistance.

Refer to the wiring diagram for your model at the end of this book when performing the following.

1. Check the wiring harness for visible signs of damage.

2. Check all of the connectors (**Figure 5**) as follows:

   a. Disconnect each electrical connector in the ignition circuit. Check for bent or damaged pins in each male connector (**Figure 6**). A bent pin will not connect to its mating receptacle in the female end of the connector. This will cause an open circuit.

   b. Check each female connector end. Make sure the metal connector on the end of each wire (**Figure 7**) is pushed all the way into the plastic connector. If not, carefully push them in with a narrow-blade screwdriver. Make sure you do not pinch or cut the wire.

   c. Check all electrical wires where they enter the individual metal connector in both the male and female plastic connectors.

   d. Make sure all electrical connectors within the connector are clean and free of corrosion. If necessary, clean connectors with a spray electrical contact cleaner.

   e. After all is checked out, push the connectors together until they "click" and make sure they are fully engaged and locked together (**Figure 8**).

   f. Never pull on the electrical wires when disconnecting an electrical connector—pull only on the connector plastic housing. See **Figure 9**.

3. Check all electrical components that are grounded to the engine for a good ground.

4. Check all wiring for disconnected wires or short or open circuits.

5. Make sure there is an adequate supply of fuel available to the engine. Make sure the oil tank is properly filled.

6. Check spark plug cable routing (**Figure 10**). Make sure the cables are properly connected to their respective spark plugs.

> *CAUTION*
> *Before removing spark plugs, blow away any dirt that has accumulated next to the spark plug base. The dirt could fall into the cylinder when the plug is removed, causing serious engine damage.*

7. Remove both spark plugs, keeping them in order. Check the condition of each plug. See Chapter Three.

8. Make the following spark test:

> *WARNING*
> *During this test do not hold the spark plug, wire or connector with fingers or a serious electrical shock may result. If necessary, use a pair of insulated pliers to hold the spark plug wire.*

   a. Open the shroud.

   b. Remove one of the spark plugs.

   c. Connect the spark plug wire and connector to the spark plug and touch the spark plug base to a good ground like the engine cylinder head. Position the spark plug so you can see the electrode.

   d. Turn the ignition switch ON and set the tether and cut-out switches to the ON position.

   e. Crank the engine over with the pull starter. A fat blue spark should be evident across the spark plug electrode.

   f. If there is no spark or only a weak one, check for loose connections at the coil. If all external wiring connections are good, the problem is most likely in the ignition system.

   g. Turn the ignition switch off.

Switch Tests

Test the following switches as described in Chapter Seven.

   a. Ignition switch.

   b. Tether cut-out switch.

   c. Emergency cut-out switch.

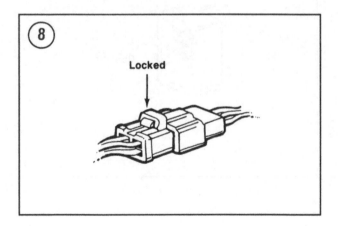

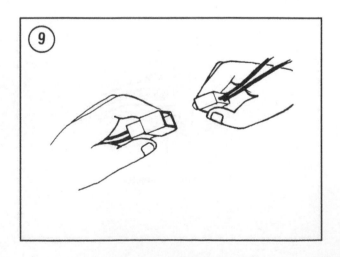

Ignition Testing with the Bombardier Ignition Tester

Before testing the ignition system, note the following:

a. The Bombardier Ignition Tester (part No. 419 0033 00) (**Figure 3**) is required for the

following tests. The tester can be purchased through Ski-Doo dealers. The Bombardier Ignition Tester will include information on the tester and its use. The procedures given should be followed only after acquainting yourself with the test equipment. If you do not have access to this test instrument, have the tests performed by a Ski-Doo dealer.

b. Perform the *Troubleshooting Preparation* procedures in this chapter.

c. The following tests must be made at cranking speed. This means that while it is not necessary to have the engine running when checking the ignition system, it is important to pull vigorously on the starter rope while reading the ignition tester.

d. Each test should be performed 3 times.

e. The ignition tester should be reset after each test by depressing the reset button on the front of the tester.

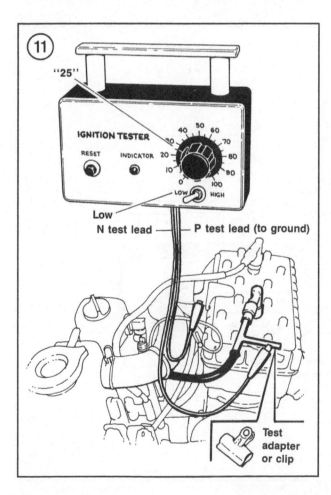

Test 1: ignition coil output

Refer to **Figure 11** for this procedure.

1. Connect the test adapter clip onto one of the spark plug cables next to the plug cap. Connect the N tester lead to the test adapter clip. Connect the P tester lead to a good ground.

2. With the ignition tester leads connected as described in Step 1, switch the tester toggle switch to LOW and the dial to position 25.

> *WARNING*
> *Do not touch any ignition component when cranking the engine in Step 3. A powerful electric shock may occur if you do so.*

3. Turn the ignition switch to ON. Set the tether and cut-out switches to the ON position. Pull the starter rope. If the engine starts, allow it to idle when performing Step 4. If the engine does not start, continue to pull the starter rope when performing Step 4.

4. Note whether or not the indicator light on the tester came on. Turn the engine off (if it started). Repeat this test twice. Interpret results as follows:

a. *Light on:* Ignition system is okay.

b. *Light off:* Perform Test 2.

5. Disconnect the ignition tester leads and detach the test adapter clip.

Test 2: CDI unit control

Refer to **Figure 12** for this procedure.

1. Disconnect the 2-prong connector at the ignition coil (**Figure 13**). This connector has 2 wires: black and white/blue.

2. Connect the ignition coil to the CDI unit with jumper wires.

3. Connect the P tester lead to the black CDI connector wire. Connect the N tester lead to the white/blue CDI connector wire.

4. With the ignition tester leads connected as described in Step 3, switch the ignition tester toggle switch to HIGH and the dial to position 55.

WARNING
Do not touch any ignition component when cranking the engine in Step 5. A powerful electric shock may occur if you do so.

5. Turn the ignition switch to ON. Set the tether and cut-out switches to the ON position. Pull the starter rope. If the engine starts, allow it to idle when performing Step 6. If the engine does not start, continue to pull the starter rope when performing Step 6.

6. Note whether or not the indicator light on the tester came on. Turn the engine off (if it started). Repeat this test twice. Interpret results as follows:

a. *Light on:* Perform Test 3.

b. *Light off:* The ignition coil may be faulty. Substitute with a known good coil and retest. If the light is still off after repeating test 3 times with a new coil, perform Test 3.

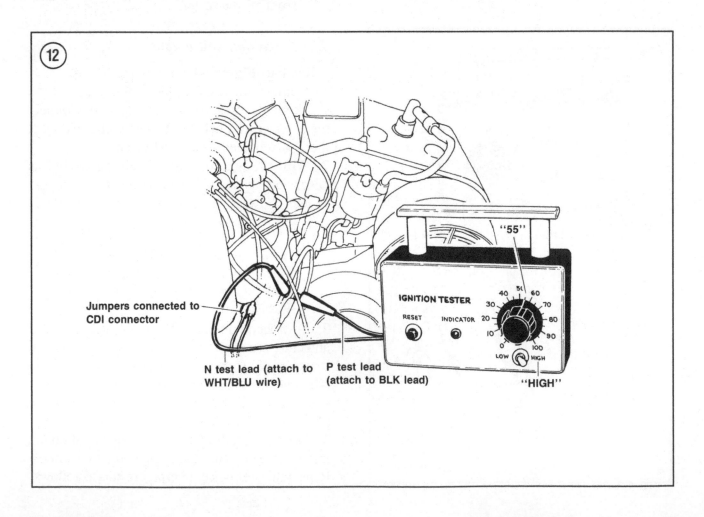

(12)

Jumpers connected to CDI connector

N test lead (attach to WHT/BLU wire)

P test lead (attach to BLK lead)

IGNITION TESTER

RESET   INDICATOR

"55"

"HIGH"

2

7. Disconnect the ignition tester leads and remove the jumper cable. Reconnect the ignition coil and CDI unit connector.

Test 3: high speed charge coil

Refer to **Figure 14** for this procedure.

1. Disconnect the 3-prong connector between the ignition module and magneto. This connector has 3 wires: black, black/red and black/white.

2. Connect the P tester lead to the black/white wire connector on the magneto harness side. Connect the N tester lead to the black/red wire connector on the magneto harness side.

3. With the ignition tester leads connected as described in Step 2, switch the ignition tester toggle switch to LOW and the dial to position 80 (1985-1988) or 70 (1989).

> **WARNING**
> *Do not touch any ignition component when cranking the engine in Step 4. A powerful electric shock may occur if you do so.*

> **WARNING**
> *Do not touch tester lead P when cranking the engine in Step 4. Do not allow either tester lead to touch any metallic object when performing Step 4.*

4. Turn the ignition switch to ON. Set the tether and cut-out switches to their OFF positions. Pull the starter rope.

5. Note whether or not the indicator light on the tester came on. Repeat this test twice. Interpret results as follows:
   a. *Light on:* High speed charge coil is working properly.
   b. *Light off:* The high speed charge coil is faulty. Replace the coil as described in Chapter Seven.

6. Disconnect the ignition tester leads. Reconnect the ignition module 3-prong connector.

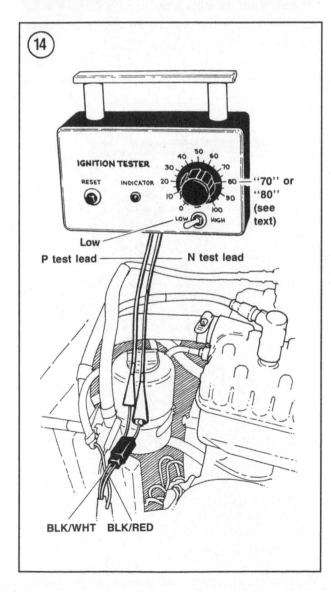

Test 4: low speed charge coil

Refer to **Figure 15** for this procedure.

1. Disconnect the 3-prong connector between the ignition module and magneto. This connector has 3 wires: black, black/red and black/white.

2. Connect the P tester lead to the black wire connector on the magneto harness side. Connect the N tester lead to the black/red wire connector on the magneto harness side.

3. With the ignition tester leads connected as described in Step 2, switch the ignition tester toggle switch to LOW and the dial to position 80.

> *WARNING*
> *Do not touch any ignition component when cranking the engine in Step 4. A powerful electric shock may occur if you do so.*

4. Turn the ignition switch to ON. Set the tether and cut-out switches to their OFF positions. Pull the starter rope.

5. Note whether or not the indicator light on the tester came on. Repeat this test twice. Interpret results as follows:
   a. *Light on:* Low speed charge coil is working properly.
   b. *Light off:* The low speed charge coil is faulty. Replace the coil as described in Chapter Seven.

6. Disconnect the ignition tester leads. Reconnect the ignition module 3-prong connector.

Test 5: lighting coil

1. Disconnect the wiring harness junction block at the engine bulkhead. See **Figure 5**.

2. Connect the P tester lead to the yellow/black wire connector on the magneto harness side. Connect the N tester lead to the yellow wire connector on the magneto harness side.

3. With the ignition tester leads connected as described in Step 2, switch the ignition tester toggle switch to LOW and the dial to position 70.

4. Pull the starter rope and note whether or not the indicator light on the tester came on. Repeat this test twice. Interpret results as follows:
   a. *Light on:* Lighting coil is working properly.
   b. *Light off:* Lighting coil is faulty. Replace the coil as described in Chapter Seven.

5. Disconnect the ignition tester leads. Reconnect the wiring harness junction block (**Figure 5**).

Ignition Component Resistance Test

An accurate ohmmeter will be required to perform the following tests. When switching between ohmmeter scales, always cross the test leads and zero the needle to assure a correct reading.

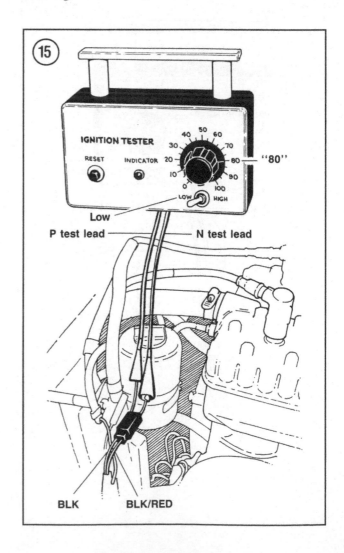

Ignition coil test

1. Open the shroud.

2. Locate the ignition coil. It is mounted on the bulkhead (**Figure 13**). Disconnect the ignition coil 2-prong connector (**Figure 16**).

3. Check ignition coil primary resistance as follows:

  a. Switch a low-reading ohmmeter to the R × 1 scale.

  b. Measure resistance between the black and white/blue coil terminals (**Figure 16**). Refer to **Table 1** for specification.

  c. Disconnect the meter leads.

4. Check ignition coil secondary resistance as follows:

  a. Remove the spark plug cap from each secondary cable (**Figure 16**).

  b. Switch an ohmmeter to the R × 100 scale.

  c. Measure resistance between each secondary lead (spark plug lead) (**Figure 16**). Refer to **Table 1** for specifications.

5. Check ignition coil insulation as follows:

  a. Switch an ohmmeter to the R × 1 scale.

  b. Measure resistance between the white/blue connector wire and each secondary lead. The meter should read infinity.

  c. Measure resistance between the white/blue wire and the ignition coil core. The meter should read infinity.

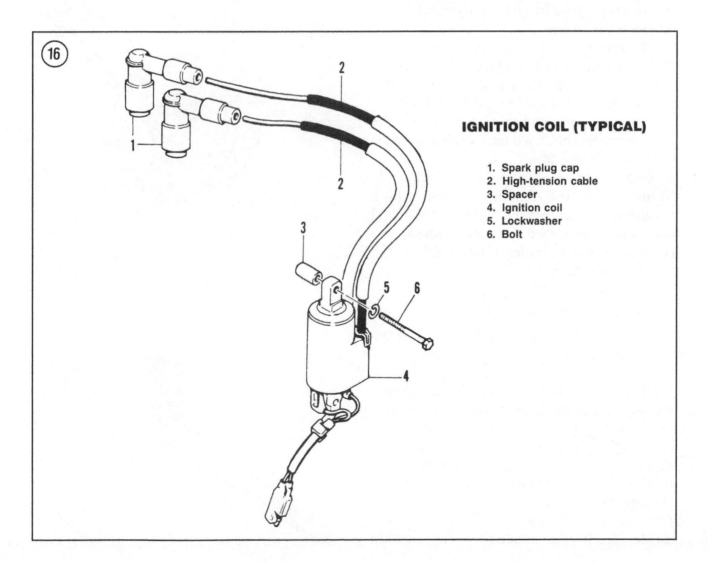

IGNITION COIL (TYPICAL)

1. Spark plug cap
2. High-tension cable
3. Spacer
4. Ignition coil
5. Lockwasher
6. Bolt

6. If the coil readings are not as specified in Steps 3-5, replace the coil. See Chapter Seven.

NOTE
Normal resistance in both the primary and secondary coil winding is not a guarantee that the unit is working properly; only an operational spark test can tell if a coil is producing an adequate spark from the input voltage. A Ski-Doo dealer may have the equipment to test the coil's output. If not, substitute a known good coil to see if the problem goes away.

High speed coil testing

The high speed coil (A, **Figure 17**) is mounted on the stator plate behind the flywheel.
1. Open the shroud.
2. Disconnect the 3-prong connector between the ignition module and magneto. This connector has 3 wires: black, black/red and black/white.
3. Switch an ohmmeter to the R × 1 scale.
4. Connect an ohmmeter between the black/white and black/red leads leading from the magneto (**Figure 18**).
5. Compare the reading to the specification in **Table 2**. If the reading is not within specifications, replace the high/low speed coil assembly as described in Chapter Seven.
6. Reconnect the 3-prong connector.
7. Close the shroud.

Low speed coil testing

The low speed coil (A, **Figure 17**) is mounted on the stator plate behind the flywheel.
1. Open the shroud.
2. Disconnect the 3-prong connector between the ignition module and magneto. This connector has 3 wires: black, black/red and black/white.
3. Switch an ohmmeter to the R × 100 scale.
4. Connect an ohmmeter between the black and black/red leads leading from the magneto (**Figure 18**).
5. Compare the reading to the specification in **Table 2**. If the reading is not within

specifications, replace the high/low speed coil assembly as described in Chapter Seven.
6. Reconnect the 3-prong connector.
7. Close the shroud.

Lighting coil testing

The lighting coil (B, **Figure 17**) is mounted on the stator plate behind the flywheel.
1. Open the shroud.
2. Disconnect the 3-prong connector between the ignition module and magneto. This connector has 3 wires: black, black/red and black/white. Disconnect the black/yellow, yellow and yellow/black wire magneto connectors.

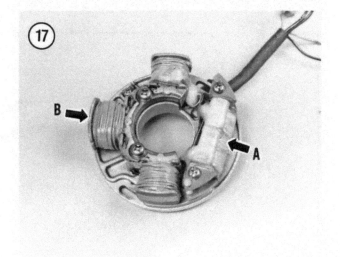

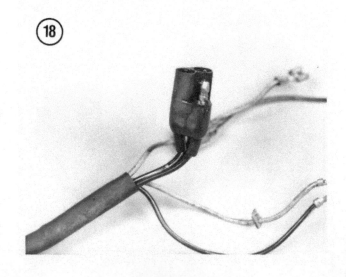

3. Switch an ohmmeter to the R × 1 scale.

4. Connect an ohmmeter between the yellow/black and yellow leads leading from the magneto (**Figure 18**).

5. Compare the reading to the specification in **Table 2**. If the reading is not within specification, replace the lighting coil assembly as described in Chapter Seven.

6. Reconnect the 3-prong connector and the 3 wire connectors.

7. Close the shroud.

VOLTAGE REGULATOR

If you are experiencing blown bulbs or if all of the lights are dim (filaments barely light), test the voltage regulator (**Figure 19**) as follows. In addition, check the bulb filament; an overcharged condition will usually melt the filament rather then break it.

1. Position the snowmobile so that the ski tips are placed against a stationary object. Raise the rear of the snowmobile so that the track is clear of the ground.

2. Open the shroud and secure it so that it cannot fall.

3. Set a voltmeter to the AC25 volt scale. Then connect the red voltmeter lead to the white/blue (1985-1986) or yellow/black (1987-1989) low beam wire at the headlight bulb connector. Connect the black voltmeter lead to a good ground.

WARNING
When performing the following steps, ensure that the track area is clear and that no one walks behind the track or serious injuries may result.

WARNING
Never lean into the snowmobile's engine compartment while wearing a scarf or other loose clothing when the engine is running or when the driver is attempting to start the engine. If the scarf or clothing should catch in the drive belt or clutch, severe injury or death could occur. Make sure the pulley guard is in place.

4. Have an assistant start the engine. When starting the engine, do not use the throttle to raise the rpm excessively.

5. Slowly increase the engine rpm and note the voltmeter reading. If the voltmeter reads more than 15 volts, replace the voltage regulator. See Chapter Seven.

6. Turn the engine off and disconnect the voltmeter.

7. Close the shroud and lower the snowmobile track to the ground.

FUEL SYSTEM

Many snowmobile owners automatically assume that the carburetor is at fault when the engine does not run properly. While fuel system problems are not uncommon, carburetor adjustment is seldom the answer. In many cases, adjusting the carburetor only compounds the problem by making the engine run worse.

Fuel system troubleshooting should start at the fuel tank and work through the system, reserving the carburetor as the final point. Most fuel system problems result from an empty fuel tank, a plugged fuel filter, malfunctioning fuel pump or sour fuel. **Figure 20** provides a series of symptoms and causes that can be useful in localizing fuel system problems.

Carburetor chokes can also present problems. A choke stuck open will show up as a hard

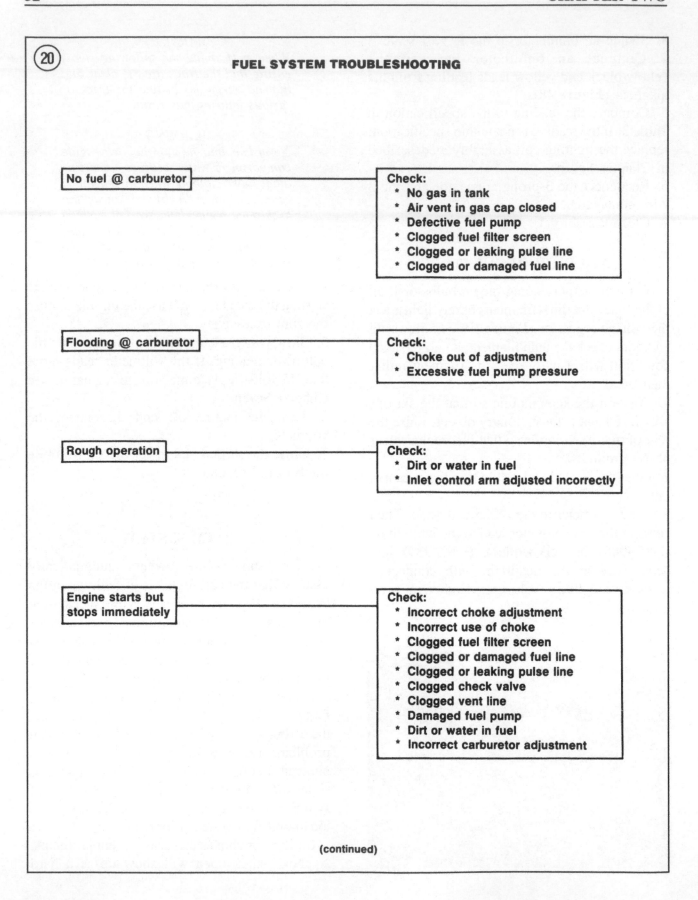

⑳   FUEL SYSTEM TROUBLESHOOTING

No fuel @ carburetor ────────────────

Check:
* No gas in tank
* Air vent in gas cap closed
* Defective fuel pump
* Clogged fuel filter screen
* Clogged or leaking pulse line
* Clogged or damaged fuel line

Flooding @ carburetor ────────────────

Check:
* Choke out of adjustment
* Excessive fuel pump pressure

Rough operation ────────────────

Check:
* Dirt or water in fuel
* Inlet control arm adjusted incorrectly

Engine starts but stops immediately ────────────────

Check:
* Incorrect choke adjustment
* Incorrect use of choke
* Clogged fuel filter screen
* Clogged or damaged fuel line
* Clogged or leaking pulse line
* Clogged check valve
* Clogged vent line
* Damaged fuel pump
* Dirt or water in fuel
* Incorrect carburetor adjustment

(continued)

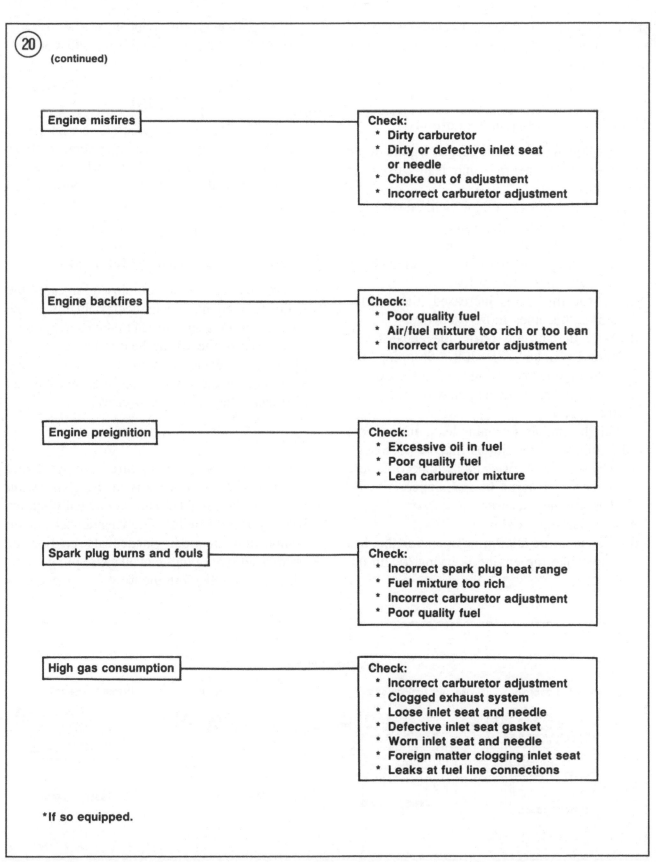

(20) (continued)

Engine misfires ——————————— Check:
* Dirty carburetor
* Dirty or defective inlet seat or needle
* Choke out of adjustment
* Incorrect carburetor adjustment

Engine backfires ——————————— Check:
* Poor quality fuel
* Air/fuel mixture too rich or too lean
* Incorrect carburetor adjustment

Engine preignition ——————————— Check:
* Excessive oil in fuel
* Poor quality fuel
* Lean carburetor mixture

Spark plug burns and fouls ——————————— Check:
* Incorrect spark plug heat range
* Fuel mixture too rich
* Incorrect carburetor adjustment
* Poor quality fuel

High gas consumption ——————————— Check:
* Incorrect carburetor adjustment
* Clogged exhaust system
* Loose inlet seat and needle
* Defective inlet seat gasket
* Worn inlet seat and needle
* Foreign matter clogging inlet seat
* Leaks at fuel line connections

*If so equipped.

2

starting problem; one that sticks closed will result in a flooding condition. Check choke operation and adjustment.

Identifying Carburetor Conditions

The following list can be used as a guide when trying to determine rich and lean carburetor conditions.

When the engine is running rich, one or more of the following conditions may be present:

a. The spark plug(s) will foul.
b. The engine will miss and run rough when it is running under a load.
c. As the throttle is increased, the exhaust smoke becomes more excessive.
d. With the throttle open, the exhaust will sound choked or dull. Stopping the snowmobile and trying to clear the exhaust with the throttle held open will not clear up the sound.

When the engine is running lean, one or more of the following conditions may be present:

a. The spark plug firing end will become very white or blistered in appearance.
b. The engine overheats.
c. Acceleration is slower.
d. Flat spots are felt during operation that feel much like the engine is trying to run out of gas.

e. Engine power is reduced.
f. At full throttle, engine rpm will not hold steady.

ENGINE

Engine problems are generally symptoms of something wrong in another system, such as ignition, fuel or starting. If properly maintained and serviced, the engine should experience no problems other than those caused by age and wear.

Overheating and Lack of Lubrication

Overheating and lack of lubrication cause the majority of engine mechanical problems. Make sure the cooling system isn't damaged. Using a spark plug of the wrong heat range can burn a piston. Incorrect ignition timing, a faulty cooling system or an excessively lean fuel mixture can also cause the engine to overheat.

Preignition

Preignition is the premature burning of fuel and is caused by hot spots in the combustion chamber (**Figure 21**). The fuel actually ignites before it is supposed to. Glowing deposits in the combustion chamber, inadequate cooling or overheated spark plugs can all cause preignition. This is first noticed in the form of a power loss

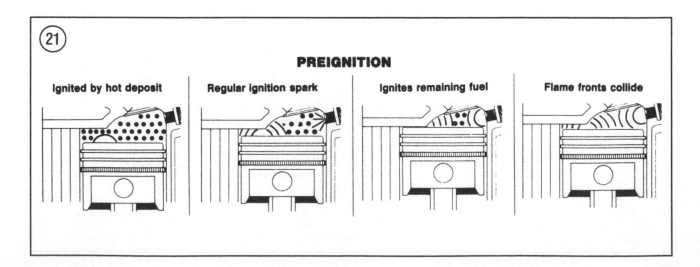

PREIGNITION

| Ignited by hot deposit | Regular ignition spark | Ignites remaining fuel | Flame fronts collide |

2

but will eventually result in extended damage to the internal parts of the engine because of higher combustion chamber temperatures.

Detonation

Commonly called "spark knock" or "fuel knock," detonation is the violent explosion of fuel in the combustion chamber prior to the proper time of combustion (**Figure 22**). Severe damage can result. Use of low octane gasoline is a common cause of detonation.

Even when high octane gasoline is used, detonation can still occur if the engine is improperly timed. Other causes are over-advanced ignition timing, lean fuel mixture at or near full throttle, inadequate engine cooling, cross-firing of spark plugs, or the excessive accumulation of deposits on piston and combustion chamber.

Since the snowmobile engine is covered, engine knock or detonation is likely to go unnoticed, especially at high engine rpm when wind noise is also present. Such inaudible detonation, as it is called, is usually the cause when engine damage occurs for no apparent reason.

Poor Idling

A poor idle can be caused by improper carburetor adjustment, incorrect timing or

ignition system malfunctions. Check the carburetor pulse and vent lines for an obstruction. Also check for loose carburetor mounting bolts or a faulty carburetor flange gasket.

Misfiring

Misfiring can result from a weak spark or a dirty spark plug. Check for fuel contamination. If misfiring occurs only under heavy load, as when accelerating, it is usually caused by a defective spark plug. Check for fuel contamination.

Water Leakage in Cylinder

The fastest and easiest way to check for water leakage in a cylinder is to check the spark plugs. Water will clean a spark plug. If one of the plugs is clean and the other is dirty, there is most likely a water leak in the cylinder with the clean plug.

To check further, install a dirty plug in each cylinder. Run the engine for 5-10 minutes. Shut the engine off and remove the plugs. If one plug is clean and the other is dirty (or if all plugs are clean), a water leak in the cylinder is the problem.

Flat Spots

If the engine seems to die momentarily when the throttle is opened and then recovers, check

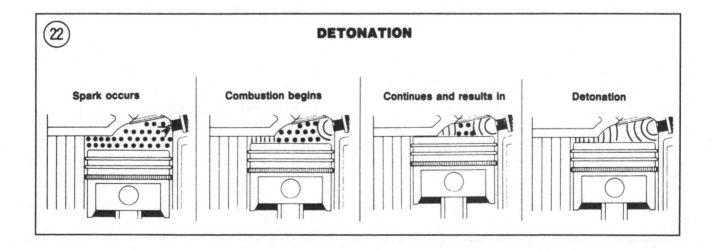

DETONATION

| Spark occurs | Combustion begins | Continues and results in | Detonation |

㉓   **LOW ENGINE POWER**

Ignition system trouble ────────────────── **Check:**
                                              * **Faulty ignition coil**
                                              * **Incorrect ignition timing**
                                              * **Incorrect spark plug heat range**
                                              * **Loose wiring connectors**

Fuel system trouble ────────────────── **Check:**
                                          * **Contaminated fuel filter**
                                          * **Contaminated fuel filter screen**
                                          * **Throttle valve does not open fully**
                                          * **Clogged high speed nozzle**
                                          * **Clogged pulse line**
                                          * **Leaking pulse line**
                                          * **Insuffient fuel supply**
                                          * **Faulty check valve diaphragm**
                                          * **Faulty regulator diaphragm**
                                          * **Faulty pulse diaphragm**

Overheating ────────────────── **Check:**
                                  * **See "Ignition system trouble"**
                                  * **See "Fuel system trouble"**
                                  * **Incorrect ignition timing**
                                  * **Excessive carbon buildup in combustion chamber**
                                  * **Incorrect fuel/oil mixture**
                                  * **Incorrect oil type**
                                  * **Incorrect fuel type**
                                  * **Incorrect carburetor adjustment**
                                  * **Clogged or leaking cooling system**
                                  * **Clogged exhaust system**

Other ────────────────── **Check:**
                            * **Dirt or water in fuel**
                            * **Clogged exhaust system**

for a dirty or contaminated carburetor, water in the fuel or an excessively lean or rich low speed mixture.

Power Loss

Several factors can cause a lack of power and speed. Look for air leaks in the fuel line or fuel pump, a clogged fuel filter or a throttle slide that does not operate properly. Dynamically (engine running) check the ignition timing at full advance. See Chapter Three. This will allow you to make sure that the ignition system is operating properly. If the ignition timing is incorrect dynamically, but was properly set with a dial indicator, there may be a problem with an ignition component.

A piston or cylinder that is galling, incorrect piston clearance or worn or sticky piston rings

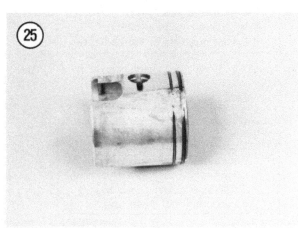

may be responsible. Look for loose bolts, defective gaskets or leaking machined mating surfaces on the cylinder head, cylinder or crankcase. Also check the crankshaft seals. Refer to *Two-stroke Pressure Testing* in this chapter.

Exhaust fumes leaking within the engine compartment can slow and even stop the engine.

Refer to **Figure 23** for a general listing of engine troubles.

Piston Seizure

Piston seizure or galling is the transfer of metal from the piston to the cylinder bore. Friction causes piston seizure. This is caused by one or more pistons with incorrect bore clearances, piston rings with an improper end gap, compression leak, incorrect type of oil, spark plug of the wrong heat range, incorrect ignition timing or an incorrectly operating oil injection pump. Overheating from any cause may result in piston seizure.

A noticeable reduction of speed may be your first sign of seizure while immediate stoppage indicates full lockup. A top end rattle should be considered as an early sign of seizure.

When diagnosing piston seizure, the pistons themselves can be used to troubleshoot and determine the failure cause. High cylinder temperatures normally cause seizure above the piston rings while seizure below the piston rings are usually caused by a lack of proper lubrication.

See **Figure 24** and **Figure 25** for examples of piston seizure.

Excessive Vibrations

Excessive vibrations may be caused by loose engine, suspension or steering mount bolts.

Engine Noises

Experience is needed to diagnose accurately in this area (**Figure 26**). Noises are difficult to differentiate and even harder to describe.

TWO-STROKE PRESSURE TESTING

Many owners of 2-stroke engines are plagued by hard starting and generally poor running, for which there seems to be no cause. Carburetion and ignition may be good, and a compression test may show that all is well in the engine's upper end.

What a compression test does *not* show is lack of primary compression. In a 2-stroke engine, the crankcase must be alternately under pressure and vacuum. After the piston closes the intake port, further downward movement of the piston causes the entrapped mixture to be pressurized so that it can rush quickly into the cylinder when the scavenging ports are opened. Upward piston movement creates a lower vacuum in the crankcase, enabling fuel/air mixture to pass in from the carburetor.

NOTE
The operational sequence of a 2-stroke engine is illustrated in Chapter One.

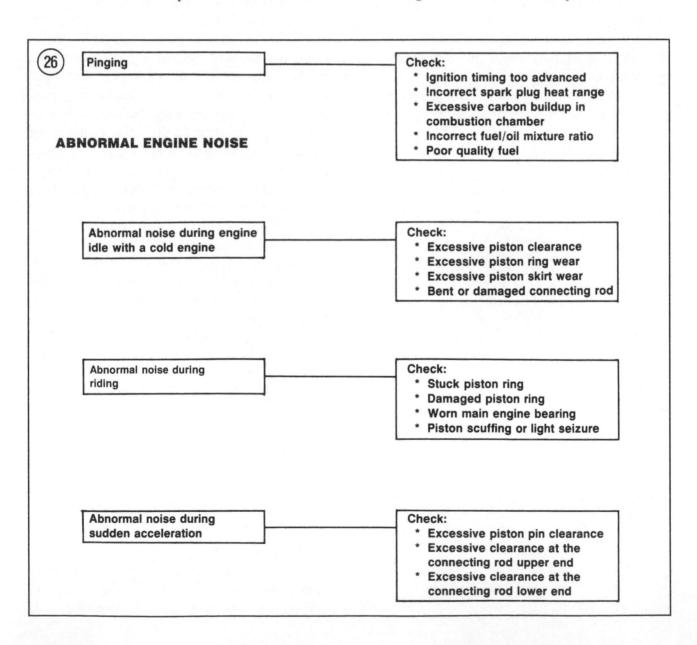

If crankcase seals or cylinder gaskets leak, the crankcase cannot hold pressure or vacuum and proper engine operation becomes impossible. Any other source of leakage such as a defective cylinder base gasket or porous or cracked crankcase castings will result in the same conditions.

It is possible, however, to test for and isolate engine pressure leaks.

The test is simple but requires special equipment. A typical 2-stroke pressure test kit is shown in **Figure 27**. Briefly, what is done is to seal off all natural engine openings, then apply air pressure. If the engine does not hold air, a leak or leaks is indicated. Then, it is only necessary to locate and repair all leaks.

The following procedure describes a typical pressure test.

NOTE
Because of the labyrinth seal on the crankshaft, the cylinders cannot be checked individually. When you pump up one cylinder you will also be pumping up the opposite cylinder. Thus, both cylinders must be blocked before applying pressure during testing.

1. Remove the carburetors as described in Chapter Six.
2. Take a rubber plug and insert it tightly in the intake manifolds.

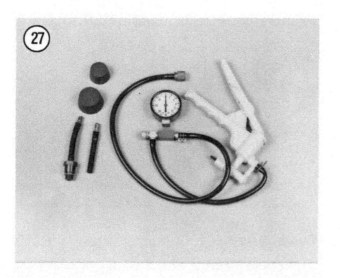

3. Remove the exhaust pipe and block off the exhaust ports using suitable adapters and fittings.
4. Remove one of the spark plugs and install the pressure gauge adapter into the spark plug hole. Connect the pressurizing lever and gauge to the pressure fitting installed where the spark plug was, then continue to squeeze the lever until the gauge indicates approximately 9 psi.
5. Observe the pressure gauge. If the engine is in good condition, the pressure should not drop more than 1 1/2 to 2 psi in several minutes. Any pressure loss of 1 psi in one minute indicates serious sealing problems.

Before condemning the engine, first be sure that there are no leaks in the test equipment or sealing plugs. If the equipment shows no signs of leakage, go over the entire engine carefully. Large leaks can be heard; smaller ones can be found by going over every possible leakage source with a small brush and soap suds solution. Possible leakage points are listed below:
  a. Crankshaft seals.
  b. Spark plug(s).
  c. Cylinder head joint.
  d. Cylinder base joint.
  e. Carburetor base joint.
  f. Crankcase joint.

POWER TRAIN

The following items provide a starting point from which to troubleshoot power train malfunctions. The possible causes for each malfunction are listed in a logical sequence.

Drive Belt Not Operating Smoothly in Primary Sheave

  a. Drive sheave face is rough, grooved, pitted or scored.
  b. Defective drive belt.

Uneven Drive Belt Wear

  a. Misaligned primary and secondary sheaves.
  b. Loose engine mounts.

Glazed Drive Belt

a. Excessive slippage caused by stuck or frozen track.
b. Engine idle speed too high.

Drive Belt Too Tight at Idle

a. Engine idle speed too high.
b. Incorrect sheave distance.
c. Incorrect belt length.

Drive Belt Edge Cord Failure

a. Misaligned primary and secondary sheaves.
b. Loose engine mounts.

Brake Not Holding Properly

a. Incorrect brake cable adjustment.
b. Worn brake pads.
c. Worn brake disc.
d. Oil saturated brake pads.
e. Sheared key on brake disc.
f. Incorrect brake adjustment.

Brake Not Releasing Properly

a. Weak or broken return spring.
b. Bent or damaged brake lever.
c. Incorrect brake adjustment.

Excessive Chaincase Noise

a. Incorrect chain tension.
b. Excessive chain stretch.
c. Worn sprocket teeth.
d. Damaged chain and/or sprockets.

Chain Slippage

a. Incorrect chain tension.
b. Excessive chain stretch.
c. Worn sprocket teeth.

Leaking Chaincase

a. Loose chaincase cover mounting bolts.
b. Damaged chaincase cover gasket.
c. Damaged chaincase oil seal(s).
d. Cracked or broken chaincase.

Rapid Chain and Sprocket Wear

a. Insufficient chaincase oil level.
b. Broken chain tensioner.
c. Misaligned sprockets.

Primary Sheave Engages Before Engagement RPM

a. Worn spring.
b. Incorrect weight.

Primary Sheave Engages After Engagement RPM

a. Incorrect spring.
b. Worn or damaged secondary sheave buttons.

Erratic Shifting

a. Worn rollers and bushings.
b. Scuffed or damaged weights.
c. Dirty primary sheave assembly.
d. Worn or damaged secondary sheave buttons.

Engine Bogs During Engagement

a. Incorrect secondary sheave width adjustment.
b. Drive belt worn too thin.
c. Incorrect sheave distance.

Primary or Secondary Sheave Sticks

a. Damaged sheave assembly.
b. Moveable sheave damaged.
c. Dirty sheave assembly.

SKIS AND STEERING

The following items provide a starting point from which to troubleshoot ski and steering malfunctions. The possible causes for each malfunction are listed in a logical sequence.

Loose Steering

a. Loose steering post bushing.

b. Loose steering post or steering column fasteners.
c. Loose tie rod ends.
d. Worn spindle bushings.
e. Stripped spindle splines.

Unequal Steering

a. Improperly adjusted tie rods.
b. Improperly installed steering arms.
c. Damaged steering components.

Rapid Ski Wear

a. Skis misaligned.
b. Worn out ski wear rods (skags).

TRACK ASSEMBLY

The following items provide a starting point from which to troubleshoot track assembly malfunctions. Also refer to *Inspection* under *Track* in Chapter Fourteen.

Frayed Track Edge

a. Incorrect track alignment.
b. Track contacts rivets in tunnel area (incorrect rivets previously installed).

Track Grooved on Inner Surface

a. Track too tight.
b. Frozen rear idler shaft bearing.

Track Drive Ratcheting

a. Track too loose.
b. Drive sprockets misaligned.
c. Damaged drive sprockets.

Rear Idlers Turning on Shaft

Frozen rear idler shaft bearings.

Table 1 IGNITION COIL TEST SPECIFICATIONS

Primary coil resistance	
All models	0.23-0.43 ohms
Secondary coil resistance	
1985-1986	2.45-4.55 K ohms
1987-on	3.85-7.15 K ohms

Table 2 MAGNETO COIL TESTING

Low speed charge coil	120-180 ohms
High speed charge coil	2.8-4.2 ohms
Lighting coil	0.21-0.31 ohms

Chapter Three

Lubrication, Maintenance And Tune-Up

This chapter covers all of the regular maintenance required to keep your snowmobile in top shape. Regular maintenance is the best guarantee of a troublefree, long lasting vehicle. Because snowmobiles are high-performance vehicles, proper lubrication, maintenance and tune-ups have become increasingly important as ways in which you can maintain a high level of performance, extend engine life and extract the maximum economy of operation. You can do your own lubrication, maintenance and tune-ups if you follow the correct procedures and use common sense. Always remember that engine damage can result from improper tuning and adjustment. In addition, where special tools or testers are called for during a particular maintenance or adjustment procedure, the tool should be used or you should refer service to a qualified dealer or repair shop.

The following information is based on recommendations from Ski-Doo that will help you keep your snowmobile operating at its peak level.

Tables 1-11 are at the end of the chapter.

NOTE
Be sure to follow the correct procedure and specifications for your specific
model and year. Also use the correct quantity and type of fluid as indicated in the tables.

PRE-RIDE CHECKS

The machine should be checked before each ride. Refer to Chapter One.

FLUID CHECKS

Vital fluids should be checked daily or before each ride to assure proper operation and prevent severe component damage. Refer to **Table 1**.

BREAK-IN PROCEDURE

Following cylinder servicing (boring, honing, new rings, etc.) and major lower end work, the engine should be broken in just as if it were new. The performance and service life of the engine depends greatly on a careful and sensible break-in.

For the first 10-15 hours of operation, no more than 3/4 throttle should be used and the speed should be varied as much as possible. Prolonged steady running at one speed, no matter how moderate, is to be avoided, as is hard acceleration. Wet snow conditions should also be avoided during break-in.

To assure adequate protection to the engine during break-in, 500 cc (16.9 oz.) of Bombardier Injection Oil (part No. 496 0133 00) should be added to the first tank of gas. This oil will be used *together* with the oil supplied by the injection system. Throughout the break-in period, check the oil injection reservoir tank to make sure the injection system is working (oil level diminishing).

After the initial 10-15 hours, all engine and chassis fasteners should be checked for tightness. If the snowmobile is going to be used in extreme conditions, you may want to increase the break-in a few hours. After break-in, retighten the cylinder head nuts as described in this chapter and perform the *10-Hour Inspection* as described in the following section.

NOTE
After the break-in is complete, install new spark plugs as described in this chapter.

10-HOUR INSPECTION

Ski-Doo lists a 10-hour inspection that is to be performed on a new snowmobile after the first 10 hours of operation or 30 days of purchase, whichever comes first. The 10-hour inspection checks are listed in **Table 2**. While this inspection has probably been performed on your machine, the engine inspection checks should be

repeated whenever the engine top- or bottom-end has been overhauled or the engine removed from the frame. Likewise, chassis and steering inspection procedures should be performed after major service has been performed to these components.

Periodic maintenance procedures are listed in **Table 3**.

LUBRICATION

WARNING
Serious fire hazards always exist around gasoline. Do not allow any smoking in areas where fuel is being mixed or while refueling your machine. Always have a fire extinguisher, rated for gasoline and electrical fires, within reach just to play it safe.

Proper Fuel Selection

Two-stroke engines are lubricated by mixing oil with the fuel. The various components of the engine are thus lubricated as the fuel/oil mixture passes through the crankcase and cylinders. All models are equipped with an oil injection system. Pre-mixing fuel is not required on any of the models covered in this manual except during engine break-in. See *Break-in Procedure* in this chapter.

Table 4 lists fuel recommendations that should be followed to prevent engine knock and assure proper operation.

Engine Oil Tank

An oil injection system is used on all models. During engine operation, oil is automatically injected into the engine at a variable ratio depending on engine rpm.

Oil capacity in the reservoir tank (A, **Figure 1**) should be checked daily and during all fuel stops.

The oil tank is equipped with an oil level sensor (B, **Figure 1**) that is wired to the injection oil level pilot lamp on the instrument panel (**Figure 2**). When the oil level in the tank reaches a specified low point, the pilot lamp will light.

> *NOTE*
> *The oil injection level pilot lamp (**Figure 2**) lights up whenever the brake lever is operated. If the lamp does not light up during brake operation, replace the lamp as described in Chapter Seven.*

When the oil level is low, perform the following.
1. Open the shroud.
2. Remove the oil tank fill cap (C, **Figure 1**) and pour in the required amount of two-stroke injection oil specified in **Table 5**. Fill the tank until the oil level is approximately 13 mm (1/2 in.) from the top of the transparent tank.
3. Reinstall the fill cap and close the shroud.

Chaincase Oil

The oil in the chaincase lubricates the chain and sprockets.

Try to use the same brand of oil. Do not mix 2 brand types at the same time as they all vary slightly in their composition.

Oil level check

The chaincase oil level is checked with the dipstick (**Figure 3**).
1. Park the snowmobile on a level surface. Open the shroud.
2. Unscrew and remove the dipstick (**Figure 3**) from the chaincase cover. Wipe the dipstick off and insert it back into the chaincase. Do not screw the dipstick back into the cover.
3. Lift the dipstick out of the chaincase and check the oil level on the dipstick. The oil level should be between the upper and lower marks (**Figure 4**).
4. If the oil level is low, top off with a chaincase oil recommended in **Table 5**. Do not overfill. Recheck the oil level.

> *NOTE*
> *If the oil level is too high, siphon some of the oil out of the chaincase through*

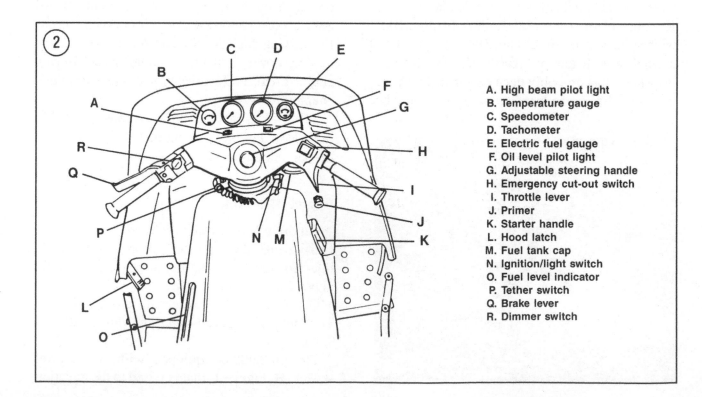

(2)

A. High beam pilot light
B. Temperature gauge
C. Speedometer
D. Tachometer
E. Electric fuel gauge
F. Oil level pilot light
G. Adjustable steering handle
H. Emergency cut-out switch
I. Throttle lever
J. Primer
K. Starter handle
L. Hood latch
M. Fuel tank cap
N. Ignition/light switch
O. Fuel level indicator
P. Tether switch
Q. Brake lever
R. Dimmer switch

the dipstick hole. A discarded plunger found in hand lotion or cleaner containers works well for this.

5. Reinstall the dipstick. Tighten it securely.

Changing

The chaincase oil should be changed once a year.

1. Park the snowmobile on a level surface.
2. Open the shroud.
3. Remove the exhaust pipe and muffler as described in Chapter Six.
4. Place a rug or a number of shop rags underneath the chaincase cover.

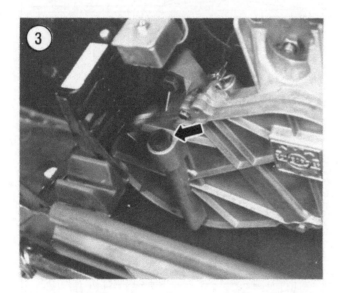

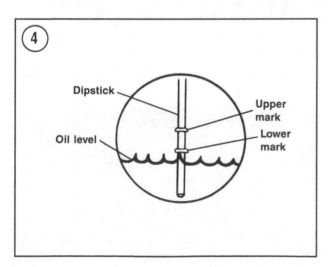

NOTE
The chaincase is filled with oil and the cover is not equipped with a drain plug. To drain the chaincase coil, the chaincase cover must be removed. Try to absorb as much of the oil on the rags as possible.

5. Remove the bolts and washers holding the chaincase cover to the chaincase. Remove the cover (**Figure 5**) and O-ring.
6. Wipe up as much oil as possible with the rags.

NOTE
Place the oil soaked rags into a suitable container until they can be cleaned.

7. Clean the chaincase cover thoroughly. Wipe up as much oil in the bottom of the chaincase as possible.
8. Replace the chaincase cover O-ring if necessary.
9. Install the chaincase cover (**Figure 5**) and O-ring. Install the cover mounting bolts and tighten securely.
10. Remove the chaincase dipstick (**Figure 3**).
11. Insert a funnel into the dipstick hole and fill the chaincase with the correct type (**Table 5**) and quantity (**Table 6**) of chaincase oil.
12. Check the oil level as described in the previous procedure.

Rotary Valve Oil Tank
(1985-1988)

A separate oil tank (**Figure 6**) supplies oil to the rotary valve system on all 1985-1988 models. The oil level in the tank should be checked frequently and filled to the maximum level line (**Figure 7**) with a rotary valve lubricant recommended in **Table 5**. Do not overfill.

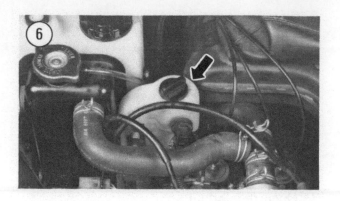

Rotary Valve Lubrication
(1989)

The rotary valve system on all 1989 models is lubricated with oil from the injection oil reservoir. The rotary valve oil tank used on 1985-1988 models has been eliminated. Always make sure the injection oil reservoir is kept full as described in this chapter.

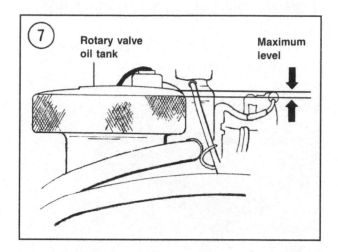

Rotary valve oil tank   Maximum level

Jackshaft Lubrication
(Brake Disc and Secondary Sheave)

The brake disc and secondary sheave should be lubricated so that they can slide freely on the jackshaft. Apply a spray penetrating lubricant at the points indicated in **Figure 8**.

> *CAUTION*
> *Do not overlubricate. Excessive lubrication can cause brake pad or drive belt contamination.*

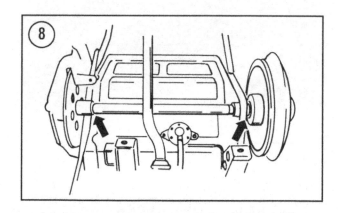

Drive Axle

Lubricate the drive axle with a low-temperature grease (**Table 5**) at the zerk fitting shown in **Figure 9**.

Steering and Suspension Lubrication

Lubricate the following components monthly or after every 40 hours of operation. If the snowmobile is operated under severe service conditions or in wet snow, perform this service more frequently. Refer to Chapter Fourteen for component removal and installation.

a. See **Figure 10**.
b. See **Figure 11**.

3

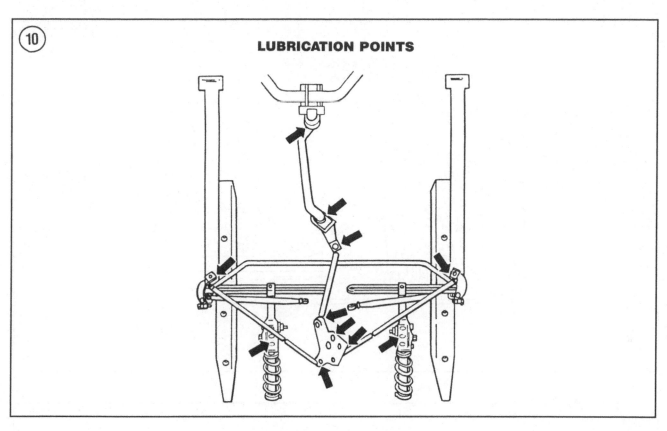

(10) **LUBRICATION POINTS**

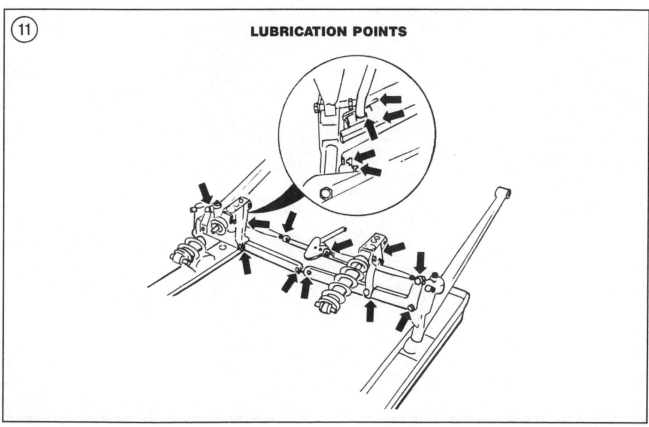

(11) **LUBRICATION POINTS**

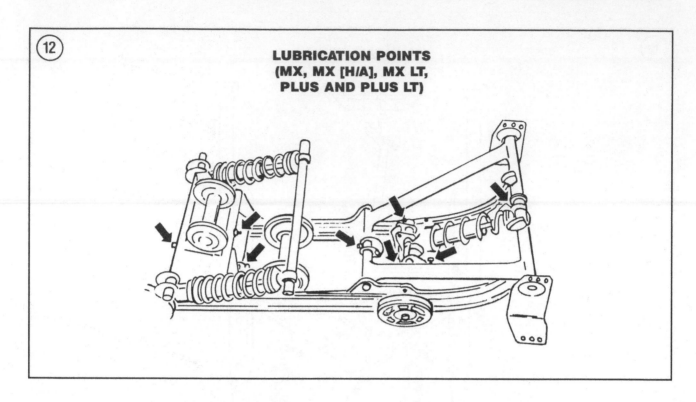

⑫ LUBRICATION POINTS
(MX, MX [H/A], MX LT,
PLUS AND PLUS LT)

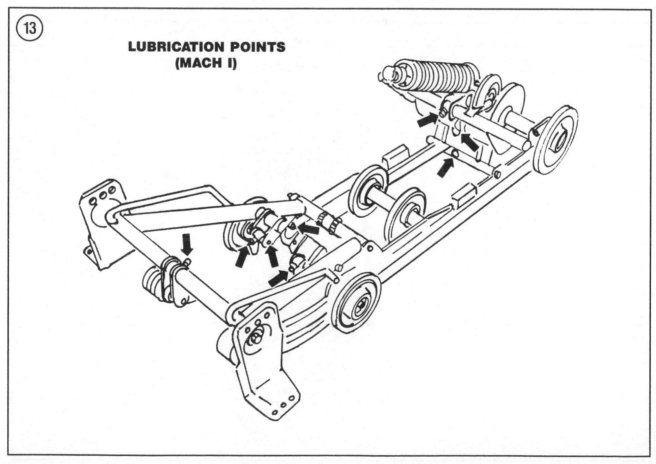

⑬ LUBRICATION POINTS
(MACH I)

Slide Suspension Lubrication

Lubricate the slide suspension with a low-temperature grease monthly or after every 40 hours of operation. If the snowmobile is operated under severe service conditions or in wet snow, perform this service more frequently. Refer to

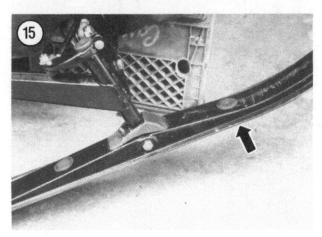

Chapter Fourteen for slide suspension removal and installation procedures.

  a. See **Figure 12** (MX, MX [H/A], MX LT, Plus and Plus LT).

  b. See **Figure 13** (Mach I).

WEEKLY OR EVERY 150 MILES (240 KM) MAINTENANCE

Maintenance intervals are specified in **Table 3**.

Drive Belt Check

Check the drive belt (**Figure 14**) for cracks, fraying or unusual wear as described in Chapter Eleven. Replace the drive belt if its width is less than specified in Chapter Eleven.

Ski and Ski Runner Check

Check the skis (**Figure 15**) for cracks, bending or other signs of damage. Raise the front of the snowmobile so that the skis clear the ground. Check ski movement by pivoting both skis up and down; skis should pivot smoothly with no signs of binding. If a ski is tight, refer to Chapter Fourteen for ski removal, installation and ski pivot bolt tightness.

Excessively worn or damaged ski runners (skags) reduce handling performance and can cause wear to the bottom of the ski. Because track and snow conditions determine runner wear, they should be inspected often. Check the ski runners for wear (**Figure 16**) and replace them if they are more than half worn at any point or cracked. Refer to Chapter Fourteen.

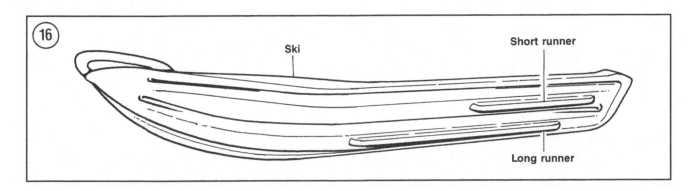

Brake Pad Wear Check

The brake pads should be replaced when the fixed brake pad measures 1 mm (1/32 in.) or less from the inner caliper half as shown in **Figure 17**. Refer to Chapter Twelve for brake service.

MONTHLY OR EVERY 500 MILES (800 KM)

Brake Adjustment

All models covered in this manual are equipped with a self-adjusting brake mechanism. Check brake operation as follows:

1. Apply the brake firmly and measure the distance from the brake lever to the handlebar grip as shown in **Figure 18**. The distance should be approximately 13 mm (1/2 in.).

2. If the distance measured in Step 1 is incorrect, squeeze the brake lever strongly several times to actuate the self-adjusting mechanism. Recheck the measurement.

3. If the distance cannot be corrected by squeezing the brake lever (Step 2), check the brake assembly as described in Chapter Twelve.

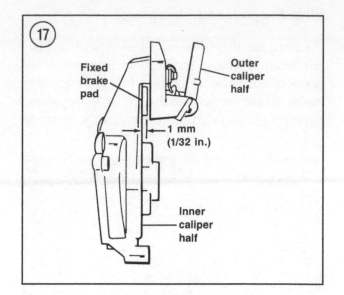

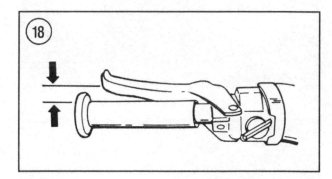

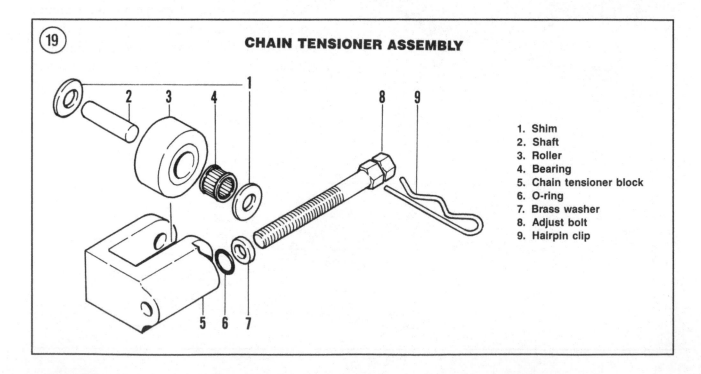

CHAIN TENSIONER ASSEMBLY

1. Shim
2. Shaft
3. Roller
4. Bearing
5. Chain tensioner block
6. O-ring
7. Brass washer
8. Adjust bolt
9. Hairpin clip

3

NOTE
If self-adjuster on brake is difficult to actuate or if you are unsure about its operation, consult a Ski-Doo dealer.

Spark Plugs

The spark plugs should be checked periodically for firing tip condition and gap. Refer to spark plug service under *Engine Tune-up* in this chapter.

Drive Chain Adjustment

The drive chain adjuster mechanism consists of a chain tensioner block, roller, adjust bolt and hair pin (**Figure 19**). Chain tension is maintained by monitoring free play at the secondary sheave (1985-1986) or brake disc (1987-on). The adjust bolt (A, **Figure 20**) threads into the chain tensioner block and is used to adjust roller (B) tension against the drive chain (C). The hair pin (D) locks the adjust bolt in position after the bolt and chaincase lock holes are aligned.

Excessive free play will cause excessive and premature chain and sprocket wear.

The drive chain can be adjusted without having to remove the chaincase cover.

1985-1986

1. Start the engine and run the snowmobile forward slightly. Turn the engine off.
2. Open the shroud.
3. Remove the drive belt guard pins and remove the guard (**Figure 21**).
4. See **Figure 22**. Lightly place a chalk mark on the drive belt. Make another chalk mark on some part of the chassis that aligns with the first mark. Turn the secondary sheave (A) *clockwise* and measure free play (B). Turn the secondary sheave *counterclockwise* and measure free play. The total movement recorded should be within 3-6 mm (1/8-1/4 in.). If the free play is incorrect, perform Step 5. If the free play is correct, perform Step 6.

5. Remove the hair pin (A, **Figure 23**) from the drive chain adjust bolt on the right-hand side of the engine. Turn adjust bolt (B, **Figure 23**) *clockwise* to tighten the drive chain. Reinstall the hair pin by inserting it through the lock hole in the case and bolt and recheck the free play as described in Step 4.

> *NOTE*
> *Figure 23 shows the chain adjuster assembly with the chaincase cover removed. It is not necessary to remove the cover when performing this procedure.*

6. Reinstall the drive belt guard (**Figure 21**) and secure it with both pins.

1987-on

1. Start the engine and run the snowmobile forward slightly. Turn the engine off.
2. Open the shroud.
3. Remove the hair pin (A, **Figure 23**) from the drive chain adjust bolt on the right-hand side of the engine. Turn the adjust bolt (B, **Figure 23**) *clockwise* by hand until it stops. Turn the adjust bolt *counterclockwise* until the hair pin can engage the locking hole in the adjust bolt and install the hair pin.

> *NOTE*
> *Figure 23 shows the chain adjuster assembly with the chaincase cover removed. It is not necessary to remove the cover when performing this procedure.*

4. Turn the brake disc (**Figure 24**) *clockwise* and measure free play. Then turn the brake disc *counterclockwise* and measure free play. The free play should be within 3-5 mm (1/8-13/64 in.). If the free play is incorrect and the adjust bolt is correctly tightened, remove the chaincase and inspect the tensioner assembly and drive chain assembly as described in Chapter Twelve.

Exhaust System

The exhaust system is a vital link to the performance and operation of the Ski-Doo engine. Check the exhaust system from the cylinder exhaust port to the muffler for damaged gaskets, loose or missing fasteners or damaged components. Refer to Chapter Six.

Air Filter Cleaning
(1987-on)

All 1987 and later models are equipped with an air filter at the air intake silencer. Service the air filter as follows:

> *CAUTION*
> *Do not operate the engine on 1987 and later models without the air filter installed or engine damage will result.*

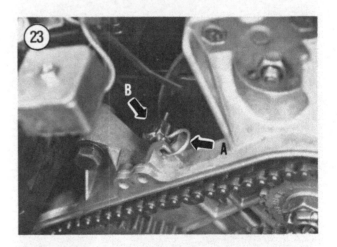

1. Open the shroud.
2. Remove the air filter (**Figure 25**) from the air intake silencer.
3. The filter should be free of all snow and water contamination. If necessary, clean the filter with solvent and blow dry with compressed air. Do not use heat to dry the air filter. If the air filter is torn or otherwise damaged, replace it.
4. When the air filter is thoroughly dry, carefully install it so that it fits snugly in the air intake silencer opening.

NOTE
Do not leave the shroud open during a snowfall where snow can collect onto and

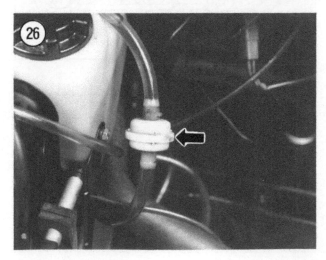

block the air filter element. If you must work in the engine compartment during these conditions, cover the air filter with some type of plate.

Oil Injection Oil Filter Inspection/Replacement

An inline oil filter (**Figure 26**) is installed between the injection oil tank and the oil pump. A clip is used on each end of the filter to prevent accidental hose separation. The oil filter will prevent contaminants from obstructing oil passages in the oil pump.

The oil filter should be inspected once a month for contamination buildup or obstruction. If the oil filter is contaminated, it should be replaced and at the same time the oil system from the reservoir tank to the oil pump should be checked for contamination and thoroughly flushed as required. After the system has been cleaned, install a new oil filter and bleed the oil pump as described in Chapter Eight.

When replacing the oil filter, note the following:
   a. Place a cloth underneath the oil filter to absorb oil when the filter is removed.
   b. Slide each hose retaining clip off the filter nipple with a pair of pliers and disconnect the hoses from the filter.
   c. Check the hose clips for fatigue or damage; install new clips as required.
   d. Slide the clips on each hose over the filter nipple and tighten the clips to ensure a tight connection.
   e. Bleed the oil pump as described in Chapter Eight.
   f. Remove and safely discard the cloth used to catch oil.

CAUTION
A contaminated oil filter will prevent oil from reaching the engine and cause engine seizure.

Wire Harness, Control Cables and Hose Lines

All wiring, cables and hoses should be inspected for proper routing. If necessary, secure loose components with cable ties. Replace damaged components as required.

Primary Sheave Adjustment

Refer to Chapter Eleven.

Track Inspection

Inspect the track as described in Chapter Fourteen.

Track Adjustment

Because the track is made of rubber and subjected to high torque loads, track stretch and alignment should be monitored on a routine schedule. Failure to periodically adjust track tension and alignment will reduce track performance and eventually cause premature wear to the track. When maintaining the track, there are 2 adjustments—tension and alignment.

Tension adjustment

Correct track tension is important because a loose track will slap on the bottom of the tunnel and wear the track, tunnel and heat exchangers. A loose track can also ratchet on the drive sprockets and damage both the track and sprockets.

A track that is too tight will rapidly wear the slide runner material and the rubber on the idler wheels. This condition will reduce performance because of increased friction and drag on the system.

> *NOTE*
> *Bombardier suggests to ride the snowmobile in snow for approximately 15-20 minutes before adjusting track tension.*

1. Support the snowmobile with a suitable lift so that the track is clear of the ground.

2. Clean ice, snow and dirt from the track and suspension.

3A. *1985-1986:* With the slide suspension extended by its own weight, measure the gap between the slider shoe and the inside of the track (**Figure 27**). The gap should be 13 mm (1/2 in.). If track tension is incorrect, proceed to Step 4.

3B. *1987-on:* Connect a spring scale to the track approximately in the middle of the slider shoe (**Figure 28**) and pull the track until the scale registers 7.3 kg. (16 lbs.). Hold the track at this point and measure the distance between the slider shoe and the inside of the track with a scale. The gap should be 13 mm (1/2 in.). If track tension is incorrect, proceed to Step 4.

4. Loosen the rear idler wheel retaining bolt (A, **Figure 29**). Loosen the axle adjuster locknut (B, **Figure 29**) and turn the adjuster (C, **Figure 29**) in or out in equal amounts to adjust the track. Recheck the adjustment.

5. Tighten the rear idler wheel bolt to the torque specification in **Table 7**.

6. After adjusting track tension, check track alignment as described in this chapter.

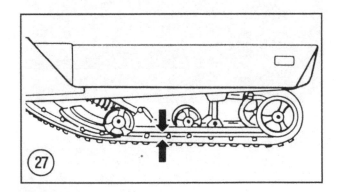

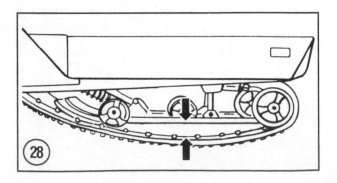

Track alignment

Track alignment is related to track tension and should be checked and adjusted when the tension is checked and adjusted. If the track is misaligned, the rear idler wheels, drive sprocket lugs and track lugs will wear rapidly. Because of resistance between the track and the sides of the wheels, performance will be greatly reduced.

1. Adjust the track tension as described in this chapter.

2. Position the machine on its skis so that the ski tips are placed against a wall or other

immovable barrier. Elevate and support the machine so that the track is completely clear of the ground and free to rotate.

> *WARNING*
> *Don't stand behind or in front of the machine when the engine is running, and take care to keep hands, feet and clothing away from the track when it is turning.*

3. Start the engine and apply just enough throttle to turn the track several complete revolutions. Shut off the engine and allow the track to coast to a stop. Don't stop it with the brake.

4. Check the alignment of the rear idler wheels and the track lugs (**Figure 30**). If the idlers are equal distance from the lugs and the openings in the track are centered with the slide runners (**Figure 30**), the alignment is correct. However, if the track is offset to one side or the other, alignment adjustment is required.

5. Loosen the rear idler wheel bolts (A, **Figure 29**).

6. If the track is offset to the left (**Figure 30**), tighten the left adjuster bolt (C, **Figure 29**) and

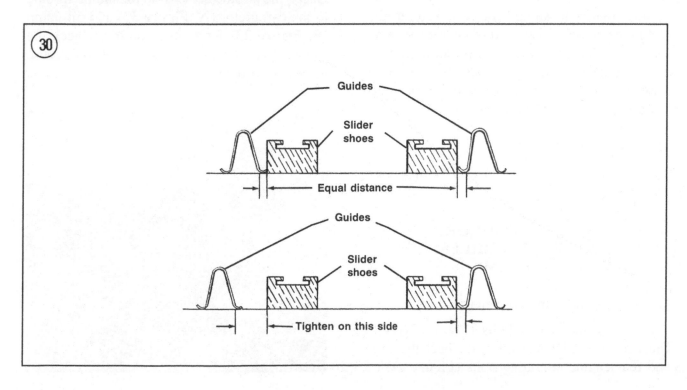

loosen the right one equal amounts until the track is centered (**Figure 30**). If the track is offset to the right (**Figure 30**), tighten the right adjuster bolt and loosen the left one in equal amounts.

7. Repeat Step 3 and Step 4 and if alignment is still not correct, repeat Step 5 and Step 6 and then Step 3 and Step 4 until the track is properly aligned.

8. Tighten the rear idler wheel bolt to the torque specification in **Table 7**.

9. Recheck track tension.

Slide Suspension Check

The suspension system should be checked for loose, damaged or missing components. Refer to Chapter Fourteen.

Steering Inspection

Because ski alignment cannot be maintained with damaged steering components or loose or missing fasteners, check the steering assembly monthly. Refer to Chapter Thirteen.

Ski Alignment

Ski alignment should be checked at the beginning of each season and whenever a steering component has been replaced. In addition, check ski alignment whenever the ski experiences a hard side impact. Refer to Chapter Thirteen for complete ski alignment procedures.

General Inspection

Refer to *General Inspection and Maintenance* in this chapter.

ONCE A YEAR OR EVERY 2,000 MILES (3,200 KM)

Engine Mounts and Fasteners

Loose engine mount bolts (**Figure 31**) will cause incorrect clutch alignment. Check the front and rear engine mounting bolts (**Figure 32**) to make sure they are tight. Check all accessible engine assembly bolts for tightness. Tighten bolts to the torque specification in **Table 7**.

NOTE
If the engine mount bolts were loose, check clutch alignment as described in Chapter Eleven.

Cylinder Head Torque

Refer to *Engine Tune-up* in this chapter.

Ignition Timing

Refer to *Engine Tune-up* in this chapter.

Carburetor Adjustment

Refer to *Engine Tune-up* in this chapter.

Throttle Cable Routing

The throttle cable is connected to the carburetors and oil pump by a junction block. Check the throttle cable from the thumb throttle to the carburetor (A, **Figure 33**) and oil pump (B, **Figure 33**) for proper routing. Check the cable ends for fraying or splitting that could cause the cable to break.

3

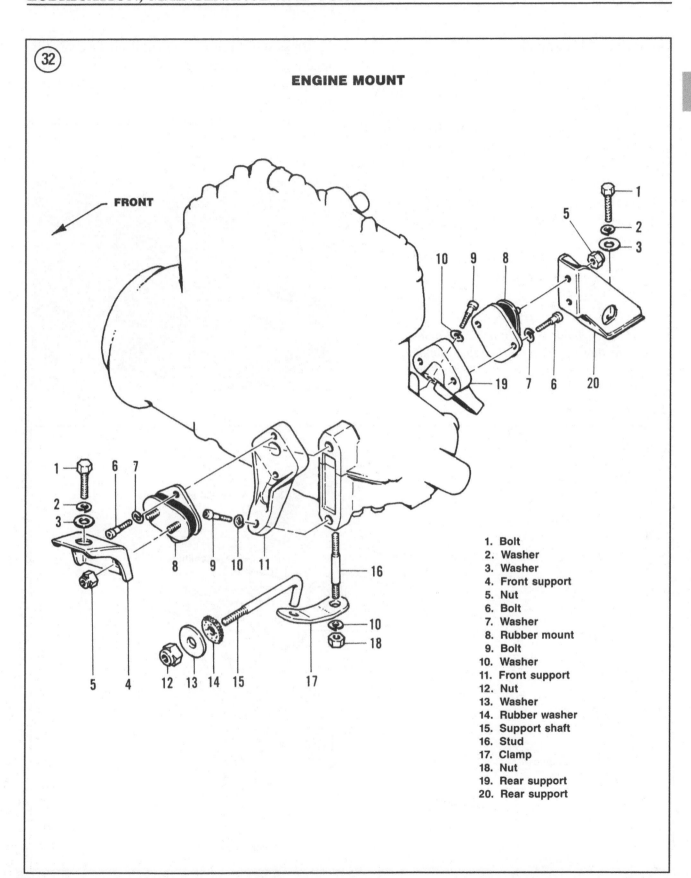

32

ENGINE MOUNT

FRONT

1. Bolt
2. Washer
3. Washer
4. Front support
5. Nut
6. Bolt
7. Washer
8. Rubber mount
9. Bolt
10. Washer
11. Front support
12. Nut
13. Washer
14. Rubber washer
15. Support shaft
16. Stud
17. Clamp
18. Nut
19. Rear support
20. Rear support

Oil Pump Adjustment

The oil pump controls oil flow to the engine. The amount of oil flow is determined by throttle position. Because of normal wear and stretch, cable adjustment is critical to engine performance. Incorrect cable adjustment can cause a delay in pump operation and result in engine seizure.

The oil pump adjustment should be checked once a year or whenever the throttle cable is disconnected or replaced.

1. Adjust the carburetor as described under *Engine Tune-up* in this chapter.

CAUTION
If the carburetor is not adjusted before adjusting the oil pump, engine damage may occur from a delay in pump opening.

2. Press the throttle lever until all free play is taken up and resistance is felt. The mark on the pump lever must align with the mark on the pump housing (**Figure 34**).

3. If the adjustment marks do not align as shown in **Figure 34**, loosen the cable adjuster locknuts

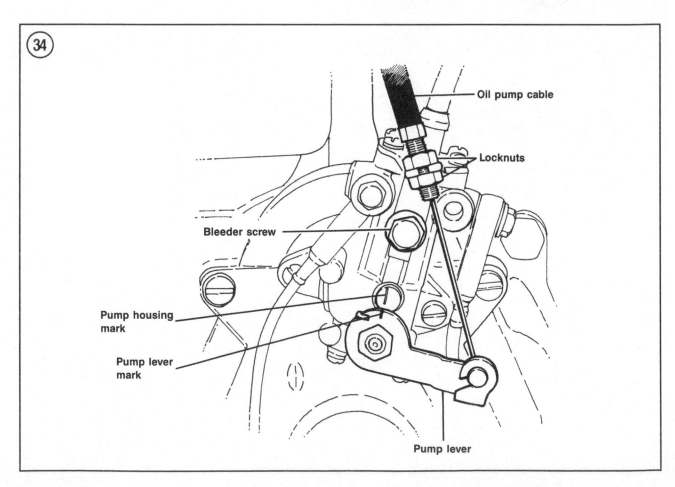

Oil pump cable

Locknuts

Bleeder screw

Pump housing mark

Pump lever mark

Pump lever

(**Figure 34**) and reposition the cable until the 2 marks align. Tighten the locknuts and recheck the adjustment.

Headlight Beam

Refer to Chapter Seven.

Cooling System Inspection

WARNING
*When performing any service work to the engine or cooling system, never remove the radiator cap (**Figure 35**), coolant*

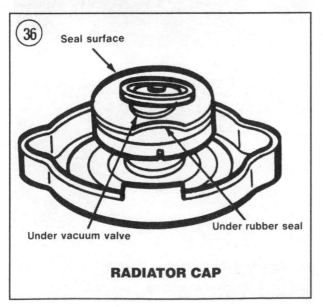

36 Seal surface

Under vacuum valve

Under rubber seal

RADIATOR CAP

drain screws or disconnect any hose while the engine is hot. Scalding fluid and steam may be blown out under pressure and cause serious injury.

Once a year, or whenever troubleshooting the cooling system, the following items should be checked. If you do not have the test equipment, the tests can be done by a Ski-Doo dealer, radiator shop or service station.

1. Loosen the radiator cap to its first detent and release the system pressure, then turn the cap to its second detent and remove it from the radiator. See **Figure 35**.

2. Check the rubber washers on the radiator cap (**Figure 36**) for tears or cracks. Check for a bent or distorted cap. Raise the vacuum valve and rubber seal and rinse the cap under warm tap water to flush away any loose rust or dirt particles.

3. Inspect the radiator cap neck seat on the coolant tank for dents, distortion or contamination. Wipe the sealing surface with a clean cloth to remove any rust or dirt.

CAUTION
Do not exceed 0.9 kg/cm² (13 psi) when performing Step 4 and Step 5 or damage to the cooling system will occur.

4. Have the radiator cap pressure tested (**Figure 37**). The specified radiator cap relief pressure

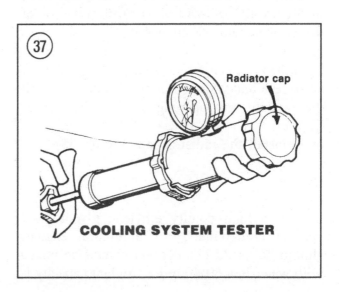

Radiator cap

COOLING SYSTEM TESTER

is 0.9 kg/cm² (13 psi). The cap must be able to sustain this pressure for 6 seconds. Replace the radiator cap if it does not hold pressure.

5. Leave the radiator cap off and have the entire cooling system pressure tested. The entire cooling system should be pressurized to 0.9 kg/cm² (13 psi). The system must be able to hold this pressure for 10 seconds. Replace or repair any components that fail this test.

6. Check all cooling system hoses for damage or deterioration. Replace any hose that is questionable. Make sure all hose clamps (**Figure 38**) are tight.

7. Check the heat exchangers (**Figure 39**) for cracks or other damage. Replace if necessary, as described in Chapter Nine.

Coolant Check

> *WARNING*
> *Do not remove the radiator cap (**Figure 35**) when the engine is hot.*

1. Park the snowmobile on level ground.
2. Open the shroud.
3. Loosen the radiator cap to its first detent and release the system pressure, then turn the cap to its second detent and remove it from the radiator. See **Figure 35**.
4. Check the level in the coolant tank. If the coolant level is not 25 mm (1.0 in.) (1985-1986) or 60 mm (2 3/8 in.) (1987-on) from the top of the coolant tank (**Figure 40**), add coolant as follows.
5. If the level is low, add a sufficient amount of antifreeze and water (in a 60:40 ratio) through the radiator cap opening as described under *Coolant*.
6. Reinstall the radiator cap (**Figure 35**).

Coolant

Only a high quality ethylene glycol-based coolant compounded for aluminum engines should be used. The coolant should be mixed with water in a 60:40 ratio. Coolant capacity is

listed in **Table 6** . When mixing antifreeze with water, make sure to use only soft or distilled water. Never use tap or salt water as this will damage engine parts. Distilled water can be purchased at supermarkets in gallon containers.

> *CAUTION*
> *Always mix coolant in the proper ratio for the coldest temperature in your area. Do not use pure antifreeze; it will freeze without water.*

Coolant Change

The cooling system should be completely drained and refilled once a year (preferably before off-season storage).

CAUTION
Use only a high quality ethylene glycol antifreeze specifically labeled for use with aluminum engines. Do not use an alcohol-based antifreeze.

The following procedure must be performed when the engine is *cold*.

CAUTION
Be careful not to spill antifreeze on painted surfaces as it will destroy the surface. Wash immediately with soapy

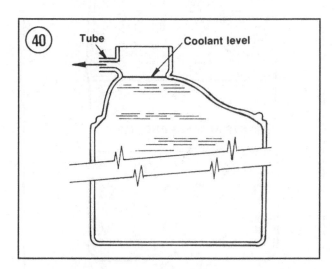

water and rinse throughly with clean water.

1. Park the snowmobile on a level surface.
2. Open the shroud.

WARNING
*Do not remove the radiator cap (**Figure 35**) when the engine is hot.*

3. Loosen the radiator cap to its first detent and release the system pressure, then turn the cap to its second detent and remove it from the radiator. See **Figure 35**.
4. Using a primer pump and a suitable length of hose, siphon the coolant from the coolant tank into a suitable container (**Figure 41**). Remove the pump when the tank is empty.

NOTE
Make sure the hose in the coolant tank is long enough to reach the bottom of the tank.

WARNING
Do not siphon coolant with your mouth and a hose. The coolant mixture is poisonous and may cause sickness. Observe warning labels on antifreeze containers. Animals are attracted to antifreeze so make sure you discard used antifreeze in a safe and suitable manner; do not store antifreeze in open containers.

WARNING
Automotive antifreeze has been classified as an environmental toxic waste by the EPA and cannot be legally disposed of by flushing down a drain or pouring onto the ground. Treat antifreeze that is to be discarded as you treat engine oil. Put it in suitable containers and dispose of it according to local regulations.

WARNING
Step 5 will make a mess. Spilled antifreeze is very slippery on cement floors. Wipe up spilled antifreeze as soon as possible.

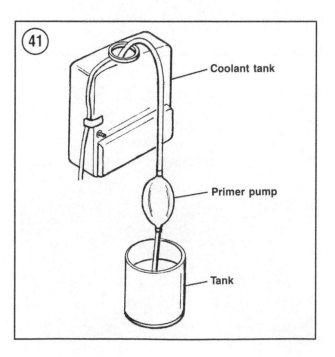

5. Remove the engine bleed screw from the top of the cylinder head (**Figure 42**). Raise the rear of the snowmobile with a jack and allow the heat exchangers to drain. After the coolant is finished draining, lower the snowmobile to the ground. Do not reinstall the bleed screw.

6. Rinse engine and engine compartment with clean water.

7. On 1987 and later models, disconnect and drain the recovery tank. Reinstall the recovery tank and reconnect the hose at the tank nipple.

8. Using a 60:40 mixture of antifreeze and distilled water, add coolant into the coolant tank until it flows out of the bleed hole on the cylinder head (**Figure 42**). Reinstall the bleed screw and tighten securely. Continue to add coolant until the level reaches 25 mm (1.0 in.) (1985-1986) or 60 mm (2 3/8 in.) (1987-on) below the top of the coolant tank (**Figure 40**).

9. Start the engine and allow it to idle until it reaches operating temperature. When the thermostat opens and the coolant begins to circulate, allow the engine to idle for a few more minutes.

10. Turn the engine off and check the coolant freezing level with a coolant checker. Be sure freezing level exceeds the coldest operating temperatures in your area.

11. If the coolant level dropped, add coolant to maintain the correct level in the filler neck (Step 8). If the coolant level dropped significantly, you may want to start the engine and allow it to idle again with the radiator cap removed. Continue to add coolant until the coolant level stabilizes. At this point, turn the engine off and install the radiator cap.

> *NOTE*
> *After flushing the cooling system, the coolant level may drop due to the displacement of entrapped air. Check the level once again before starting the engine to prevent operating the engine with a low coolant level.*

12. Close and secure the shroud.

GENERAL INSPECTION AND MAINTENANCE

Recoil Starter

Pull out the starter rope (**Figure 43**) and inspect it for fraying. If its condition is

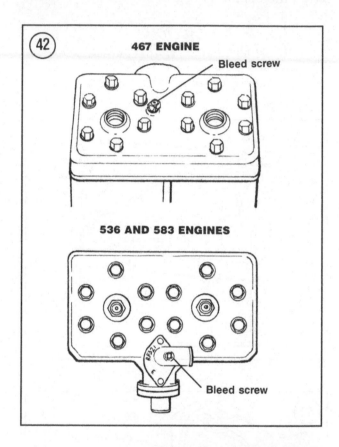

(42)   467 ENGINE
Bleed screw

536 AND 583 ENGINES

Bleed screw

(43)

questionable, replace the rope as described in Chapter Ten.

Check the action of the starter. It should be smooth, and when the rope is released, it should return all the way. If the starter action is rough or if the rope does not return, service the starter as described in Chapter Ten.

Body Inspection

Damaged body panels should be repaired or replaced.

Body Fasteners

Tighten any loose body bolts. Replace loose rivets by first drilling out the old rivet and then installing a new one with a pop-riveter. This tool, along with an assortment of rivets, is available

through many hardware and auto parts stores. Follow the manufacturer's instructions for installing rivets.

Welded joints should be checked for cracks or other damage. Damaged welded joints should be repaired by a competent welding shop.

Drive Assembly

Refer to Chapter Four for drive assembly tuning and adjustment. Refer to Chapter Eleven for drive assembly service.

Guide Wheel Inspection

Inspect the rubber on the guide wheels for wear and damage (**Figure 44**). Replace the wheels if they are in poor condition. Refer to Chapter Fourteen.

Drive Axle Sprockets

Inspect the teeth on the drive axle sprockets for wear and damage (**Figure 45**). If the sprockets are damaged, replace them as described in Chapter Fourteen.

Fuel Tank and Lines

Remove the seat and inspect the fuel tank for cracks and abrasions. If the tank is damaged and leaking, replace it. See Chapter Six.

Fuel Tank Cleaning

The fuel tank should be removed and thoroughly flushed once a season. Refer to Chapter Six.

Oil Tank

Inspect the oil tank for cracks and abrasions that are leaking or may soon be and replace the tank if its condition is in doubt.

Electrical System

All of the switches should be checked for proper operation. Refer to Chapter Seven.

Electrical Connectors

Inspect the high-tension electrical leads to the spark plugs (**Figure 46**) for cracks and breaks in the insulation and replace the leads if they are less than perfect; breaks in the insulation allow the spark to arc to ground and will impair engine performance.

Check primary ignition wiring and lighting wiring for damaged insulation. Usually minor damage can be repaired by wrapping the damaged area with electrical insulating tape. If insulation damage is extensive, the damaged wires should be replaced.

Abnormal Engine Noise

> **WARNING**
> *Never lean into the snowmobile's engine compartment while wearing a scarf or other loose clothing when the engine is running or when the driver is attempting to start the engine. If the scarf or clothing should catch in the drive belt or clutch, severe injury could occur. Make sure the belt guard is in place.*

Open the shroud. Start the engine and listen for abnormal noises. This could be a rattle indicating a loose fastener or a loud damaging engine sound. Periodic inspection for abnormal engine noises can prevent engine failure later on.

Oil and Fuel Lines

Inspect the oil and fuel lines for loose connections and damage. Tighten all connections and replace any lines that are damaged or cracked.

Pulse Hose

The fuel pump-to-engine pulse hose should be inspected at the beginning of each season for cracks or other damage that results from age or secondary damage. Worn or damaged hoses will cause air leaks that result in intermittent operating problems. Check the pulse hose from the fuel pump to the engine crankcase for loose connections or damage. Replace the pulse hose when necessary.

ENGINE TUNE-UP

The number of definitions of the term "tune-up" is probably equal to the number of people defining it. For the purposes of this book, a tune-up is general adjustment and maintenance to insure peak engine performance.

The following paragraphs discuss each facet of a proper tune-up which should be performed in the order given.

Have the new parts on hand before you begin.

To perform a tune-up on your snowmobile, you will need the following tools and equipment:
 a. 14 mm spark plug wrench.
 b. Socket wrench and assorted sockets.
 c. Phillips head screwdriver.
 d. Spark plug wire feeler gauge and gapper tool.
 e. Dial indicator.
 f. Flywheel puller.
 g. Compression gauge.

Cylinder and Cylinder Head Nuts

The engine must be at room temperature for this procedure.

1. Open the shroud.

2. Tighten each cylinder head nut equally in a crisscross pattern to the tightening torque in **Table 7**. Refer to **Figure 47**.

Cylinder Compression

A cylinder cranking compression check is one of the quickest ways to check the internal condition of the engine: rings, head gasket, etc. It's a good idea to check compression at each tune-up, write it down, and compare it with the reading you get at the next tune-up. This will help you spot any developing problems.

1. Elevate and support the machine so that the track is completely clear of the ground and free to rotate. Start and run the engine until it warms up to normal operating temperature. Turn the engine off.

> *WARNING*
> *Don't stand behind or in front of the machine when the engine is running, and take care to keep hands, feet and clothing away from the track when it is turning.*

> *CAUTION*
> *Before removing spark plugs, blow away any dirt that has accumulated next to the spark plug base. The dirt could fall into the cylinder when the plug is removed, causing serious engine damage.*

2. Remove both spark plugs (**Figure 46**). Insert the plugs in the caps and ground both plugs to the exhaust pipe.

> *CAUTION*
> *If the plugs are not grounded during the compression test, the CDI ignition could be damaged.*

3. Screw a compression gauge into one spark plug hole or, if you have a press-in type gauge, hold it firmly in position.

4. Check that the emergency cut-out switch is in the OFF position.

5. Hold the throttle wide open and crank the engine several revolutions until the gauge gives its highest reading. Record the reading. Remove the pressure tester and relieve the pressure valve.

6. Repeat for the opposite cylinder.

7. There should be no more than a 10% difference in compression between cylinders.

8. If the compression is very low, it's likely that a ring is broken or there is a hole in the piston.

Correct Spark Plug Heat Range

The proper spark plug is very important in obtaining maximum performance and reliability. The condition of a used spark plug can tell a trained mechanic a lot about engine condition and carburetion.

Select plugs of the heat range designed for the loads and conditions under which the snowmobile will be run. Use of incorrect heat ranges can cause a seized piston, scored cylinder wall, or damaged piston crown.

In general, use a hot plug for low speeds and low temperatures. Use a cold plug for high speeds, high engine loads and high temperatures. The plugs should operate hot enough to burn off unwanted deposits, but not so hot that they burn themselves or cause preignition. A spark plug of the correct heat range will show a light tan color on the portion of the insulator within the cylinder after the plug has been in service. See **Figure 48**.

SPARK PLUG CONDITIONS

NORMAL USE

OIL FOULED

CARBON FOULED

OVERHEATED

GAP BRIDGED

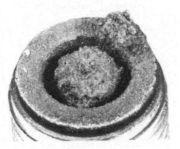

SUSTAINED PREIGNITION

WORN OUT

3

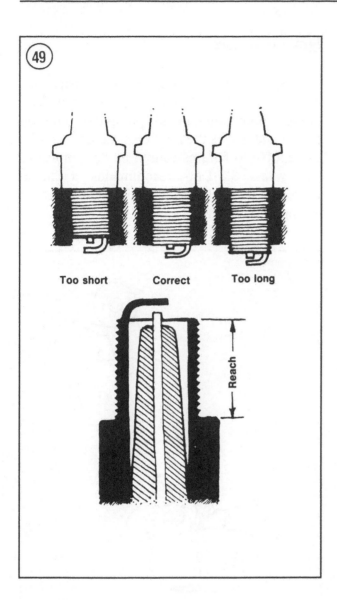

Too short   Correct   Too long

Reach

The reach (length) of a plug is also important. A shorter than normal plug will cause hard starting, reduced engine performance and carbon buildup on the exposed cylinder head threads. A longer than normal plug could interfere with the piston or cause overheating; both conditions result in permanent and severe engine damage. Refer to **Figure 49**.

The standard heat range spark plug for the various models is listed in **Table 8**.

Spark Plug Removal/Cleaning

1. Grasp the spark plug leads as near the plug as possible and pull it off the plug. If it is stuck to the plug, twist it slightly to break it loose. See **Figure 46**.

2. Blow away any dirt that has accumulated next to the spark plug base.

> *CAUTION*
> *The dirt could fall into the cylinder when the plug is removed, causing serious engine damage.*

3. Remove the spark plug with a 14 mm spark plug wrench.

> *NOTE*
> *If the plug is difficult to remove, apply penetrating oil, like WD-40 or Liquid Wrench, around the base of the plug and let it soak in about 10-20 minutes.*

4. Inspect the plug carefully. Look for a broken center porcelain, excessively eroded electrodes, and excessive carbon or oil fouling. See **Figure 48**.

Gapping and Installing the Plug

A new spark plug should be carefully gapped to ensure a reliable, consistent spark. You must use a special spark plug gapping tool and a wire feeler gauge.

1. Insert a wire feeler gauge between the center and side electrode (**Figure 50**). The correct gap is listed in **Table 8**. If the gap is correct, you will

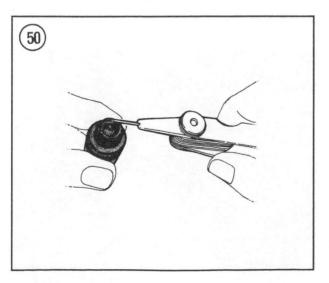

feel a slight drag as you pull the wire through. If there is no drag, or the gauge won't pass through, bend the side electrode with a gapping tool (**Figure 51**) to set the proper gap.

> *NOTE*
> *Never try to close the spark plug gap by tapping the spark plug on a solid surface. This can damage the plug internally. Always use the special tool to open or close the gap.*

2. Apply anti-seize to the plug threads before installing the spark plug.

> *NOTE*
> *Anti-seize can be purchased at most automotive parts stores.*

3. Screw the spark plug in by hand until it seats. Very little effort is required. If force is necessary, you have the plug cross-threaded. Unscrew it and try again.

4. Use a spark plug wrench and tighten the plug an additional 1/4 to 1/2 turn after the gasket has made contact with the head. If you are installing an old, regapped plug and reusing the old gasket, only tighten an additional 1/4 turn.

> *CAUTION*
> *Do not overtighten. This will only squash the gasket and destroy its sealing ability. This could cause compression leakage around the base of the plug.*

5. Install the spark plug wires. Make sure they snap onto the top of the plugs tightly.

> *CAUTION*
> *Make sure the spark plug wire is pulled away from the exhaust pipe.*

Reading Spark Plugs

Much information about engine and spark plug performance can be determined by careful examination of the spark plugs. Refer to Chapter Four.

Ignition Timing

These models are equipped with a capacitor discharge ignition (CDI). This system uses no breaker points and greatly simplifies ignition timing and makes the ignition system much less susceptible to failures caused by dirt, moisture and wear. Ski-Doo recommends a 2-step procedure when checking ignition timing—static timing and dynamic timing.

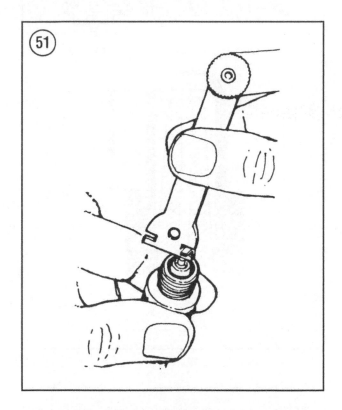

Static timing of Bombardier engines requires the use of an accurate dial indicator to determine top dead center (TDC) of the piston before making any timing adjustment. TDC is determined by removing the spark plug and

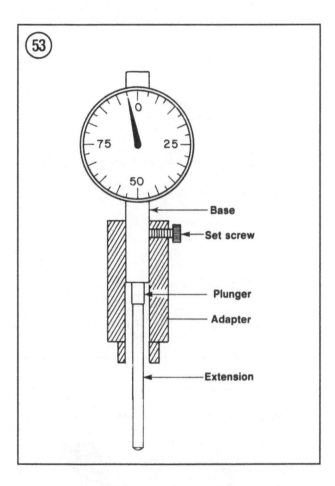

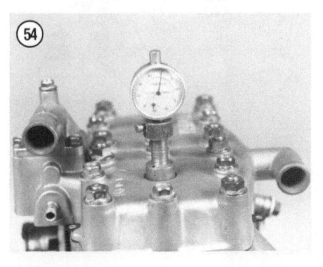

installing the dial indicator in the plug opening. The static timing method is used solely to verify the flywheel timing marks before using a timing light to check ignition timing. The static timing procedure can be used for the following:

a. Verify factory timing marks.
b. Detect a broken or missing flywheel Woodruff key.
c. Detect a twisted crankshaft.
d. To scribe timing marks on a new flywheel.

Dynamic engine timing uses a timing light connected to the MAG side spark plug lead. As the engine is cranked or operated, the light flashes each time the spark plug fires. When the light is pointed at the moving flywheel, the mark on the flywheel appears to stand still. The flywheel mark should align with the stationary timing pointer on the engine.

Static Timing Check

1. Open the shroud.
2. Remove the drive belt as described in Chapter Eleven.
3. Remove both spark plugs as described in this chapter.
4. Remove the crankcase inspection plug (**Figure 52**).
5. Install and position a dial indicator as follows:
   a. Screw the extension onto a dial indicator and insert the dial indicator into the adapter (**Figure 53**).
   b. Screw the dial indicator adapter into the cylinder head (**Figure 54**) on the MAG side. Do not lock the dial indicator in the adapter at this time.
   c. Rotate the flywheel (by turning the primary sheave) until the dial indicator rises all the way up in its holder (piston is approaching top dead center). Slide the indicator far enough into the holder to obtain a reading.
   d. Lightly tighten the set screw on the dial indicator adapter to secure the dial gauge.
   e. Rotate the flywheel until the dial on the gauge stops and reverses direction. This is

top dead center. Zero the dial gauge by aligning the zero with the dial (**Figure 55**).

   f. Tighten the set screw on the dial indicator adapter securely.

6. Rotate the crankshaft clockwise (viewed from the right-hand side) until the gauge needle has made approximately 3 revolutions. Then carefully turn the crankshaft clockwise until the gauge indicates the timing shown in **Table 9**. View the timing marks through the hole in the crankcase (**Figure 56**).

7. Check the static timing marks on the stator and flywheel. They should line up (**Figure 56**). If they do not, perform the following:

   a. Scribe a new mark on the flywheel that aligns with the crankcase center mark shown in **Figure 56**. This mark will be used as the reference when using the timing light.

   b. Repeat the timing procedure to check the accuracy of the new mark.

8. Remove the gauge and adapter. Install the spark plugs and connect the high-tension leads.

9. Reinstall the drive belt (Chapter Eleven).

Dynamic Timing Check

1. Open the shroud.

2. Perform the *Static Timing Check* in this chapter.

3. Hook up a stroboscopic timing light according to the manufacturer's instructions on the MAG side spark plug lead (**Figure 57**).

4. Connect a tachometer according to the manufacturer's instructions.

5. Position the machine, on its skis, so the tips of the skis are against a wall or other immovable barrier. Elevate and support the machine so the track is completely clear of the ground and free to rotate.

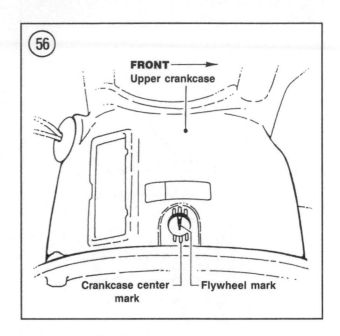

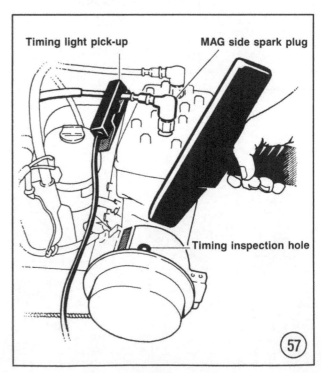

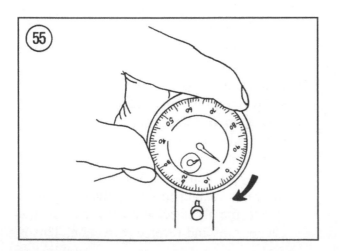

3

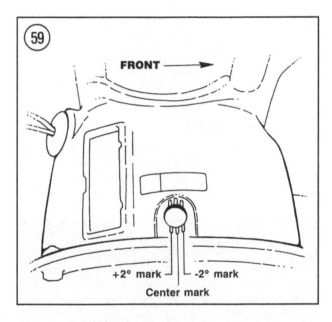

FRONT ——→

+2° mark —| |— -2° mark
Center mark

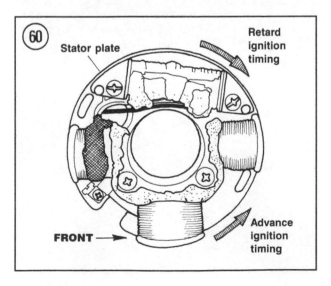

Stator plate

Retard ignition timing

Advance ignition timing

FRONT ——→

WARNING
Don't allow anyone to stand behind or in front of the machine when the engine is running, and take care to keep hands, feet and clothing away from the track when it is running.

NOTE
Because ignition components are temperature sensitive, ignition timing must be checked when the engine is cold.

6. Start the engine and turn the headlight on.
7. Allow the engine to idle for approximately 10-15 seconds, then run the engine idle up to 6,000 rpm and point the timing light at the crankcase inspection hole (**Figure 57**). The light will flash as the flywheel timing mark aligns with the center crankcase pointer if ignition timing is correct. Turn the engine off. Lower the snowmobile to the ground. Note the following:
  a. If the ignition timing is incorrect, proceed to Step 8.
  b. If the ignition timing is correct, proceed to Step 11.
8. Remove the recoil starter assembly as described in Chapter Ten.
9. Remove the starter pulley and flywheel as described in Chapter Seven.
10. Ignition timing is changed by moving the stator plate (A, **Figure 58**). Note the following:
  a. Each mark on either side of the center crankcase mark represents 2° of crankshaft rotation. See **Figure 59**.
  b. If the flywheel mark was to the right-hand side of the crankcase mark (**Figure 59**), ignition timing is retarded.
  c. If the flywheel mark was to the left-hand side of the crankcase mark (**Figure 59**), ignition timing is advanced.
  d. To change ignition timing, loosen the 2 stator plate Allen screws (B, **Figure 58**). Turn the stator plate counterclockwise to advance or clockwise to retard ignition timing (**Figure 60**).
  e. Tighten the stator plate Allen screws.

f. Reinstall the flywheel, starter pulley and recoil starter assembly.

g. Repeat Step 6 to recheck ignition timing. Repeat procedure until ignition timing is correct.

11. When ignition timing is correct, remove the timing light and tachometer.

12. Lower the snowmobile to the ground and close the shroud.

Carburetor Adjustment

Check the starter cable free play before performing these procedures as described in this chapter.

Pilot air screw

1. Open the shroud.

2. Locate the pilot air screws on the side of each carburetor (**Figure 61**).

3. Turn both pilot air screws in until they lightly seat. Then back the screws out the number of turns specified in **Table 10**.

> *CAUTION*
> *Do not use the pilot air screws to attempt to set engine idle speed. Pilot air screws must be set as specified in **Table 10** or a "too lean" mixture and subsequent engine damage may result.*

Synchronization

For maximum engine performance, both cylinders must work equally. If one cylinder's throttle valve opens earlier, that cylinder will be required to work harder. This will cause poor acceleration, rough engine performance and engine overheating. For proper carburetor synchronization, both throttle valves must lift at the same time.

Because of throttle cable stretch, carburetor synchronization should be checked at each tune-up or whenever the engine suffers from reduced performance.

1. Open the shroud.

2. Remove the air intake silencer (**Figure 62**).

3. Back out both carburetor throttle stop screws (**Figure 63**) so that the carburetor throttle valves drop all the way to the bottom.

4. Turn one carburetor throttle stop screw (**Figure 63**) clockwise until it just touches the carburetor slide. Then turn the throttle stop screw 2 turns clockwise.

5. Repeat Step 4 for the opposite carburetor.

6. Use a strong rubber band and clamp the throttle lever to the handlebar grip in the wide-open throttle position.

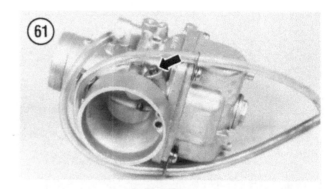

NOTE
Make sure the throttle lever is held in the wide-open position. This will ensure full mixture flow during engine operation.

7. Loosen the locknut securing the throttle valve adjuster at the top of each carburetor (**Figure 64**). Feel inside the carburetor bore and turn the adjuster (**Figure 64**) until the cut-out portion on the throttle valve is flush with the inside of the carburetor bore as shown in **Figure 65**. If necessary, turn the adjuster as required to position the throttle valve flush with the carburetor bore. Tighten the locknut and recheck the slide position.

8. Repeat Step 7 for the opposite carburetor.

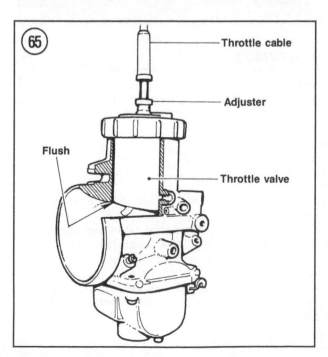

9. Remove the rubber band clamp from handlebar and allow the throttle to return to the idle position.

10. Operate the throttle lever a few times, then hold the throttle lever at the wide-open throttle position. Check that there is 1.5 mm (1/16 in.) amount of free play between the carburetor cap and the top of the throttle valve. Check free play by moving the slide with your fingers. If necessary, loosen the locknut and turn the adjuster (**Figure 64**) to obtain correct free play. Recheck synchronization.

CAUTION
Maintaining throttle cable free play is critical to prevent throttle cable damage.

11. After completing carburetor synchronization, perform the *Oil Pump Adjustment* procedure in this chapter.

12. Reinstall the air intake silencer.

13. Start the engine and warm to operating temperature. Check idle speed (see **Table 11**). If necessary, turn both throttle stop adjustment screws (**Figure 63**) in equal amounts to obtain specified idle speed.

CAUTION
*Do not use the pilot air screws to attempt to set engine idle speed. Pilot air screws must be set as specified in **Table 10** or a "too lean" mixture and subsequent engine damage may result.*

Idle speed

1. Open the shroud.

2. Connect a tachometer according to the manufacturer's instructions.

NOTE
When turning the throttle stop screw in Step 3, turn the left- and right-hand screws the same amount of turns.

3. Set the idle speed by turning the throttle stop screw (**Figure 63**) in to increase or out to decrease idle speed. Refer to **Table 11** for the correct idle speed for your model.

Table 1 WEEKLY INSPECTION

Check coolant level
Check injection system oil level
Check chaincase oil level
Check rotary valve oil level (1985-1988)

Table 2 INITIAL 10 HOUR INSPECTION

Check coolant level
Check rotary valve oil level (1985-1988)
Check injection system oil level
Check chaincase oil level
Check drive chain tension
Retorque cylinder head
Check engine mount tightness
Check exhaust system fasteners
Check ignition timing
Check spark plug condition and gap
Check carburetor adjustment
Check oil injection pump adjustment
Check brake pad wear and brake performance
Check ski alignment and adjustment
Retighten the handlebar bolts
Retighten the steering arm bolts
Retighten rear axle screw
Visually inspect the drive belt
Check sheave alignment
Check driven sheave preload
Inspect track for wear or damage
Check track tension and alignment
Perform all steering, suspension and drive axle lubrication procedures
Check light operation
Check all switches for proper operation

Table 3 PERIODIC MAINTENANCE

Weekly or every 150 miles (240 km)	Check drive belt condition Visually inspect skis and ski runners for wear or damage Check brake pad wear and brake performance
Monthly or every 500 miles (800 km)	Check brake adjustment Check spark plug condition and gap Adjust drive chain tension Clean the air filter [1] Visually inspect the oil injection oil filter for contamination; replace oil filter when necessary Visually inspect all wiring harness, control cable and hose assemblies for incorrect routing and missing fasteners Check the primary sheave for correct adjustment Visually check track condition Check track tension and alignment and adjust if necessary Check all suspension components for missing fasteners and excessive play (continued)

Table 3 PERIODIC MAINTENANCE (continued)

Monthly or every 500 miles (800 km) (continued)	Check all steering components for missing fasteners and excessive play
	Check steering adjustments and adjust if necessary
	Check exhaust system for missing or damaged components and fasteners
	Check all fasteners for tightness
	Check all wiring for chafing or other damage
	Check all electrical connectors for looseness or damage
	Perform general inspection
Once a year or every 2000 miles (3,200 km)	Check engine mount bolt tightness
	Retorque cylinder head
	Check all control cables for routing and damage
	Check ignition timing
	Adjust carburetors
	Adjust oil injection pump (2)
	Check headlight beam aim
	Check cooling system hoses for looseness or damage
	Check antifreeze condition
	Replace coolant

1. 1987 and later models only.
2. The carburetors must always be adjusted before adjusting the oil injection pump.

Table 4 FUEL RECOMMENDATIONS

1985-1987	Regular—leaded or unleaded
1988	
MX and MX LT	Regular—leaded or unleaded
Plus	Super
1989	Regular unleaded

Table 5 RECOMMENDED LUBRICANTS

Item	Lubricant type
Drive shaft bearing, hub bearings, bogie wheels, ski legs, idler bearings, leaf spring cushion pads, etc.	A
Oil seal interior lips	A
Engine injection oil	B
Chaincase	C
Rotary valve lubricant (1985-1988)	D

Lubricant legend:
A. Bombardier bearing grease or equivalent multi-purpose lithium base grease for use through a temperature range of −40° to 95° C (−40° to 200° F). This grease will be referred to as a 'low temperature grease' throughout this manual.
B. Bombardier injection oil or equivalent. Injection oil must flow at −40° C (−40° F).
C. Bombardier chaincase oil or equivalent. Make sure equivalent oil provides lubrication at low temperatures.
D. Bombardier injection oil or equivalent.
* WD-40 can be used as a general lubricant.

Table 6 APPROXIMATE REFILL CAPACITY

Rotary valve reservoir	
1985-1988	455 cc (16 oz.)
1989	N.A.
Chaincase	
1985-1988	256 cc (9 oz.)
1989	200 cc (7 oz.)
Oil injection reservoir	
All models	2.9 L (98 oz.)
Cooling system	
All models	4.2 L (142 oz.)
Fuel tank	
All models	28.6 L (7.6 gal.)

Table 7 MAINTENANCE TIGHTENING TORQUES

	N·m	ft.-lb.
Cylinder head		
1985-1988	20	15
1989	22	16
Engine mounts		
1985-1987	11	8
1988-on	38	28
Rear idler wheel bolt	48	35

Table 8 SPARK PLUGS

Model	Plug type	Gap mm (in.)
1985	NGK BR9ES	0.40 (0.016)
1986		
Formula MX	NGK BR10ES	0.40 (0.016)
Formula MX (H/A)	NGK BR10ES	0.40 (0.016)
Formula Plus	NGK BR9ES	0.40 (0.016)
1987-1988		
All models	NGK BR9ES	0.40 (0.016)
1989		
Formula Mach I	NGK BR9ES	0.45 (0.018)
All other models	NGK BR9ES	0.40 (0.016)

Table 9 IGNITION TIMING* (WITH DIAL INDICATOR)

	mm	in.
1985		
Formula MX	2.50	0.098
Formula Plus	1.76	0.069
1986		
Formula MX	2.50	0.098
Formula MX (H/A)	2.50	0.098
Formula Plus	1.75	0.069
1987		
Formula MX	2.50	0.098
Formula MX LT	2.50	0.098
Formula Plus	1.75	0.069
1988		
Formula MX	2.51	0.099
Formula MX LT	2.51	0.099
Formula Plus	1.75	0.069
1989		
Formula MX	2.51	0.099
Formula MX LT	2.51	0.099
Formula Plus	2.18	0.086
Formula Plus LT	2.18	0.086
Formula Mach I	1.75	0.069

*All measurements before top dead center (BTDC).

Table 10 CARBURETOR PILOT AIR SCREW ADJUSTMENT

	Turns out*
1985	
Formula Plus	2.0
All other models	1 1/2
1986	
Formula Plus	1.0
All other models	1 1/2
1987-1988	
Formula Plus	1.0
All other models	1 1/2
1989	
All models	1 1/2

* (± 1/8 turn)

Table 11 CARBURETOR IDLE SPEED

All models	1,800-2,000 rpm

Chapter Four

High-Altitude and Rear Suspension Adjustment

If the snowmobile is to deliver its maximum efficiency and peak performance, the engine and chassis must be properly adjusted. This chapter describes high altitude tuning and rear suspension adjustment.

Basic tune-up procedures are described in Chapter Three. **Tables 1-22** are found at the end of the chapter.

HIGH-ALTITUDE CARBURETOR TUNING

Ski-Doo snowmobiles are tuned at the factory for sea level conditions. However, when the snowmobile is operated at a higher altitude, engine performance will drop because of a change in air density. At sea level, the air is quite a bit denser than the air at 10,000 feet. You should figure on a 3% loss of power output for every 1,000 feet of elevation change (increase). This decrease in power is caused by a drop in cylinder pressure and a change in the fuel:air ratio. For example, an engine that produces 40 horsepower at sea level will produce approximately 38.8 horsepower at 1,000 feet. At 10,000 feet, the engine produces 29.5 horsepower. With sea level jetting, the engine would run extremely rich at 10,000 feet.

Air temperature must also be considered when jetting the carburetor. For example, the carburetors are set at the factory to run at temperatures of 32-minus 4° F (0-minus 20° C) at sea level. If the snowmobile is to be operated under conditions other than those specified, the carburetors must be adjusted accordingly.

Figure 1 illustrates the different carburetor circuits and how they overlap during engine operation. The charts in **Tables 1-8** show tuning requirements for variations in elevation.

Before adjusting the carburetor, make the following adjustments as described in Chapter Three:

a. Carburetor adjustment (including synchronization).
b. Oil pump adjustment.

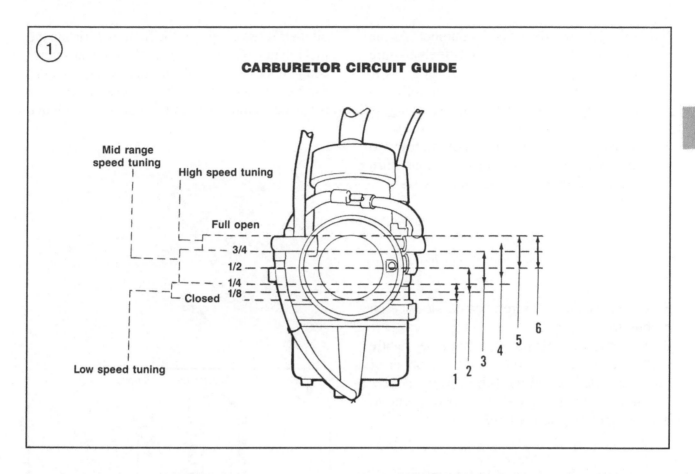

Figure 1 — CARBURETOR CIRCUIT GUIDE

NOTE

Changes in port shape and smoothness, expansion chamber, carburetor, etc., will also require jetting changes because these factors alter the engine's ability to breathe. When installing aftermarket equipment or when the engine has been modified, equipment manufacturers often include a tech sheet listing suitable jetting changes to correspond to their equipment or modification. This information should be taken into account along with altitude and temperature conditions previously mentioned.

NOTE

It is important to note that the following jetting guidelines should be used as guidelines only. Individual adjustments will vary because of altitude, temperature and snow conditions. The condition of the spark plugs should be used as the determining factor when changing jets and adjusting the carburetor.

Carburetor Adjustment

When the snowmobile is going to be run at a high altitude, refer to the tuning chart (**Tables 1-8**) for your model and replace jets as required. Refer to Chapter Six for carburetor removal and disassembly.

Low speed tuning

The pilot jet and pilot air screw setting control the fuel mixture from 0 to about 1/4 throttle (**Figure 1**). In addition, the pilot air screw controls mixture adjustment when the throttle is opened from idle to the full open position quickly and when the engine is run at half-throttle. Note the following when adjusting the pilot air screw:

a. Turning the pilot air screw clockwise leans the fuel mixture.

b. Turning the pilot air screw counterclockwise richens the fuel mixture.

c. Pilot jets are identified by number. As the pilot numbers increase, the fuel mixture richens.

d. When operating the snowmobile in relatively warm weather or at a higher altitude, turn the pilot air screw clockwise. When operating the snowmobile in excessively cold weather conditions, turn the pilot air screw counterclockwise.

1. Open the shroud.

NOTE
Figure 2 shows the carburetor removed for clarity.

2. Locate the pilot air screws on the side of each carburetor (**Figure 2**).

3. Turn both pilot air screws in until they lightly seat. Then back the screws out the number of turns specified in **Tables 1-8** for your model.

4. Start the engine and allow it to warm up to normal operating temperature.

NOTE
Figure 3 shows the carburetor removed for clarity.

5. Adjust the throttle stop screw (**Figure 3**) until the engine is idling at the rpm listed in **Tables 1-8** for your model. Slowly turn the pilot air screws (**Figure 2**) counterclockwise. Turning the

pilot air screws counterclockwise will richen the fuel mixture. Continue to turn the pilot air screws counterclockwise until the highest engine rpm can be reached. When the highest engine rpm is reached, adjust the throttle stop screw (**Figure**

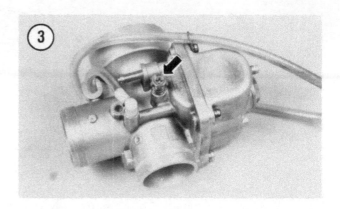

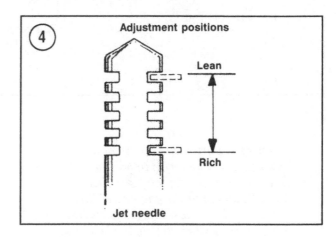

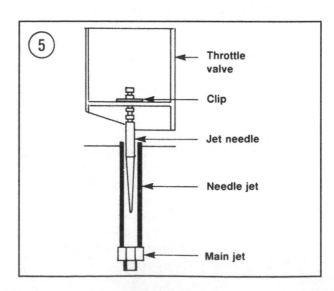

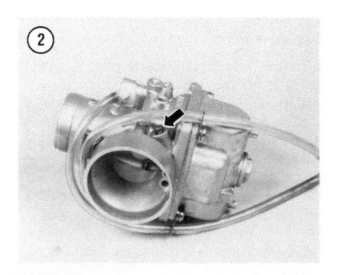

3) to set the idle speed to the specification listed in **Tables 1-8**.

6. Operate the engine and check performance. If the engine performance is off at high altitudes or in extremely cold areas or if the off-idle pickup is poor, install larger pilot jets. Refer to

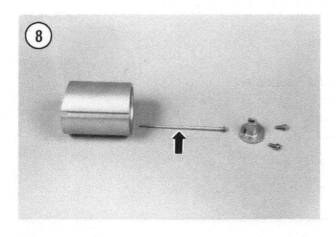

Chapter Six for carburetor removal and disassembly.

7. After replacing the pilot jets, repeat Steps 2-6.

Mid-range tuning

The jet needle controls the mixture at medium speeds, from approximately 1/4 to 3/4 throttle (**Figure 1**). The jet needle has 2 operating ends. The top of the needle has 5 evenly spaced circlip grooves (**Figure 4**). The bottom half of the needle is tapered; this portion extends into the needle jet. While the jet needle is fixed into position by the circlip, fuel cannot flow through the space between the needle jet and jet needle until the throttle valve is raised approximately 1/4 open. As the throttle valve is raised, the jet needle's tapered portion moves out of the needle jet (**Figure 5**). The grooves permit adjustment of the mixture ratio. If the clip is raised (thus dropping the needle deeper into the jet), the mixture will be leaner; lowering the clip (raising the needle) will result in a rich mixture.

1. Open the shroud.

> *NOTE*
> *Before removing the top cap, throughly clean the area around it so that no dirt can fall into the carburetor.*

2. Unscrew and remove the carburetor top cap (**Figure 6**) and pull the throttle valve assembly (**Figure 7**) out of the carburetor.

3. Remove the jet needle (**Figure 8**) from the throttle valve.

> *NOTE*
> *Some models have a washer installed on the jet needle. Don't lose it when removing the jet needle.*

4. Note the position of the clip (**Figure 4**) before removing it. Remove the clip and reposition it on the jet needle. Make sure the clip seats in the needle groove completely.

5. Reverse to install. Make sure the O-ring in the cap is positioned correctly (**Figure 9**) before installing the cap.

High-speed tuning

The main jet controls the mixture from 3/4 to full throttle and has some effect at lesser throttle openings (**Figure 1**). Each main jet is stamped with a number. Larger numbers provide a richer mixture, smaller numbers a leaner mixture. Refer to **Tables 1-8** for main jet sizes at different altitudes.

When operating the snowmobile in relatively warm weather or at a higher altitude, a smaller main jet should be used. When operating the snowmobile in excessively cold weather conditions, install a larger main jet.

CAUTION
The information given in Step 1 for determining main jet sizes should be used as a guideline only. Because of variables that exist with each individual machine, the spark plug condition should be used as the determining factor. When in doubt, always jet on the rich side.

CARBURETOR MAIN JET CORRECTION CHART

FT/METER	°F/°C							
	−60/ −50	−40/ −40	−20/ −30	0/ −20	20/ −5	40/ 5	60/ 15	80/ 25
0	111.10	107.40	103.70	% 100.00	96.30	92.60	88.90	85.20
2000/ 600	105.77	102.07	98.37	94.67	90.97	87.27	83.57	79.87
4000/ 1200	100.43	96.73	93.03	89.33	85.63	81.93	78.23	74.53
6000/ 1800	95.10	91.40	87.70	84.00	80.30	76.60	72.90	69.20
8000/ 2400	89.77	86.07	82.37	78.67	74.97	71.27	67.57	63.27
10,000/ 3000	84.44	80.74	77.04	73.34	69.64	65.94	62.24	58.54

1A. *1985-1988:* After determining the altitude and temperature range that the snowmobile will be operated in, refer to **Figure 10** and compute the necessary jet changes as follows:

a. Refer to **Tables 1-6** and determine the stock main jet for your model.

b. Refer to **Figure 10** to determine the temperature and altitude range the snowmobile will be operated in. Cross-reference the altitude and temperature to obtain your correction percentage number.

c. When the correction number has been determined, multiply the stock main jet number times the correction number for your model to determine the new main jet size.

d. For example, with a stock 300 main jet, a temperature of 20° F (-5° C) and an altitude of 6000 ft. (1800 meters): 300 × 80.30% = 240 main jet.

e. When the answer does not equal an existing main jet size, always select the next highest jet.

1B. *1989:* After determining the altitude and temperature range that the snowmobile will be operated in, compute the necessary jet changes as follows:

a. Refer to **Figure 11**, **Figure 12** or **Figure 13** for your model and find the temperature and altitude range the snowmobile will be operated in. Cross-reference the altitude and temperature to obtain the new main jet size.

**MAIN JET CHART
(1989 MX AND MX LT)**

TEMPERATURE		MAIN JET						
	°F (°C)	−40 (−40)	−20 (−30)	0 (−20)	14 (-10)	32 (0)	50 (10)	70 (20)
ALTITUDE FEET (METERS)								
Sea level 0	PTO	240	230	220	210	210	200	190
	MAG	260	250	240	230	230	220	210
4000 (1200)	PTO	220	210	200	195	190	180	170
	MAG	240	230	220	210	210	200	190
6000 (1800)	PTO	200	195	185	180	175	165	155
	MAG	220	210	200	195	185	180	170
8000 (2400)	PTO	190	185	175	170	165	155	145
	MAG	210	200	190	185	175	170	160
10,000 (3000)	PTO	180	175	165	160	155	145	135
	MAG	195	185	175	170	160	155	145

b. For example, refer to **Figure 11**. When the operating temperature is 0° F (-20° C) at an altitude of 6000 ft. (1800 meters), the new main jet size would be 185 for the PTO side and 200 for the MAG side.

2. Refer to Chapter Six for carburetor removal and installation. Replace main jets as required.

3. Reinstall the carburetor as described in Chapter Six.

CAUTION
Do not run the engine without the air intake silencer installed as engine seizure may result.

WARNING
If you are taking spark plug readings, the engine will be HOT! Use caution as the

fuel in the float bowl will spill out when the main jet cover is removed from the bottom of the float bowl. Have a fire extinguisher and an assistant standing by when performing this procedure.

4. Run the snowmobile at high speed, then stop the engine.

5. Open the shroud and remove the spark plugs. Read the spark plugs as described in this chapter.

6. Reinstall the spark plugs.

7. If it is necessary to change main jets, perform the following:

a. Remove the carburetor as described in Chapter Six.

b. Remove the float bowl (**Figure 14**).

c. Remove and replace the main jet (**Figure 15**).

MAIN JET CHART
(1989 PLUS AND PLUS LT)

TEMPERATURE		MAIN JET						
ALTITUDE FEET (METERS)	°F (°C)	−40 (−40)	−20 (−30)	0 (−20)	14 (−10)	32 (0)	50 (10)	70 (20)
Sea level 0	PTO	270	260	250	240	240	230	220
	MAG	270	260	250	240	240	230	220
4000 (1200)	PTO	240	230	220	210	210	200	190
	MAG	240	230	220	210	210	200	190
6000 (1800)	PTO	230	220	210	200	195	190	180
	MAG	230	220	210	200	195	190	180
8000 (2400)	PTO	220	210	200	195	185	175	165
	MAG	220	210	200	195	185	175	165
10,000 (3,000)	PTO	200	190	180	175	165	160	150
	MAG	200	190	180	175	165	160	150

⑬

MAIN JET CHART
(1989 MACH I)

TEMPERATURE		MAIN JET						
°F (°C)		−40 (−40)	−20 (−30)	0 (−20)	14 (−10)	32 (0)	50 (10)	70 (20)
ALTITUDE FEET (METERS)								
Sea level 0	PTO	280	270	260	250	250	240	230
	MAG	310	300	290	280	270	260	250
4000 (1200)	PTO	260	250	240	230	230	220	210
	MAG	280	270	260	250	240	230	220
6000 (1800)	PTO	230	220	210	200	200	190	180
	MAG	250	240	230	220	210	200	195
8000 (2400)	PTO	210	200	190	185	175	170	160
	MAG	230	220	210	200	195	185	175
10,000 (3000)	PTO	200	190	180	175	165	160	150
	MAG	220	210	200	195	185	175	165

4

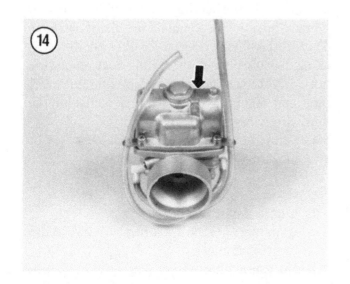

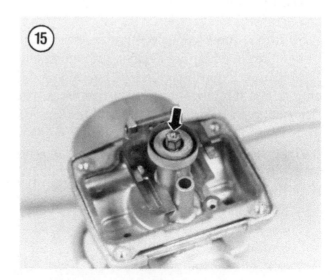

d. Reinstall the float bowl. Make sure the float bowl gasket is in place and not torn or damaged.
e. Repeat for the opposite carburetor.
f. Reinstall the carburetors as described in Chapter Six.
g. Make sure the throttle cables work smoothly before starting the engine

Reading Spark Plugs

Because the firing end of a spark plug operates in the combustion chamber, it reflects the operating condition of the engine. Much information about engine and spark plug performance can be determined by careful examination of the spark plug. Refer to **Figure 16**.

Normal condition

If the plug has a light tan-or gray-colored deposit and no abnormal gap wear or erosion, good engine, carburetion and ignition condition are indicated. The plug in use is of the proper heat range and may be serviced and returned to use.

Carbon fouled

Soft, dry, sooty deposits covering the entire firing end of the plug are evidence of incomplete combustion. Even though the firing end of the plug is dry, the plug's insulation decreases. An electrical path is formed that lowers the voltage from the ignition system. Engine misfiring is a sign of carbon fouling. Carbon fouling can be caused by one or more of the following:
a. Too rich fuel mixture (incorrect jetting).
b. Spark plug heat range too cold.
c. Over-retarded ignition timing.
d. Ignition component failure.
e. Low engine compression.

Oil fouled

The tip of an oil fouled plug has a black insulator tip, a damp oily film over the firing end and a carbon layer over the entire nose. The electrodes will not be worn. Common causes for this condition are:
a. Too much oil in the fuel (incorrect jetting or incorrect oil pump adjustment).
b. Wrong type of oil.
c. Ignition component failure.
d. Spark plug heat range too cold.
e. Engine still being broken in.
Oil fouled spark plugs may be cleaned in an emergency, but it is better to replace them. It is important to correct the cause of fouling before the engine is returned to service.

Gap bridging

Plugs with this condition exhibit gaps shorted out by combustion deposits between the electrodes. If this condition is encountered, check for an improper oil type, excessive carbon in combustion chamber or a clogged exhaust port and pipe. Be sure to locate and correct the cause of this condition.

Overheating

Badly worn electrodes and premature gap wear are signs of overheating, along with a gray or white "blistered" porcelain insulator surface. The most common cause for this condition is using a spark plug of the wrong heat range (too hot). If you have not changed to a hotter spark plug and the plug is overheated, consider the following causes:
a. Lean fuel mixture (incorrect main jet or incorrect oil pump adjustment).
b. Ignition timing too advanced.
c. Cooling system malfunction.
d. Engine air leak.
e. Improper spark plug installation (overtightening).
f. No spark plug gasket.

4

SPARK PLUG CONDITIONS

NORMAL USE

OIL FOULED

CARBON FOULED

OVERHEATED

GAP BRIDGED

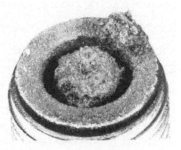

SUSTAINED PREIGNITION

WORN OUT

Worn out

Corrosive gases formed by combustion and high voltage sparks have eroded the electrodes. Spark plugs in this condition require more voltage to fire under hard acceleration. Replace with a new spark plug.

Preignition

If the electrodes are melted, preignition is almost certainly the cause. Check for carburetor mounting or intake manifold leaks and overadvanced ignition timing. It is also possible that a plug of the wrong heat range (too hot) is being used. Find the cause of the preignition before returning the engine into service.

HIGH-ALTITUDE CLUTCH TUNING

When the snowmobile is operated at an altitude of more than 4,000 feet (1,200 meters), the clutch should be changed to the specifications listed in **Tables 9-20** (for your model) to compensate for engine power loss. See *High Altitude Carburetor Tuning* in this chapter. If the clutch is not adjusted for altitude, engine bogging at the point of belt engagement will occur. In addition, the engine is easier to bog when running in deep snow. Both conditions can cause premature drive belt failure.

Refer to Chapter Eleven for complete clutch service procedures.

GEARING

Depending upon altitude, snow and track conditions, a different gear ratio may be required. Snow conditions that offer few rough sections require a higher gear ratio. Less optimum snow conditions or more rugged terrain require a lower gear ratio. Refer to the gear ratio chart in **Table 21** and the chain length chart in **Table 22**. Replacement sprockets and chains can be purchased through Ski-Doo dealers. Refer to Chapter Twelve for sprocket and chain replacement procedures.

SUSPENSION ADJUSTMENT

The suspension can be adjusted to accommodate rider weight and snow conditions.

Correct suspension adjustment is arrived at largely through a matter of trial-and-error "tuning." There are several fundamental points that must be understood and applied before the suspension can be successfully adjusted to your needs.

Ski pressure—the load on the skis relative to the load on the track—is the primary factor controlling handling performance. If the ski

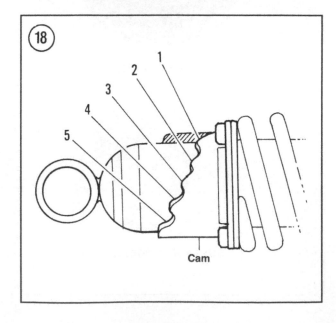

Cam

pressure is too light, the front of the machine tends to float and steering control becomes vague, with the machine tending to drive straight ahead rather than turn and wander when running straight at steady throttle.

On the other hand, if ski pressure is too heavy, the machine tends to plow during cornering and the skis dig in during straight line running rather than stay on top of the snow.

Ski pressure for one snow condition is not necessarily good for another condition. For instance, if the surface is very hard and offers little steering traction, added ski pressure—to permit the skis to "bite" into the snow—is desirable. Also, the hard surface will support the skis and not allow them to penetrate when the machine is running in a straight line under power.

On the other hand, if the surface is soft and tacky, lighter ski pressure is desirable to prevent the skis from sinking into the snow. Also, the increased traction afforded by the snow will allow the skis to turn with light pressure.

Good suspension adjustment involves a thorough analysis relating to ski pressure versus conditions. The suspension has been set at the factory to work in most conditions encountered by general riding. However, when the snowmobile is operated in varying or more difficult conditions, the suspension should be adjusted. It is important to remember that

suspension tuning is a compromise. An adjustment that works well in one situation may not work as well in another.

Front spring preload

The front spring (**Figure 17**) is provided with 5 preload positions. See **Figure 18**. The No. 1 position is soft and the No. 5 position hard. The No. 1 position works well when riding over small bumps and provides more positive steering. The No. 5 position reduces bottoming over large bumps but steering is reduced. The spring preload is changed by rotating the cam at the end of the spring. To adjust the front shock preload, remove the front shock as described in Chapter Thirteen. Use the adjustment key to turn the cam.

Rear spring preload

The rear spring (**Figure 19**) is provided with 5 preload positions. See **Figure 18**. The No. 1 position is soft and the No. 5 position hard. Always adjust both shocks to the same preload position. Initial preload adjustment can be set according to rider weight. Note the following:

a. Up to 140 lb. (64 kg): Position No. 1.
b. 140 lb. (64 kg) to 160 lb. (73 kg): Position No. 2.
c. 160 lb. (73 kg) to 180 lb. (82 kg): Position No. 3.
d. 180 lb. (82 kg) and up: Position No. 4 or No. 5.

The suspension should collapse 38 mm (1 1/2 in.) with a rider on the seat. After adjusting spring preload according to rider weight, consider track and speed conditions. Flat track conditions and low speeds require a smaller preload position. High terrain and fast speed conditions require a higher preload position.

If you are having difficulty adjusting the suspension with the stock spring, different spring rates are available from Ski-Doo dealers.

The rear spring preload is changed by rotating the cams (**Figure 19**) at the end of the spring.

Limiter Screw Adjustment (1986-on)

The limiter screw (**Figure 20**) controls weight transfer during acceleration. Adjustment is provided by changing the length of the limiter screw with the adjust nut. Note the following:

a. To provide less ski pressure and more speed, adjust the nut so that it is *further away* from the cotter pin (or end or screw).

b. To provide more ski pressure and more steering stability, adjust the nut so that it is *closer* to the cotter pin (or end of screw).

Table 1 CARBURETOR SPECIFICATIONS—1985 MX

	Sea level	4,000 ft.	6,000 ft.	8,000 ft.	10,000 ft.
Main jet					
PTO	240	220	180	170	160
MAG	250	220	210	200	180
Pilot jet	40	40	40	40	40
Needle jet	159 P4	159 P4	159 P4	159 P4	159 P4
Jet needle	6DH7	6DH7	6DH7	6DH7	6DH7
Clip position	3	3	3	3	3
Throttle valve	2.5	2.5	2.5	2.5	2.5
Pilot air screw	1 1/2	1/2	1 1/2	1 1/2	1 1/2
Idle speed	N/A	N/A	N/A	N/A	N/A

Table 2 CARBURETOR SPECIFICATIONS—1986 MX and MX (H/A)

	Sea level	4,000 ft.	6,000 ft.	8,000 ft.	10,000 ft.
Main jet					
PTO	220	195	185	175	160
MAG	240	210	200	190	175
Pilot jet	40	40	40	40	40
Needle jet	159 P4	159 P4	159 P4	159 P4	159 P4
Jet needle	6DH7	6DH7	6DH7	6DH7	6DH7
Clip position	3	3	3	3	3
Throttle valve	2.5	2.5	2.5	2.5	2.5
Pilot air screw	1 1/2	1 1/2	1 1/2	1 1/2	1 1/2
Idle speed	N/A	N/A	N/A	N/A	N/A

4

Table 3 CARBURETOR SPECIFICATIONS—1987-ON MX AND MX LT

	Sea level	4,000 ft.	6,000 ft.	8,000 ft.	10,000 ft.
Main jet					
PTO	220	200	185	175	165
MAG	240	220	200	190	175
Pilot jet	40	40	40	45	45
Needle jet	159 P4	159 P4	159 P4	159 P4	159 P4
Jet needle	6DH7	6DH7	6DH7	6DH7	6DH7
Clip position	3	3	3	3	3
Throttle valve	2.5	2.5	2.5	2.5	2.5
Pilot air screw	1 1/2	1 1/2	1 1/2	1	1
Idle speed	1800-2000	1800-2000	2400-2600	2400-2600	2400-2600

Table 4 CARBURETOR SPECIFICATIONS—1985 PLUS

	Sea level	4,000 ft.	6,000 ft.	8,000 ft.	10,000 ft.
Main jet					
PTO	300	270	250	240	220
MAG	350	320	300	280	260
Pilot jet	40	40	40	40	40
Needle jet	224 AA 5	224 AA 5	224 AA 5	224 AA 5	224 AA 5
Jet needle	7DH2	7DH2	7DH2	7DH2	7DH2
Clip position	2	2	2	1	1
Throttle valve	2.5	2.5	2.5	2.5	2.5
Pilot air screw	1	1	1	1	1
Idle speed	1800-2000	2800-3000	2800-3000	2800-3000	2800-3000

Table 5 CARBURETOR SPECIFICATIONS—1986 PLUS

	Sea level	4,000 ft.	6,000 ft.	8,000 ft.	10,000 ft.
Main jet					
PTO	330	290	280	260	240
MAG	350	310	290	280	260
Pilot jet	40	40	40	40	40
Needle jet	224 AA 5	224 AA 5	224 AA 5	224 AA 5	224 AA 5
Jet needle	7DH2	7DH2	7DH2	7DH2	7DH2
Clip position	2	2	2	2	2
Throttle valve	2.5	2.5	2.5	2.5	2.5
Pilot air screw	1	1	1	1	1
Idle speed	1800-2000	2800-3000	2800-3000	2800-3000	2800-3000

Table 6 CARBURETOR SPECIFICATIONS—1987-1988 PLUS

	Sea level	4,000 ft.	6,000 ft.	8,000 ft.	10,000 ft.
Main jet					
PTO	330	290	280	260	240
MAG	350	310	290	280	260
Pilot jet	40	40	40	60	60
Needle jet	224 AA 5	224 AA 5	224 AA 5	224 AA 5	224 AA 5
Jet needle	7DH2	7DH2	7DH2	7DH2	7DH2
Clip position	2	2	2	3	3
Throttle valve	2.5	2.5	2.5	2.5	2.5
Pilot air screw	1	1	1	2	2
Idle speed	1800-2000	1800-2000	2800-3000	2800-3000	2800-3000

Table 7 CARBURETOR SPECIFICATIONS—1989 PLUS AND PLUS LT

	Sea level	4,000 ft.	6,000 ft.	8,000 ft.	10,000 ft.
Main jet					
PTO	250	220	210	200	180
MAG	250	220	210	200	180
Pilot jet	30	30	30	30	30
Needle jet	159 Q 4	159 Q 4	159 Q 4	159 Q 4	159 Q 4
Jet needle	6FJ6	6FJ6	6FJ6	6FJ6	6FJ6
Clip position	3	3	3	3	3
Throttle valve	2.0	2.0	2.0	2.0	2.0
Pilot air screw	1 1/2	1 1/2	1 1/2	1 1/2	1 1/2
Idle speed	1800-2000	1800-2000	2400-2600	2400-2600	2400-2600

Table 8 CARBURETOR SPECIFICATIONS—1989 MACH I

	Sea level	4,000 ft.	6,000 ft.	8,000 ft.	10,000 ft.
Main jet					
PTO	260	240	210	190	180
MAG	290	260	230	210	200
Pilot jet	45	45	45	45	45
Needle jet	480 P 2	480 P 2	480 P 2	480 P 2	480 P 2
Jet needle	6DH8	6DH8	6DH8	6DH8	6DH8
Clip position	2	2	1	1	1
Throttle valve	2.5	2.5	2.5	2.5	2.5
Pilot air screw	1 1/2	1 1/2	3/4	3/4	3/4
Idle speed	1800-2000	1800-2000	2400-2600	2400-2600	2400-2600

4

Table 9 PRIMARY SHEAVE SPECIFICATIONS (1985-1986 MX AND MX [H/A])

	SEA LEVEL	4000 FT.	6000 FT.	8000 FT.	10,000 FT.
RETURN SPRING Color Part No.	Black 414 4784 00	Orange 414 4065 00	Orange 414 4065 00	Orange 414 4065 00	Orange 414 4065 00
CENTRIFUGAL LEVER ARM I.D. Part No.	A8S 860 4181 00	A8S 860 4181 00	A8S 860 4181 00	A8S 860 4181 00	A8S 860 4181 00
CALIBRATION WASHERS	6	0	0	0	0
GOVERNOR CUP OR RAMP All 1985 MX Part No.	M 504 0574 00	M 504 0574 00	M 505 0574 00	M 504 0574 00	M 505 0574 00
1986 MX and MX (H/A) Part No.	M 504 0574 00	MI 504 2543 00	MI 504 2543 00	MI 504 2543 00	MI 504 2543 00

Table 10 PRIMARY SHEAVE SPECIFICATIONS (1987-ON MX AND MX LT)

	SEA LEVEL	4000 FT.	6000 FT.	8000 FT.	10,000 FT.
RETURN SPRING Color Part No.	Blue/Yellow 420 4380 93	Blue/Yellow 420 4380 93	Blue/Yellow 420 4380 93	Blue/Yellow 420 4380 93	Blue/Yellow 420 4380 93
CENTRIFUGAL LEVER ARM	420 4481 96	420 4481 96	420 4481 96	420 4481 96	420 4481 96
CALIBRATION WASHERS	420 4291 40	420 4291 40	420 4291 40	420 4291 40	420 4291 40
GOVERNOR CUP OR RAMP I.D. Part No.	"140" 420 4801 40	"145" 420 4801 45	"145" 420 4801 45	"145" 420 4801 45	"145" 420 4801 45
CALIBRATION SCREW POSITION	3	3	3	3	5

Table 11 PRIMARY SHEAVE SPECIFICATIONS (1985 PLUS)

	SEA LEVEL	4000 FT.	6000 FT.	8000 FT.	10,000 FT.
RETURN SPRING Color Part No.	Orange 414 4065 00	White 414 4471 00	White 414 4471 00	White 414 4471 00	White 414 4471 00
CENTRIFUGAL LEVER ARM I.D. Part No.	A8S 860 4181 00	A8S 860 4181 00	A8S 860 4181 00	A8S 860 4181 00	A8S 860 4181 00
CALIBRATION WASHERS	6	0	0	0	0
GOVERNOR CUP OR RAMP	P1 504 0585 00	504 0589 00	505 0589 00	504 0589 00	505 0589 00

Table 12 PRIMARY SHEAVE SPECIFICATIONS (1986-1988 PLUS)

	SEA LEVEL	4000 FT.	6000 FT.	8000 FT.	10,000 FT.
RETURN SPRING Color Part No.	Blue/Yellow 420 4380 93	Black/Yellow 414 6055 00	Black/Yellow 414 6055 00	Black/Yellow 414 6055 00	Black/Yellow 414 6055 00
CENTRIFUGAL LEVER ARM	420 4481 95	420 4481 95	420 4481 95	420 4481 95	420 4481 95
CALIBRATION WASHERS	N.A	N.A.	N.A.	N.A.	N.A.
GOVERNOR CUP OR RAMP I.D. Part No.	"144" 420 4801 44	"145" 420 4801 45	"145" 420 4801 45	"145" 420 4801 45	"145" 420 4801 45
CALIBRATION SCREW POSITION	3	3	3	4	4

Table 13 PRIMARY SHEAVE SPECIFICATIONS (1989 PLUS AND PLUS LT)

	SEA LEVEL	4000 FT.	6000 FT.	8000 FT.	10,000 FT.
RETURN SPRING Color Part No.	Blue/Orange 420 4381 33	Yellow/Orange 420 4380 98	Yellow/Orange 420 4380 98	Yellow/Orange 420 4380 98	Yellow/Orange 420 4380 98
CENTRIFUGAL LEVER ARM	420 4481 96	420 4481 96	420 4481 96	420 4481 96	420 4481 96
CALIBRATION WASHERS	420 4291 40	420 4291 40	420 4291 40	420 4291 40	420 4291 40
GOVERNOR CUP OR RAMP I.D. Part No.	"149" 420 4801 49	"DA9" 504 2594 00	"DA9" 504 2594 00	"DA9" 504 2594 00	"DA9" 504 2594 00
CALIBRATION SCREW POSITION	3	3	3	3	3

Table 14 PRIMARY SHEAVE SPECIFICATIONS (1989 MACH1)

	SEA LEVEL	4000 FT.	6000 FT.	8000 FT.	10,000 FT.
RETURN SPRING Color Part No.	Red/Blue 420 4380 95	Yellow/Purple 414 6784 00	Yellow/Purple 414 6784 00	Yellow/Purple 414 6784 00	Yellow/Purple 414 6784 00
CENTRIFUGAL LEVER ARM	420 4481 96	420 4481 96	420 4481 96	420 4481 96	420 4481 96
CALIBRATION WASHERS	420 4291 40	420 4291 40	420 4291 40	420 4291 40	420 4291 40
GOVERNOR CUP OR RAMP I.D. Part No.	"147" 420 4801 47	"147" 420 4801 47	"147" 420 4801 47	"147" 420 4801 47	"147" 420 4801 47
CALIBRATION SCREW POSITION	3	4	4	5	5

Table 15 SECONDARY SHEAVE SPECIFICATIONS (1985-1986 MX AND MX [H/A])

	Sea level	4,000 ft.	6,000 ft.	8,000 ft.	10,000 ft.
Spring preload	13.5 lb.	15.5 lb.	15.5 lb.	15.5 lb.	15.5 lb.
Cam	Std.	Std.	Std.	Std.	Std.

Table 16   SECONDARY SHEAVE (1987-ON MX AND MX LT)

	SEA LEVEL	4000 FT.	6000 FT.	8000 FT.	10,000 FT.
SPRING PRELOAD	14 lbs. ± 2	16 lbs. ± 2	16 lbs. ± 2	16 lbs. ± 2	16 lbs. ± 2
CAM					
Angle	44°	44°	44°	36°	36°
Part No.	504 1282 00	504 1282 00	504 1282 00	504 1303 00	504 1303 00

Table 17 SECONDARY SHEAVE SPECIFICATIONS (1985 PLUS)

	Sea level	4,000 ft.	6,000 ft.	8,000 ft.	10,000 ft.
Spring preload	19 lb.	19 lb.	19 lb.	19 lb.	19 lb.
Cam	Std.	Std.	Std.	Std.	Std.

Table 18   SECONDARY SHEAVE (1986-1988 PLUS)

	SEA LEVEL	4000 FT.	6000 FT.	8000 FT.	10,000 FT.
SPRING PRELOAD	14 lbs. ± 2	16 lbs. ± 2	16 lbs. ± 2	16 lbs. ± 2	16 lbs. ± 2
CAM					
Angle	44°	44°	44°	36°	36°
Part No.	504 1282 00	504 1282 00	504 1282 00	504 1303 00	504 1303 00

Table 19   SECONDARY SHEAVE (1989 PLUS AND PLUS LT)

	SEA LEVEL	4000 FT.	6000 FT.	8000 FT.	10,000 FT.
SPRING PRELOAD	14 lbs. ± 2	16 lbs. ± 2	16 lbs. ± 2	16 lbs. ± 2	16 lbs. ± 2
CAM					
Angle	44°	36°	36°	36°	36°
Part No.	504 1348 00	504 1355 00	504 1355 00	504 1355 00	504 1355 00

Table 20   SECONDARY SHEAVE (1989 MACH I)

	SEA LEVEL	4000 FT.	6000 FT.	8000 FT.	10,000 FT.
SPRING PRELOAD	16 lbs. ± 2	18 lbs. ± 2	18 lbs. ± 2	18 lbs. ± 2	18 lbs. ± 2
CAM Angle Part No.	50° 504 1351 00	44° 504 1348 00	44° 504 1348 00	44° 504 1348 00	44° 504 1348 00

Table 21 CHAINCASE GEARING

	Sea level	4,000 ft.	6,000 ft.	8,000 ft.	10,000 ft.
MX, MX (H/A) and MX LT					
1985-1986	26/40	26/44	26/44	26/44	26/44
1987-on	22/44	20/44	20/44	20/44	20/44
Plus and Plus LT					
1985	26/38	26/44	26/44	26/44	26/44
1986-1988	20/38	20/44	20/44	20/44	20/44
1989	20/38	22/44	22/44	22/44	22/44
Mach I					
1989	22/40	22/44	22/44	22/44	22/44

Table 22 CHAIN LENGTH

Upper sprocket No. of teeth	Lower sprocket No. of teeth		
	38	40	44
20	68	68	72
22	70	70	72
26	70	72	74

Chapter Five

5

Engine

All of the Ski-Doo snowmobiles covered in this manual are equipped with a Rotax rotary valve, water-cooled, 2-stroke parallel twin engine. The Rotax engines are equipped with ball-type main crankshaft bearings and needle bearings on both ends of the connecting rods. Crankshaft components are available as individual parts. However, other than to replace the outer seals, it is recommended that the crankshaft work be entrusted to a dealer or other competent engine specialist.

This chapter covers information to provide routine top-end service as well as crankcase disassembly and crankshaft service.

Work on the snowmobile engine requires considerable mechanical ability. You should carefully consider your own capabilities before attempting any operation involving major disassembly of the engine.

Much of the labor charge for dealer repairs involves the removal and disassembly of other parts to reach the defective component. Even if you decide not to tackle the entire engine

overhaul after studying the text and illustrations in this chapter, it can be cheaper to perform the preliminary operations yourself and take the engine to your dealer. Since dealers have lengthy waiting lists for service (especially during the fall and winter season), this practice can reduce the time your unit is in the shop. If you have done much of the preliminary work, your repairs can be scheduled and performed much quicker.

Engine identification numbers are listed in **Table 1**. General engine specifications are listed in **Table 2**. Engine service specifications are listed in **Tables 3-6**. Engine tightening torques are listed in **Table 7**. **Tables 1-7** are found at the end of the chapter.

ENGINE NUMBER IDENTIFICATION

Bombardier uses a series of 3 numbers to identify Ski-Doo snowmobiles engines. Refer to **Table 1** for model listing and engine number identification.

ENGINE LUBRICATION

Engine lubrication is provided by the fuel/oil mixture used to power the engine. Refer to Chapter Eight for oil pump service.

SERVICE PRECAUTIONS

Whenever you work on your Ski-Doo, there are several precautions that should be followed to help with disassembly, inspection and reassembly.

1. In the text there is frequent mention of the left-hand and right-hand side of the engine. This refers to the engine as it is mounted in the frame, not as it sits on your workbench. See **Figure 1**.

2. Always replace a worn or damaged fastener with one of the same size, type and torque requirements. Make sure to identify each bolt before replacing it with another. Bolt threads should be lubricated with engine oil, unless otherwise specified, before torque is applied. If a tightening torque is not listed in **Table 7** (end of this chapter), refer to the torque and fastener information in Chapter One.

3. Use special tools where noted. In some cases, it may be possible to perform the procedure with makeshift tools, but this procedure is not recommended. The use of makeshift tools can damage the components and may cause serious personal injury. Where special snowmobile tools are required, these may be purchased through any Ski-Doo dealer. Other tools can be purchased through your dealer, or from a motorcycle or automotive accessory store. When purchasing tools from automotive or motorcycle accessory stores, remember that all threaded parts that screw into the engine must have metric threads.

4. Before removing the first bolt and to prevent frustration during assembly, get a number of boxes, plastic bags and containers and store the parts as they are removed (**Figure 2**). Also have on hand a roll of masking tape and a permanent, waterproof marking pen to label each part or assembly as required. If your snowmobile was purchased second hand and it appears that some of the wiring may have been changed or replaced, label each electrical connection before disconnecting it.

5. Use a vise with protective jaws to hold parts. If protective jaws are not available, insert wooden blocks on either side of the part(s) before clamping them in the vise.

6. Remove and install pressed-on parts with an appropriate mandrel, support and hydraulic press. **Do not** try to pry, hammer or otherwise force them on or off.

7. Refer to **Table 7** at the end of the chapter for torque specifications. Proper torque is essential to assure long life and satisfactory service from snowmobile components.

8. Discard all O-rings and oil seals during disassembly. Apply a small amount of heat durable grease to the inner lips of each oil seal to prevent damage when the engine is first started.

9. Keep a record of all shims and where they came from. As soon as the shims are removed,

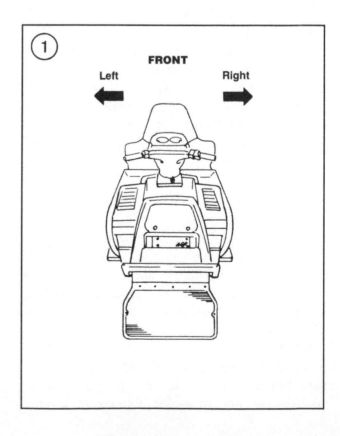

①

FRONT

Left ← → Right

inspect them for damage and write down their thickness and location.

10. Work in an area where there is sufficient lighting and room for component storage.

SERVICING ENGINE IN FRAME

Some of the components can be serviced while the engine is mounted in the frame:

a. Cylinder head.
b. Cylinder.
c. Piston.
d. Carburetor.
e. Magneto.
f. Oil pump.
g. Recoil starter.
h. Clutch assembly.

ENGINE REMOVAL

Engine removal and crankcase separation is required for repair of the "bottom end" (crankshaft, connecting rod and bearings).

1. Open the shroud all the way. Unplug the harness from the headlight and free the harness from the shroud. Disconnect the shroud restraining cable and with assistance, support the shroud and remove the bolts that attach the shroud hinge to the chassis. Remove the shroud

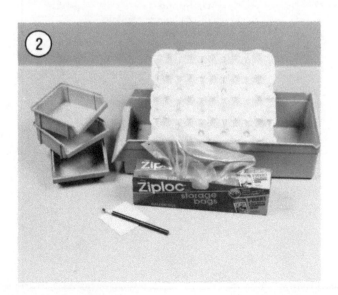

and set it well out of the way to prevent it from being damaged.

2. See **Figure 3** or **Figure 4**. Remove the exhaust assembly as follows:

a. Disconnect the tailpipe springs (A, **Figure 5**) at the muffler and exhaust pipe.
b. Remove the tailpipe and muffler (B, **Figure 5**).
c. Disconnect the long exhaust pipe spring (**Figure 6**).
d. Disconnect the 2 exhaust pipe springs at the manifold (**Figure 7**) and remove the exhaust pipe.

3. Remove the drive belt as described in Chapter Eleven.

4. Remove the primary sheave as described in Chapter Eleven.

5. Drain the cooling system as described under *Coolant Change* in Chapter Three.

6. Remove the recoil starter as described in Chapter Ten.

7. Remove the air silencer as described in Chapter Six.

8. Unscrew the carburetor cap and remove the throttle valves (slides) from the carburetors. Be careful that you do not damage the jet needles. Disconnect the throttle cable (**Figure 8**) from the slide. Reinstall the slide and cap. Repeat for the opposite carburetor.

9. Disconnect the oil pump cable at the oil pump (**Figure 9**).

10. Disconnect and plug the oil hose at the oil pump.

11. Remove the carburetors as described in Chapter Six.

12. Disconnect the coolant hose at the cylinder head. A, **Figure 10** shows the hose connection for 467 engines; other models are similar.

13. *1985-1988:* These models are equipped with a rotary valve oil tank (B, **Figure 10**). It will be easier (and less messy) to remove the oil tank with the engine. Remove the bolts holding the oil tank and allow it to hang in place during engine removal.

③

**EXHAUST SYSTEM
(1985-ON MX, MX [H/A], MX LT,
PLUS AND PLUS LT)**

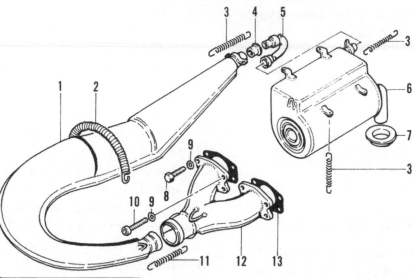

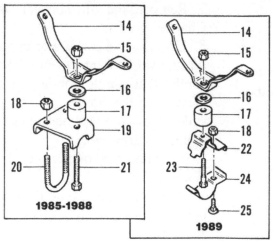

1985-1988

1989

1. Exhaust pipe
2. Spring
3. Spring
4. Ring
5. Tail pipe
6. Muffler
7. Grommet
8. Bolt
9. Lockwasher
10. Bolt
11. Spring
12. Exhaust manifold
13. Gasket
14. Support
15. Nut
16. Asbestos washer
17. Rubber spacer
18. Nut
19. Clamp bracket
20. U-clamp
21. Bolt
22. Clamp bracket
23. Bolt
24. Clamp bracket
25. Bolt

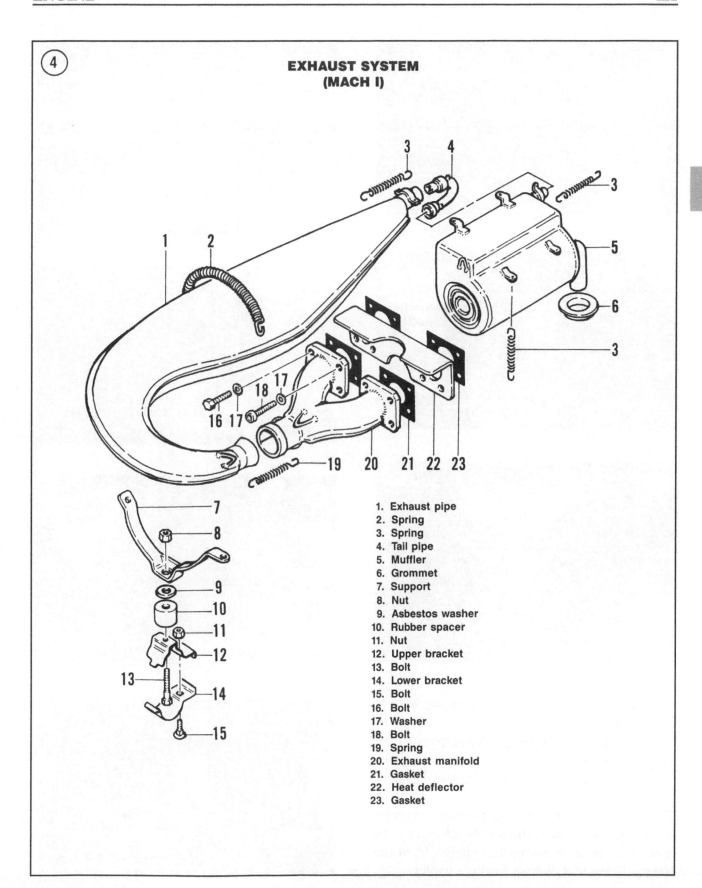

④ **EXHAUST SYSTEM**
(MACH I)

1. Exhaust pipe
2. Spring
3. Spring
4. Tail pipe
5. Muffler
6. Grommet
7. Support
8. Nut
9. Asbestos washer
10. Rubber spacer
11. Nut
12. Upper bracket
13. Bolt
14. Lower bracket
15. Bolt
16. Bolt
17. Washer
18. Bolt
19. Spring
20. Exhaust manifold
21. Gasket
22. Heat deflector
23. Gasket

14. Disconnect the fuel pump impulse hose at the crankcase.

15. Disconnect the CDI electrical connectors. See **Figure 11**, typical.

16. Disconnect the spark plug caps at the spark plugs.

17. If necessary, the engine top end (cylinder head, pistons and cylinder blocks) can be removed before removing the engine from the frame. Refer to *Cylinder* in this chapter.

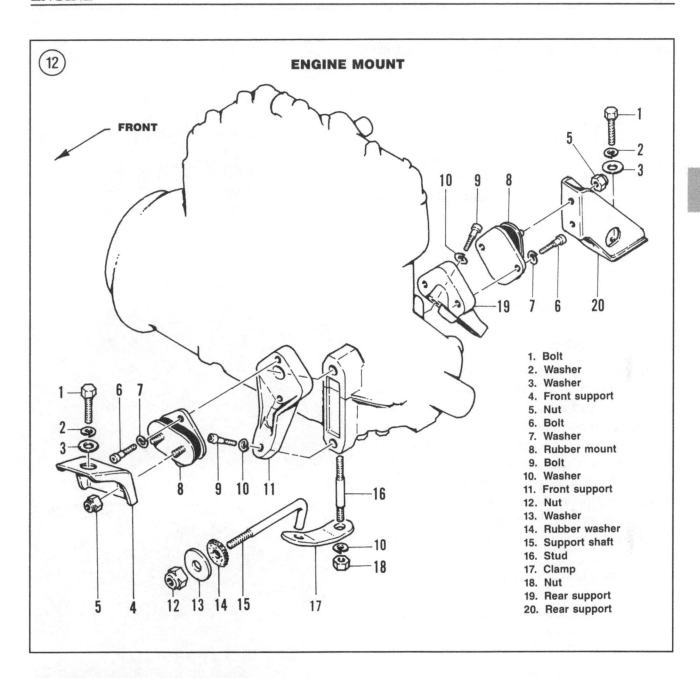

⑫   **ENGINE MOUNT**

FRONT

1. Bolt
2. Washer
3. Washer
4. Front support
5. Nut
6. Bolt
7. Washer
8. Rubber mount
9. Bolt
10. Washer
11. Front support
12. Nut
13. Washer
14. Rubber washer
15. Support shaft
16. Stud
17. Clamp
18. Nut
19. Rear support
20. Rear support

NOTE
Because of the number of bushings, washers and rubber dampers used on the engine bracket assemblies, refer to Figure 12 when performing Step 18 and Step 19.

18. Loosen and remove the engine torque rod nut and disconnect the torque rod from the engine clamp.

19. See **Figure 12**. Loosen the front (**Figure 13**) and rear engine bracket nuts. Remove the nut,

washer and rubber damper from the front engine brackets. Remove the nut and washer from the rear bracket.

20. Check to make sure all of the wiring and hoses have been disconnected from the engine.

21. With at least one assistant, lift the engine up and remove it from the frame. Take it to a workbench for further disassembly.

22. Make sure the engine (**Figure 14**) is placed on a sturdy workbench.

ENGINE MOUNTS

Removal/Installation

The engine mounts and supports are a critical part of the snowmobile drive train. Damaged or loose engine mounts will allow the engine to shift or pull out of alignment during operation. This condition will cause primary and secondary sheave misalignment that will result in drive belt wear and reduced performance. The engine mount and support assembly should be inspected carefully whenever the engine is removed from the frame or if clutch misalignment becomes a problem.

Refer to **Figure 12** when performing this procedure.

1. To ease installation, mark each of the engine mounts, supports and rubber mounts for position during removal. In addition, different length bolts are used to hold the mounts and supports. Identify bolts before removal and store accordingly.

2. Remove the front engine mounts, rubber mounts and engine supports (**Figure 15**) as follows:

   a. Loosen and remove the front engine mount bolts (A, **Figure 16**) and remove the engine mount (B).

   b. Remove the Allen bolt (C, **Figure 16**) holding the rubber mount to the front support and remove the rubber mount.

c. Remove the front support bolts (D, **Figure 16**) and remove the support.

3. Remove the engine torque rod clamp (**Figure 17**) from the left-hand side of the engine.

4. Remove the rear engine mounts, rubber mounts and engine supports (**Figure 18**) as follows:

    a. Loosen and remove the rear engine mount bolts (A, **Figure 19**) and remove the engine mount (B).

    b. Remove the Allen bolt (C, **Figure 19**) holding the rubber mount to the front support and remove the rubber mount.

    c. Remove the front support bolts (D, **Figure 19**) and remove the support.

5. Installation is the reverse of these steps. Note the following:

    a. Check the engine mounts, rubber mounts and engine supports for cracks or damage. Check the rubber mounts for separation or other damage.

    b. Check all of the engine and support mount bolts for thread damage. Replace damaged bolts with the same grade bolt. Weaker bolts will loosen and allow the engine to slip.

    c. Inspect the washers for splitting or other damage.

    d. Check the support plate tapped holes in the crankcase (**Figure 20**) for stripped threads or debris buildup. Clean threads with a suitable size metric tap. If Loctite was previously used, make sure to remove all traces of Loctite residue before reinstalling bolts.

    e. Check the front and rear engine mount brackets in the frame (A, **Figure 21**) for cracks or other signs of damage. Check the bracket tapped holes for thread damage or debris buildup. Clean threads as described in sub-step d.

    f. Replace worn or damaged parts as required.

    g. Tighten the engine mount, rubber mount and support bolts to the torque specification in **Table 7**.

ENGINE INSTALLATION

1. Wash the engine compartment with clean water.

2. Spray all of the exposed electrical connectors with a spray electrical contact cleaner.

3. Before installing the engine, now is a good time to check components which are normally inaccessible for visual inspection. Note the following:

   a. Check the inlet coolant hose (**Figure 22**) for damage. Check the hose connection at the radiator (B, **Figure 21**) for looseness or damage. Make sure the water clamp is tight. If you are replacing the inlet coolant hose, make sure it is positioned underneath the tie rod as shown in **Figure 22**.

   b. Check the outlet (A, **Figure 23**) and inlet (B, **Figure 23**) cooling hoses as described in sub-step a.

   c. Check the steering shaft clamp (A, **Figure 24**) for tightness.

   d. Check all of the steering component tightening torques as described in Chapter Thirteen.

   e. Check the ignition coil mounting bracket for tightness.

   f. Check the fuel pump (B, **Figure 24**) mounting bolts for tightness.

   g. Check the fuel pump outlet and pulse hoses for age deterioration, cracks or other damage; replace damaged hoses as required.

4. Examine the engine mounts and supports as described in this chapter.

5. Check inside the frame for tools or other objects that may interfere with engine installation. Make sure all wiring harnesses are routed and secured properly.

6. Install the front and rear engine supports and mounts onto the engine as described in this chapter.

7. With an assistant, place the engine part way in the frame.

NOTE
To ease installation of the coolant hoses in the following steps, coat the inside of the hose where it slides on the mating joint with antifreeze.

8. It will be easier to connect the inlet coolant hose (**Figure 22**) at the water pump before the engine is lowered all the way in the frame. Attach the coolant hose to the water pump and secure the end of the hose with the hose clamp. Make sure the hose is not twisted and that the hose clamp will not contact the frame when the engine is lowered in the frame. When the hose is securely attached, lower the engine onto the

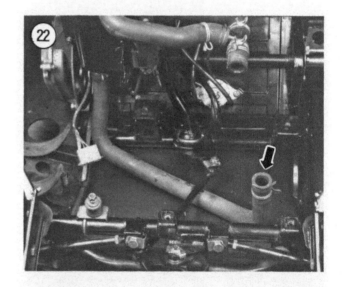

frame brackets. Check the hose clamp to make sure it is free of the frame.

9. Connect the engine torque rod at the clamp on the left-hand side of the engine. See **Figure 12**.

10. Install the engine mount bolts (**Figure 12**). Do not tighten the bolts at this point.

11. Install the primary sheave and drive belt as described in Chapter Eleven.

12. Align the engine/clutch assembly as described in Chapter Eleven. Tighten the engine mount bolts to the torque specifications in **Table 7**.

13. If the engine top end components were previously removed, install them as described in this chapter.

14. Reconnect the spark plug caps.

15. Reconnect all electrical connectors.

16. Reconnect the pulse hose at the crankcase nozzle and secure the hose with the metal clamp. The pulse hose is the center hose mounted on the fuel pump as shown in C, **Figure 24**.

17. *1985-1988:* Align the rotary valve oil tank with its mounting bracket; install and tighten the tank mounting bolt.

18. Reconnect the coolant hoses at the cylinder head. Position and secure the hose clamp with a pair of pliers.

19. Reconnect the oil injection hose at the oil pump.

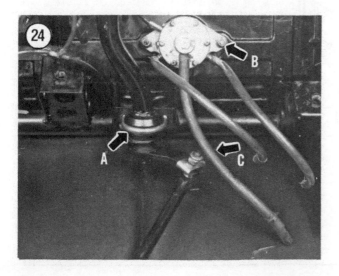

20. Reconnect the oil pump cable at the oil pump (**Figure 9**).

> *CAUTION*
> *Do not adjust the oil pump cable until the carburetors have been synchronized as described in Chapter Three.*

21. Install the carburetors as described in Chapter Six.

> *NOTE*
> *While the left- and right-hand carburetor assemblies are identical, their jetting is different. Make sure to follow the identification and installation information given in Chapter Six.*

22. Perform the carburetor synchronization procedure as described in Chapter Three.

23. Adjust the oil pump cable as described in Chapter Three.

24. Bleed the oil pump as described in Chapter Eight.

25. Install the air silencer as described in Chapter Six.

26. Install the recoil starter assembly as described in Chapter Ten. Check starter operation.

27. Refill the cooling system as described in Chapter Nine.

28. Install the exhaust pipe and muffler. See **Figure 3** or **Figure 4**. Make sure all springs are properly attached.

29. Install the engine shroud. Reconnect the headlight electrical connector.

ENGINE TOP END

The engine "top end" consists of the cylinder head, cylinder blocks, pistons, piston rings, piston pins and the connecting rod small-end bearings.

The engine top end can be serviced with the engine installed in the frame. However, the following service procedures are shown with the engine removed for clarity.

Refer to illustration for your model when servicing the engine top end.

a. **Figure 25**: Engine 467 (all 1985-on MX, MX [H/A] and MX LT).

b. **Figure 26**: Engine 537 (1985-1988 Plus).

c. **Figure 27**: Engine 536 (1989 Plus and Plus LT).

d. **Figure 28**: Engine 583 (1989 Mach I).

Cylinder Head
Removal/Installation

CAUTION
To prevent warpage and damage to any component, remove the cylinder head

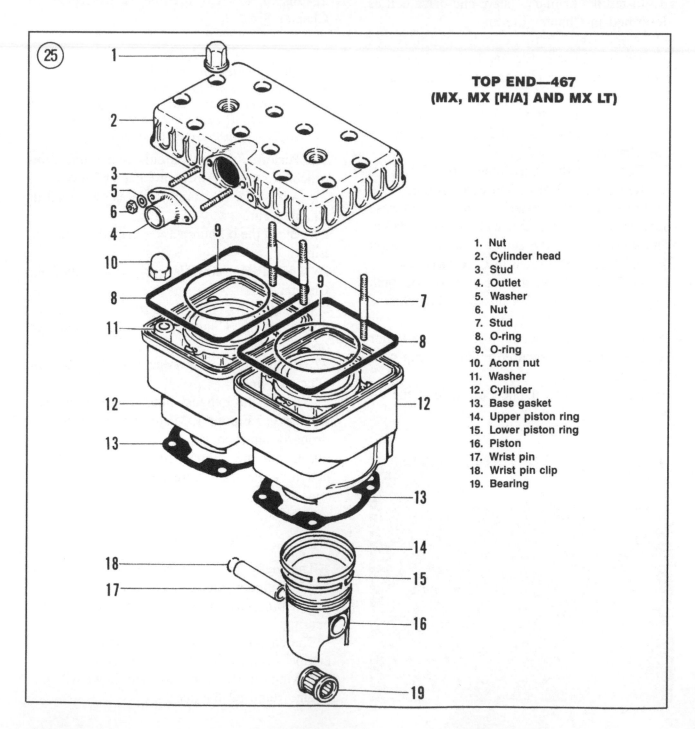

TOP END—467
(MX, MX [H/A] AND MX LT)

1. Nut
2. Cylinder head
3. Stud
4. Outlet
5. Washer
6. Nut
7. Stud
8. O-ring
9. O-ring
10. Acorn nut
11. Washer
12. Cylinder
13. Base gasket
14. Upper piston ring
15. Lower piston ring
16. Piston
17. Wrist pin
18. Wrist pin clip
19. Bearing

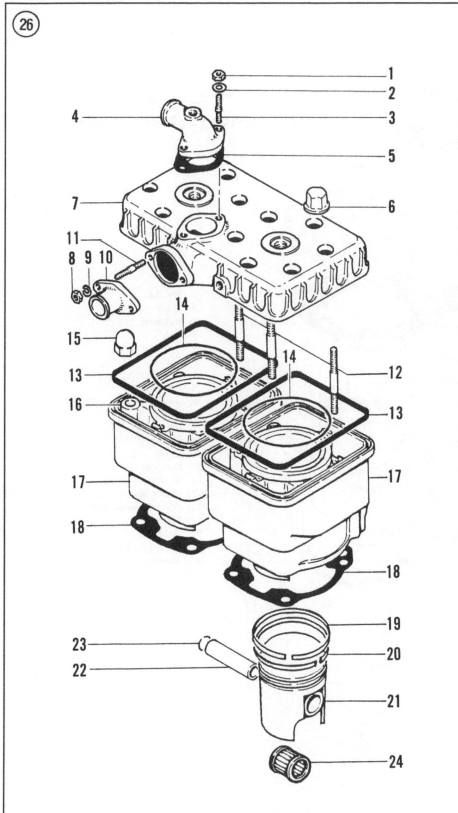

**TOP END—537
(1985-1988 PLUS)**

1. Nut
2. Washer
3. Stud
4. Housing
5. Gasket
6. Nut
7. Cylinder head
8. Nut
9. Washer
10. Housing
11. Stud
12. Stud
13. O-ring
14. O-ring
15. Nut
16. Washer
17. Cylinder
18. Base gasket
19. Upper piston ring
20. Lower piston ring
21. Piston
22. Wrist pin
23. Wrist pin clips
24. Bearing

26

5

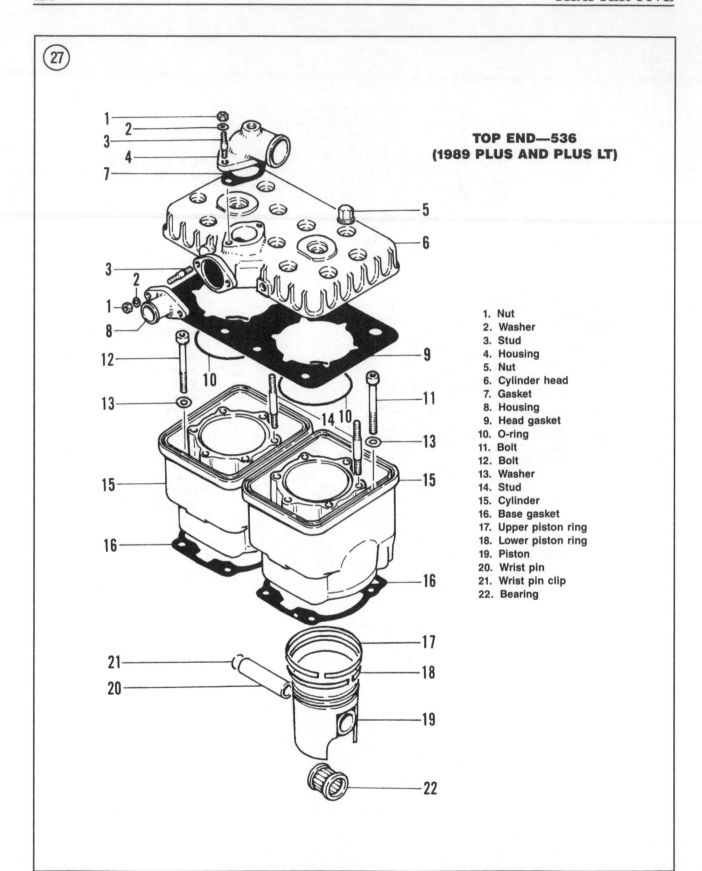

**TOP END—536
(1989 PLUS AND PLUS LT)**

1. Nut
2. Washer
3. Stud
4. Housing
5. Nut
6. Cylinder head
7. Gasket
8. Housing
9. Head gasket
10. O-ring
11. Bolt
12. Bolt
13. Washer
14. Stud
15. Cylinder
16. Base gasket
17. Upper piston ring
18. Lower piston ring
19. Piston
20. Wrist pin
21. Wrist pin clip
22. Bearing

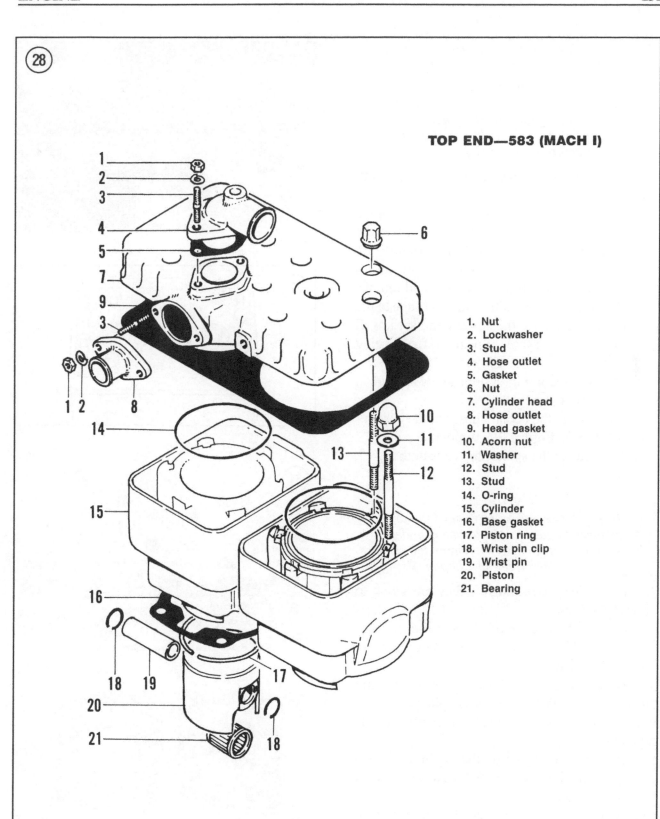

TOP END—583 (MACH I)

1. Nut
2. Lockwasher
3. Stud
4. Hose outlet
5. Gasket
6. Nut
7. Cylinder head
8. Hose outlet
9. Head gasket
10. Acorn nut
11. Washer
12. Stud
13. Stud
14. O-ring
15. Cylinder
16. Base gasket
17. Piston ring
18. Wrist pin clip
19. Wrist pin
20. Piston
21. Bearing

only when the engine is at room temperature.

NOTE
*If the engine is being disassembled for inspection procedures, check the compression as described in Chapter Three **before** disassembly.*

1. If the engine is mounted in the frame, perform the following:
   a. Open the engine shroud all the way. Unplug the harness from the headlight and free the harness from the shroud. Disconnect the shroud restraining cable and with assistance, support the shroud and remove the bolts that attach the shroud hinge to the chassis. Remove the shroud and set it well out of the way to prevent it from being damaged.
   b. Drain the cooling system as described in Chapter Three.
   c. Disconnect the coolant hoses at the cylinder head. See **Figure 29**, typical.
   d. Disconnect the spark plug caps at the spark plugs.

NOTE
It may be necessary to remove the air silencer to gain access to one cylinder head mounting nut. If so, remove the air silencer as described in Chapter Six.

2. Loosen the spark plugs if they are going to be removed later.

3. Referring to **Figure 30**, loosen the cylinder head nuts and bolts in two to three steps following a crisscross pattern.

4. Loosen the cylinder head by tapping around the perimeter with a rubber or plastic mallet. Remove the cylinder head. See **Figure 31**, typical.

5. Remove and discard the cylinder head gasket(s). See **Figures 25-28** for gasket removal for your model.

6. Lay a rag over the cylinder block to prevent dirt from falling into the cylinders.

7. Inspect the cylinder head as described in this chapter.

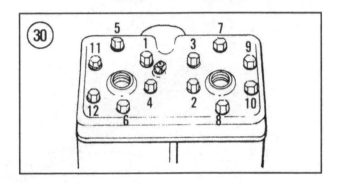

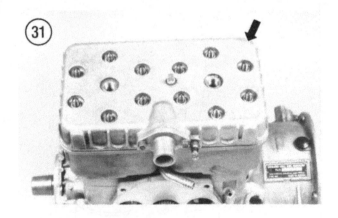

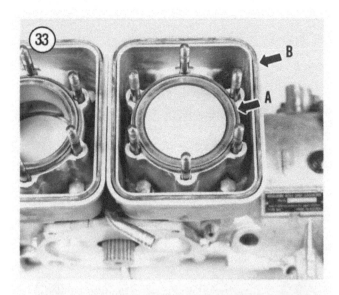

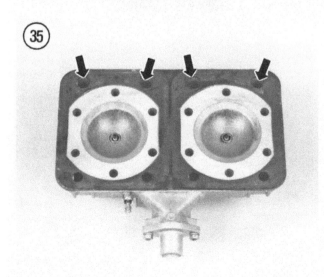

NOTE
While the cylinder head is removed, check the cylinder studs (Figure 32) for stripped threads, looseness or other damage. If necessary, remove the cylinder and replace damaged studs as described in this chapter.

8. Install the head gasket(s) as follows:
   a. *467 and 537:* See **Figure 33**. Install the inner (A) and outer (B) O-rings into the cylinder O-ring grooves. Check that each O-ring seats squarely in the groove.
   b. *536 and 583:* Install the inner O-ring in the cylinder groove (**Figure 34**). The head gasket is made with a series of 4 large holes and 4 small holes along the top and bottom edges of the gasket. Install the head gasket so that the large holes are on the exhaust side of the cylinder head. See **Figure 35**.

9. Install the cylinder head so that the thermostat mounting area faces to the rear of the engine. See **Figures 25-28**.

10. Lightly lubricate the cylinder studs and install the Acorn nuts (**Figure 36**) finger-tight.

NOTE
*If you replaced the cylinder studs and one or more Acorn nuts bottom out against a stud, the stud was not installed deep enough or the wrong end of the stud was threaded into the cylinder. Refer to **Cylinder Head Stud Replacement** in this*

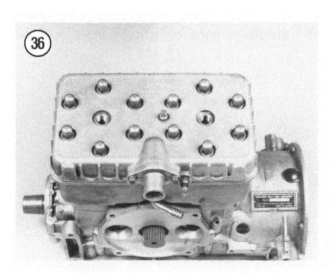

chapter for the correct procedure. If you are still having problems, install a washer underneath the nut until you can reposition or replace the stud.

11. Tighten the nuts in two to three steps following a crisscross pattern (**Figure 30**) to the torque specification in **Table 7**.

12. If the spark plugs were removed, gap and install them as described in Chapter Three.

13. If the engine is installed in the frame, note the following:

    a. Connect the spark plug caps.

    b. Connect the coolant hoses at the cylinder head (**Figure 29**, typical).

    c. Refill the cooling system as described in Chapter Three.

    d. Reinstall the engine shroud.

Inspection

1. Wipe away any soft deposits on the cylinder head (**Figure 37**). Hard deposits should be removed with a wire brush mounted in a drill or drill press or with a soft metal scraper (**Figure 38**). Be careful not to gouge the aluminum surfaces. Burrs created from improper cleaning will cause preignition and heat erosion.

> *NOTE*
> *Always use an aluminum thread fluid or Kerosene on the thread chaser and cylinder head threads when performing Step 2.*

2. With the spark plug removed, check the spark plug threads (**Figure 39**) in the cylinder head for

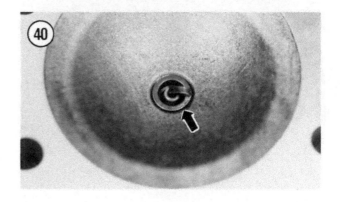

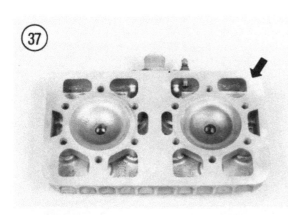

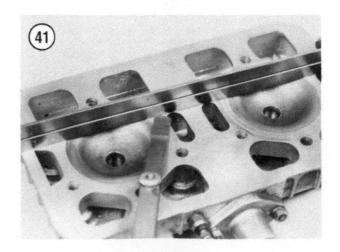

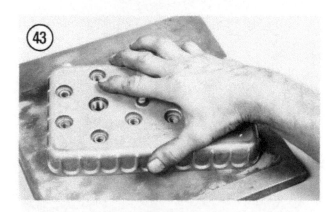

any signs of carbon buildup or cracking. The carbon can be removed with a 14 mm spark plug chaser. After cleaning the threads, reinstall the spark plug and make sure it can be installed all the way (**Figure 40**).

3. Use a straightedge and feeler gauge and measure the flatness of the cylinder head (**Figure 41**). If the cylinder head shows signs of minor warpage, resurface the cylinder head as follows:

a. Tape a piece of 400-600 grit wet emery sandpaper onto a piece of thick plate glass or surface plate (**Figure 42**).

b. Slowly resurface the head by moving it in figure-eight patterns on the sandpaper (**Figure 43**).

c. Rotate the head several times to avoid removing too much material from one side. Check progress often (**Figure 44**) with the straightedge and feeler gauge. See **Figure 41**.

d. If the cylinder head warpage still exceeds the service limit, it will be necessary to have the head resurfaced by a machine shop familiar with snowmobile and motorcycle service. Note that removing material from the cylinder head mating surface will change the compression ratio. Consult with the machinist on how much material was removed.

4. Check the cylinder head water passages (**Figure 45**) for coolant residue and sludge buildup. Clean passages thoroughly with solvent and dry thoroughly.

5. Wash the cylinder head in hot soapy water and rinse thoroughly before installation.

CYLINDER

An aluminum cylinder block is used with a cast iron liner pressed into the block. When severe wear is experienced, the cylinder liner can be bored oversize and new pistons and rings installed.

Refer to illustration for your model when servicing the cylinder.

a. **Figure 25**: Engine 467 (all 1985-on MX, MX [H/A] and MX LT).
b. **Figure 26**: Engine 537 (1985-1988 Plus).
c. **Figure 27**: Engine 536 (1989 Plus and Plus LT).
d. **Figure 28**: Engine 583 (1989 Mach I).

Removal

1. Remove the cylinder head as described in this chapter.

2. If the engine is installed in the frame, remove the muffler and exhaust pipe as described in Chapter Six.

3A. *Engines 467, 537 and 536:* Remove the bolts holding the exhaust manifold (**Figure 46**) onto the cylinders. Remove the exhaust manifold (**Figure 46**) and both gaskets.

3B. *Engine 583:* See **Figure 4**. Remove the bolts holding the exhaust manifold onto the cylinders. Remove the exhaust manifold, gaskets and heat deflector.

4. Gradually loosen the 4 cylinder Acorn nuts in a crisscross pattern. Remove the nuts (**Figure 47**) and washers (**Figure 48**).

> *NOTE*
> *Mark the left-hand cylinder with PTO before removal. Both cylinders must be reinstalled in their original mounting positions.*

5. Rotate the engine so that the piston is at the bottom of its stroke. Pull the cylinder block (**Figure 49**) straight up and off the crankcase studs.

6. Repeat for the opposite side.

7. Remove the cylinder base gasket and discard it.

8. Stuff clean rags around the connecting rods to keep dirt and loose parts from entering the crankcase.

9. *Engine 583:* The 583 cylinders are equipped with the Rotax Automatic Variable Exhaust (RAVE) system. If necessary, remove and service the RAVE assembly as described in this chapter.

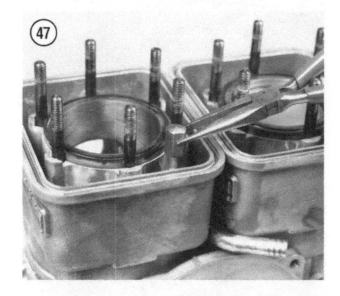

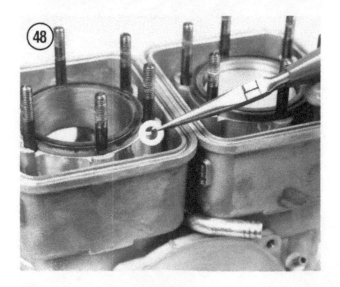

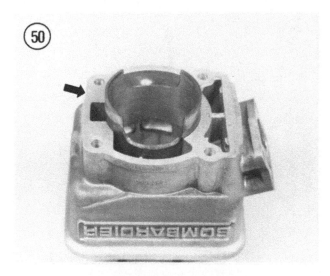

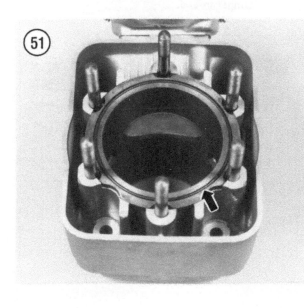

Inspection

Cylinder measurement requires a precision inside micrometer or bore gauge. If you don't have the right tools, have your dealer or a machine shop take the measurements.

1. Remove all gasket residue from the bottom (**Figure 50**) gasket surface.

2. Clean the cylinder O-ring groove (**Figure 51**) with a scribe or other sharp tool.

3. Clean the exhaust manifold sealing area on the cylinder block (A, **Figure 52**) of all gasket residue.

NOTE
On 583 model engines, remove the RAVE valve before cleaning the exhaust port in Step 4. Remove the RAVE valve as described in this chapter.

4. Using a soft scraper or a wire wheel mounted on a drill or hand grinder, remove all carbon deposits from the exhaust port (B, **Figure 52**). If you don't have access to a drill or hand grinder, the port can be cleaned with a medium to coarse grade emery paper. You can either work the emery paper by hand or wrap it around a length of wooden dowel.

CAUTION
When cleaning the exhaust port in Step 4, do not allow the grinder to slip inside the cylinder and damage the cylinder lining. Work carefully when using power equipment.

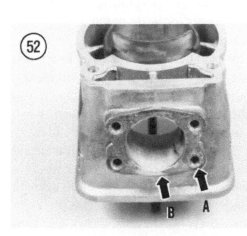

5. Wash the cylinder in solvent to remove any oil and carbon particles. The cylinder bore must be cleaned thoroughly before attempting any measurement as incorrect readings may be obtained.

6. Measure the cylinder bore diameter as described under *Piston/Cylinder Clearance* in this chapter.

> *NOTE*
> *If the cylinder diameters are still within specifications, it is possible to buy new pistons without reboring. New pistons will take up some of the excessive piston-to-cylinder clearance. However, do not install new pistons in a cylinder that is worn past the wear limit.*

7. If the cylinder is not worn past the service limit, check the bore carefully for scratches or gouges. The bore may require reconditioning.

> *NOTE*
> *If the engine experienced a slight seizure, there may be bits of aluminum stuck to the cylinder wall. A machine shop may be able to remove the aluminum without having to rebore the cylinder. However, if the surface of the cylinder wall has been scarred by the seizure, the cylinder will require reboring.*

8. Check the threaded holes in the cylinder block for thread damage. Minor damage can be cleaned up with a suitable metric tap. Refer to Chapter One for information pertaining to threads, fasteners and repair tools. If damage is severe, a thread insert should be installed.

9. Check the cylinder block studs (**Figure 53**) for stripping or other damage. If necessary, replace studs as described in this chapter.

10. After the cylinder has been serviced, wash the bore in hot soapy water. This is the only way to clean the cylinder wall of the fine grit material left from the boring or honing job. After washing the cylinder wall, run a clean white cloth through it. The cylinder wall should show no traces of grit or other debris. If the rag shows any sign

of debris, the cylinder wall is not clean and must be rewashed. After the cylinder is thoroughly cleaned, lubricate the cylinder walls with clean engine oil to prevent the cylinder liners from rusting.

> *CAUTION*
> *A combination of soap and water is the only solution that will completely clean the cylinder wall. Solvent and kerosene cannot wash fine grit out of cylinder crevices. Grit left in the cylinder will act as a grinding compound and cause premature wear to the new rings.*

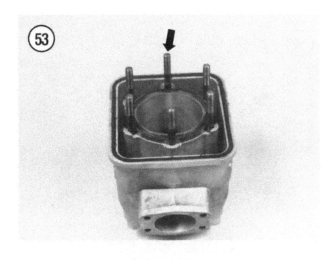

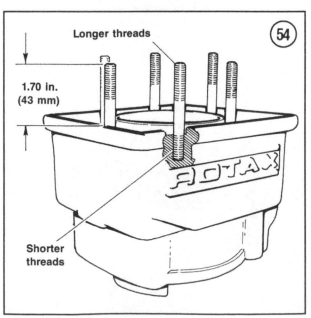

Cylinder Stud Replacement

Damaged cylinder studs will allow combustion gases to escape, a condition that will reduce engine performance and cause engine damage.

Because Acorn nuts are used to tighten the cylinder head, a specified stud length (**Figure 54**) must be maintained to prevent the nut from bottoming out before it is properly torqued. Replace damaged studs as follows:

A tube of Loctite 242 (blue), 2 nuts, 2 wrenches and a new stud will be required during this procedure. Replace damaged studs as follows:

1. Thread 2 nuts onto the damaged cylinder stud. Tighten the 2 nuts against each other so that they are locked.

NOTE
If the threads on the damaged stud do not allow installation of the 2 nuts, you

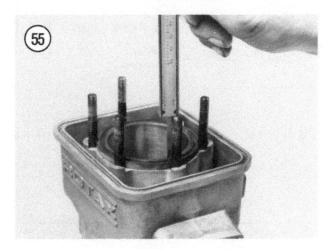

will have to remove the stud with a pair of Vise Grips.

2. Turn the bottom nut counterclockwise and unscrew the stud.
3. Clean the threads with solvent or electrical contact cleaner and allow to dry thoroughly.

NOTE
When installing the new stud, install it so that the end with the shorter threads face down (Figure 54).

4. Install 2 nuts on the top half of the new stud as in Step 1. Make sure they lock securely.
5. Coat the bottom half of a new stud with Loctite 242 (blue).
6. Turn the top nut clockwise and thread the new stud securely. Measure the portion of the installed stud from the cylinder surface to the top of the stud (**Figure 55**). The stud height should be 43 mm (1.70 in.). If not, reposition the stud as necessary.
7. Remove the nuts and repeat for each stud as required.
8. Follow Loctite's directions on cure time before assembling the component.

Installation

The cylinders must be properly aligned during installation so that the cylinder head stud holes will align with the cylinder studs. Ski-Doo offers special aligning tools for this purpose. The following procedures will describe cylinder installation and alignment with and without the special tools. If you want to purchase the aligning tools, make sure to purchase the correct tool for your engine.

1. *583 engines:* Install the RAVE assembly, if previously removed, as described in this chapter.
2. Clean the cylinder bore as described under *Inspection* in this chapter.
3. Check that the top surface of the crankcase and the bottom cylinder surface are clean prior to installation.
4. Install new base gaskets (**Figure 56**).

NOTE
Check the pistons to be sure the wrist pin clips are installed and correctly positioned.

5. Make sure the end gaps of the piston rings are lined up with the locating pins in the ring grooves (**Figure 57**). Lightly oil the piston rings and the inside of the cylinder bores with injection oil.

6. Place a piston holding tool under one piston and turn the crankshaft until the piston is down firmly against the tool. This will make cylinder installation easier. You can make this tool out of wood as shown in **Figure 58**.

NOTE
When installing the cylinders, make sure to install the cylinder previously marked

PTO on the left-hand side. If the cylinders were rebored, make sure to match the cylinder with its individual piston.

7. Align the cylinder with the piston so that the exhaust port will face toward the front of the engine. Install the cylinder block on the crankcase studs, compressing each piston ring with your fingers as the cylinder starts to slide over it. When both rings are in the cylinder, slide the cylinder all the way down. If the rings are hard to compress, you can use a large hose clamp as an effective compressor. Make sure the

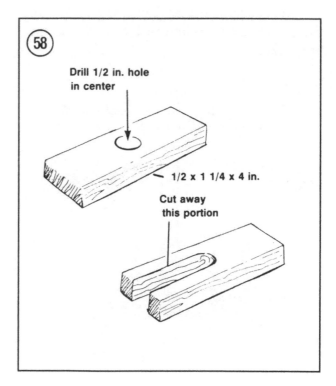

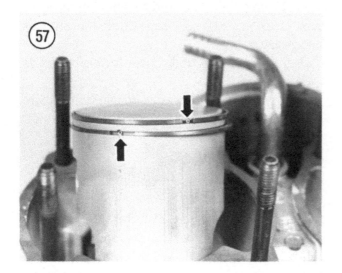

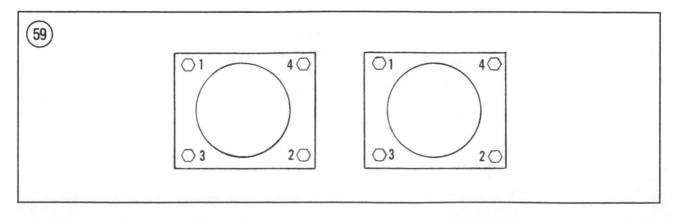

cylinder block is fully seated on the crankcase. Remove the piston holding tool and hose clamp.

8. Repeat for the opposite cylinder.

9A. *Cylinder alignment and installation without special tool:* Align the cylinders as follows:

a. Install the cylinder washers (**Figure 48**) and nuts (**Figure 47**). Install nuts finger-tight.

b. Turn crankshaft one revolution to align cylinders. Install the cylinder head (upside down with spark plugs removed) and turn the crankshaft again.

c. Install the exhaust manifold and secure it with its mounting bolts finger-tight.

d. Turn the crankshaft once again to align cylinders.

NOTE
When tightening the cylinder nuts in sub-step e, do not hold onto the cylinder

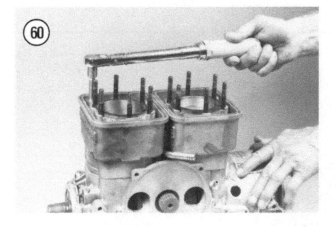

blocks for support. Hold onto the crankcase.

e. Remove the cylinder head and tighten the cylinder nuts in a crisscross pattern (**Figure 59**) to the torque specification in **Table 7**. See **Figure 60**.

f. Check cylinder alignment with the cylinder head.

g. Remove the exhaust manifold.

9B. *Cylinder alignment and installation with Ski-Doo special tool:*

a. Install the cylinder washers (**Figure 48**) and nuts (**Figure 47**). Install nuts finger-tight.

b. Install the Ski-Doo cylinder aligning tools onto the cylinders as shown in **Figure 61**.

c. Install the exhaust manifold and secure it with its mounting bolts finger-tight.

NOTE
When tightening the cylinder nuts in sub-step d, do not hold onto the cylinder blocks for support. Hold onto the crankcase.

d. Tighten the cylinder nuts in a crisscross pattern (**Figure 59**) to the torque specification in **Table 7**. See **Figure 60**.

e. Remove the aligning tools.

f. Remove the exhaust manifold.

NOTE
Before installing the exhaust manifold, check the manifold ports for carbon

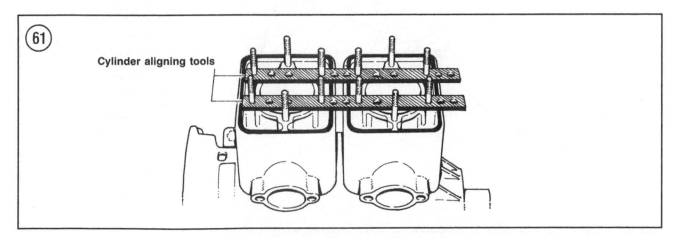

Cylinder aligning tools

buildup (*Figure 62*). *If carbon buildup is severe, clean the ports as described under* **Exhaust System** *in Chapter Six.*

10A. *Engines 467, 537 and 536:* Install the exhaust manifold with 2 new gaskets. Install the manifold bolts and washers and tighten to the torque specification in **Table 7**.

10B. *Engine 583:* See **Figure 4**. Install the exhaust manifold assembly in the following order:

a. Gaskets.
b. Install the heat deflector so that the bend faces away from the cylinders as shown in **Figure 4**.
c. Gaskets.
d. Exhaust manifold.
e. Install the manifold bolts and washers and tighten to the torque specification in **Table 7**.

11. Install the cylinder head as described in this chapter.

12. Install the exhaust pipe and muffler as described in Chapter Six.

13. If new components were installed or if the cylinders were bored or honed, the engine must be broken in as if it were new. Refer to *Break-in* in Chapter Three.

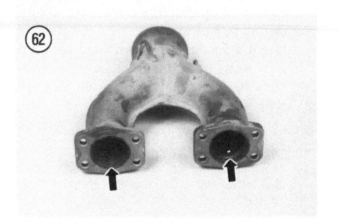

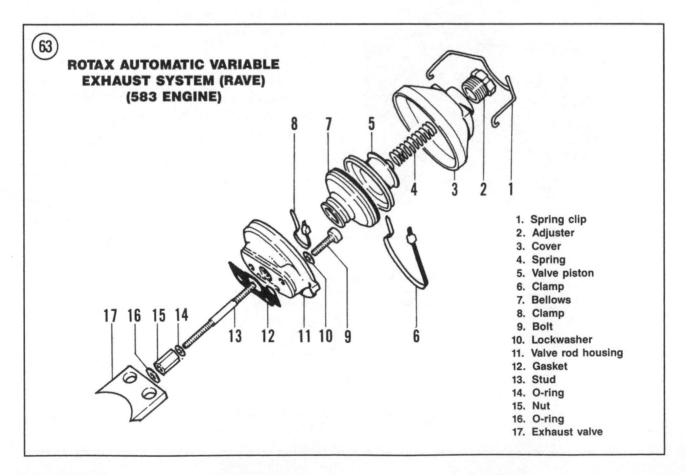

ROTAX AUTOMATIC VARIABLE
EXHAUST SYSTEM (RAVE)
(583 ENGINE)

1. Spring clip
2. Adjuster
3. Cover
4. Spring
5. Valve piston
6. Clamp
7. Bellows
8. Clamp
9. Bolt
10. Lockwasher
11. Valve rod housing
12. Gasket
13. Stud
14. O-ring
15. Nut
16. O-ring
17. Exhaust valve

ROTAX AUTOMATIC VARIABLE EXHAUST (RAVE) (583 ENGINES)

The 583 engine is equipped with the Rotax Automatic Variable Exhaust (RAVE). Refer to **Figure 63** when performing procedures in this section.

RAVE Assembly
Removal

The RAVE assembly can be removed with the cylinders mounted in the snowmobile. This procedure shows the cylinders removed for clarity.

1. Pry the spring clip (A, **Figure 64**) away from the cover (B, **Figure 64**) and remove the cover.
2. Remove the spring (**Figure 65**).
3. Unscrew and remove the valve piston (**Figure 66**).
4. Remove the screws holding the valve rod housing onto the cylinder block (**Figure 67**). Remove the valve rod housing (**Figure 68**).

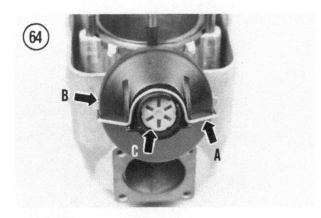

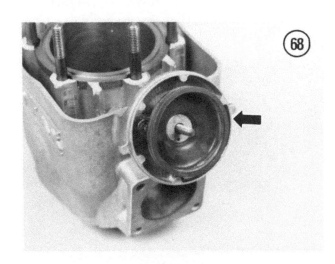

5. Remove the gasket (**Figure 69**).

6. Remove the exhaust valve (**Figure 70**).

RAVE Assembly
Inspection

> *CAUTION*
> *Before cleaning plastic or rubber components, make sure the cleaning agent is compatible with these materials. Some types of solvents can cause permanent damage.*

1. Remove all carbon residue from the exhaust valve and the valve port in the cylinder. Clean the exhaust valve in solvent and dry thoroughly.

2. Inspect the valve rod housing assembly (A, **Figure 71**) as follows:

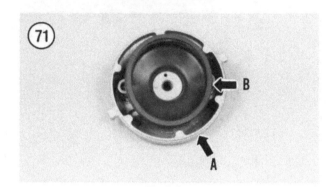

a. If there is evidence of oil draining from the drain hole on the back of the valve rod housing (A, **Figure 72**), check for a loose or damaged bellows (**Figure 63**).

b. Check the bellows (B, **Figure 71**) for splitting or other damage. If the bellows is damaged, replace it by removing it from the valve rod housing. When installing the bellows, secure it with a new cable tie (**Figure 63**).

c. Check the valve rod housing (B, **Figure 72**) and cylinder (**Figure 73**) passages for residue buildup. Clean with a piece of wire or other tool that won't enlarge or otherwise damage the passage.

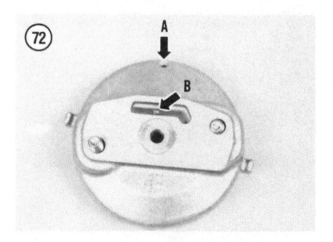

3. Visually check the exhaust valve (**Figure 74**) for cracks, deep scoring, excessive wear, heat

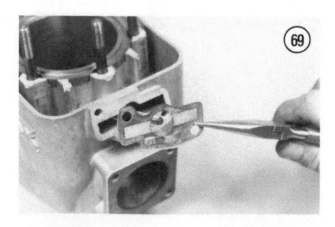

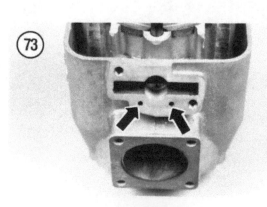

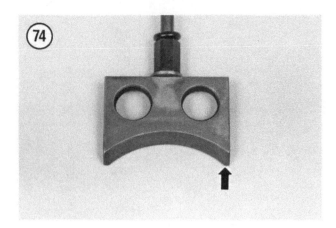

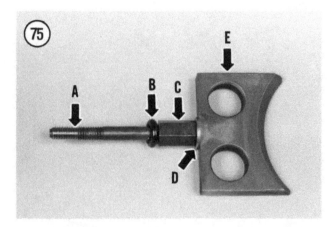

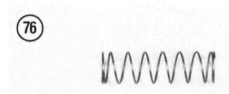

discoloration or other damage. Check the O-ring (B, **Figure 75**) for heat deterioration that shows up as cracks or splitting. If necessary, service the exhaust valve as follows (**Figure 75**):

    a. Slide the O-ring (B) off of the valve rod.

    b. Secure the exhaust valve (E) in a vise with soft jaws. Unscrew the valve rod (C) from the exhaust valve and remove the washer (D). Separate the nut from the valve rod.

    c. Replace worn or damaged parts. Remove all Loctite residue from the valve rod, nut and exhaust valve.

    d. Align the notch on the nut (C) with the portion of the valve rod (A) with the longer threads and turn the nut onto the valve rod until it stops. Slide the washer (D) over the valve rod until it rests against the nut (C). Apply Loctite 242 (blue) onto the valve rod threads that mate with the exhaust valve threads (E). Screw the valve rod (A) into the exhaust valve (E) until the rod bottoms out. Secure the exhaust valve in a vise with soft jaws and tighten the nut (C) securely against the exhaust valve (E). Install a new O-ring (B) over the valve rod.

4. Inspect the spring (**Figure 76**) for fatigue or breakage.

5. Inspect the valve piston (**Figure 77**) for wear, cracks, distortion or other damage.

6. Inspect the cover (A, **Figure 78**) for wear, cracks, distortion or other damage. Check the adjustment screw (B, **Figure 78**) for cracks or damage at the adjustment slots.

7. Replace worn or damaged parts.

RAVE Assembly
Assembly

Prior to assembly, perform the *RAVE Assembly Inspection* procedure to make sure all worn or defective parts have been repaired or replaced. All parts should be thoroughly cleaned before installation or assembly.

1. Insert the exhaust valve into the cylinder port so that the tapered side of the valve faces down (**Figure 70**). Push the valve in all the way (**Figure 79**).

2. Install a new exhaust valve gasket (**Figure 69**). Install the gasket so that the holes in the gasket align with the passage in the bottom of the valve rod housing. See **Figure 80**.

> *NOTE*
> *Make sure the bolts installed in Step 3 do not damage the gasket.*

3. Align the valve rod housing (**Figure 68**) with the gasket (**Figure 69**) and install the 2 housing bolts. Tighten the bolts securely (**Figure 67**).

4. Thread the valve piston (**Figure 66**) onto the valve rod until the edge of the valve piston is approximately 17 mm (43/64 in.) from the edge of the valve rod housing as shown in **Figure 81**.

5. Install the spring (**Figure 65**) and cover (B, **Figure 64**). Secure the cover with the spring clip (A, **Figure 64**).

6. Turn the adjustment screw (C, **Figure 64**) by hand until it bottoms.

PISTON, WRIST PIN AND PISTON RINGS

The piston is made of an aluminum alloy. The wrist pin is a precision fit and is held in place by a clip at each end. A caged needle bearing is used on the small end of the connecting rod.

Refer to illustration for your model when servicing the piston assembly.

a. **Figure 25**: Engine 467 (all 1985-on MX, MX [H/A] and MX LT).
b. **Figure 26**: Engine 537 (1985-1988 Plus).
c. **Figure 27**: Engine 536 (1989 Plus and Plus LT).
d. **Figure 28**: Engine 583 (1989 Mach I).

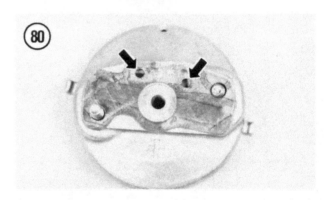

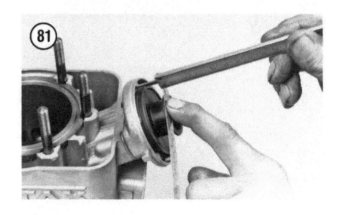

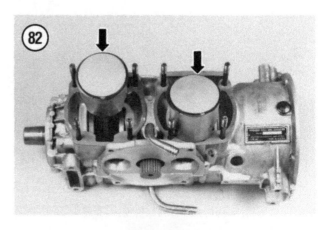

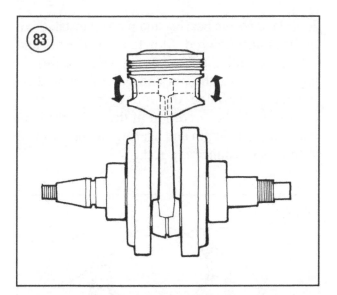

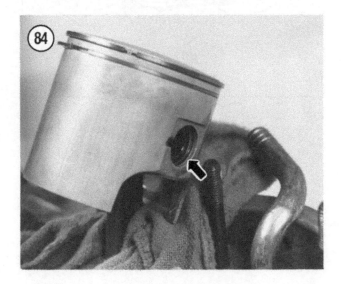

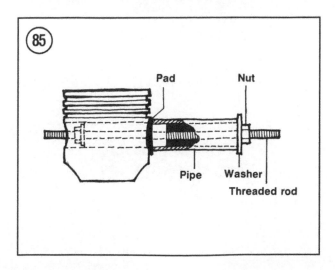

Piston and Piston Ring Removal

1. Remove the cylinder head and cylinder as described in this chapter.

2. Identify the pistons by scratching PTO (left-hand side) and MAG (right-hand side) on the piston crowns (**Figure 82**). In addition, keep each piston together with its own pin, bearing and piston rings to avoid confusion during reassembly.

3. Before removing the piston, hold the rod tightly and rock the piston as shown in **Figure 83**. Any rocking motion (do not confuse with the normal sliding motion) indicates wear on the wrist pin, needle bearing, wrist pin bore or, more likely, a combination of all three.

NOTE
Wrap a clean shop cloth under the piston so that the clip will not fall into the crankcase.

WARNING
Safety glasses should be worn when performing Step 4.

4. Remove the wrist pin clip from the outside of the piston (**Figure 84**) with needlenose pliers. Hold your thumb over one edge of the clip when removing it to prevent it from springing out.

5. Use a proper size wooden dowel or socket extension and push out the wrist pin.

CAUTION
If the engine ran hot or seized, the wrist pin may be difficult to remove. However, do not drive the wrist pin out of the piston. This will damage the piston, needle bearing and connecting rod. If the wrist pin will not push out by hand, remove it as described in Step 6.

6. If the wrist pin is tight, fabricate the tool shown in **Figure 85**. Assemble the tool onto the piston and pull the wrist pin out of the piston. Make sure to install a pad between the piston and piece of pipe to prevent from scoring the side of the piston.

7. Lift the piston off the connecting rod.

8. Remove the needle bearing from the connecting rod (**Figure 86**).

9. Repeat for the opposite piston.

10. If the pistons are going to be left off for some time, place a piece of foam insulation tube, or shop cloth, over the end of each rod to protect it.

NOTE
Always remove the top piston ring first.

11. Models can be equipped with 1 or 2 piston rings. Remove the upper ring by spreading the ends with your thumbs just enough to slide it up over the piston (**Figure 87**). Repeat for the lower ring.

Wrist Pin and Needle Bearing Inspection

1. Clean the needle bearing (**Figure 88**) in clean solvent to prevent contamination and dry thoroughly. Using a magnifying glass, inspect the bearing cage for cracks at the corners of the needle slots and inspect the needles themselves for cracking. If any cracks are found, the bearing must be replaced.

2. Check the wrist pin (**Figure 89**) for severe wear, scoring or chrome flaking. Also check the wrist pin for cracks along the top and side. Replace the wrist pin if necessary.

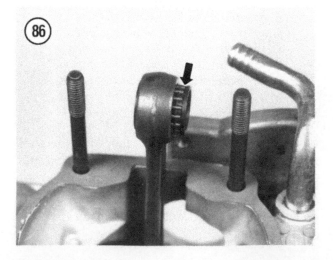

3. Oil the needle bearing and pin and install them in the connecting rod. Slowly rotate the pin and check for radial and axial play (**Figure 90**). If any play exists, the pin and bearing should be replaced, providing the rod bore is in good condition. If the condition of the rod bore is in question, the old pin and bearing can be checked with a new connecting rod.

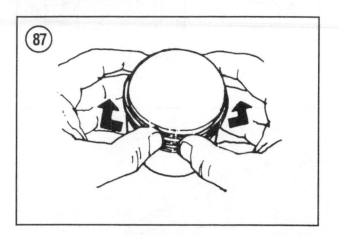

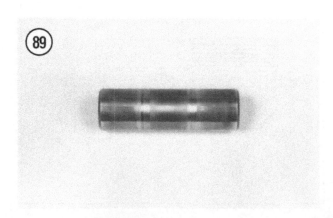

4. Oil the wrist pin and install it in the wrist pin hole (**Figure 91**). Check for up and down play between the pin and piston. There should be no noticeable play. If play is noticeable, replace the wrist pin and/or piston.

> *CAUTION*
> *If there are signs of piston seizure or overheating, replace the wrist pins and bearings as a set (**Figure 92**). These parts have been weakened from excessive heat and may fail later.*

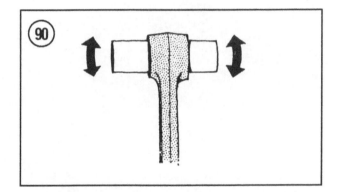

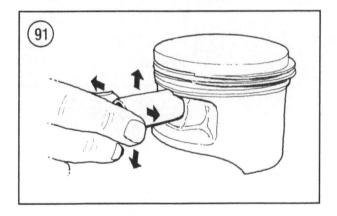

Connecting Rod Inspection

1. Wipe the wrist pin bore in the connecting rod with a clean rag and check it for galling, scratches or any other signs of wear or damage. If any of these conditions exist, replace the connecting rods as described in this chapter.

2. Check the connecting rod big end axial play. You can make a quick check by simply rocking the connecting rod back and forth (**Figure 93**). If there is more than a very slight rocking motion (some side-to-side sliding is normal), you should measure the connecting rod axial play with a feeler gauge. Measure between the side of the crankshaft and the washer. **Figure 94** shows the measurement being taken with the crankshaft

5

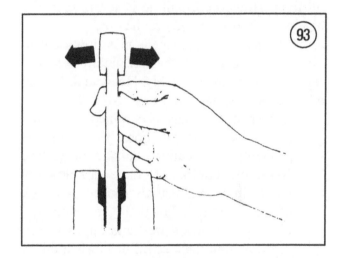

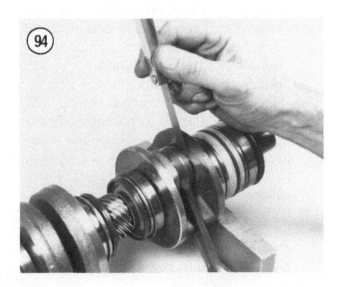

removed for clarity. If the play exceeds the wear limit specified in **Table 3**, **Table 4** or **Table 5**, the crankshaft will have to be rebuilt.

Piston and Ring Inspection

1. Carefully check the piston for cracks at the top edge of the transfer cutaways (**Figure 95**) and replace if found. Check the piston skirt (**Figure 96**) for brown varnish buildup. More than a slight amount is an indication of worn or sticking rings which should be replaced.

2. Check the piston skirt for galling and abrasion which may have resulted from piston seizure. If light galling is present, smooth the affected area with No. 400 emery paper and oil or a fine oilstone. However if galling is severe or if the piston is deeply scored, replace it.

3. If the piston is damaged, it is important to pinpoint the cause so that the failure will not repeat after engine assembly. Note the following when checking damaged pistons:

   a. If the piston damage is confined to the area above the wrist pin bore, the engine is probably overheating. Seizure or galling conditions confined to the area below the wrist pin bore is usually caused by a lack of lubrication, rather than overheating.

   b. If the piston has seized and appears very dry (apparent lack of oil or lubrication on the piston), a lean fuel mixture probably caused the overheating. Overheating can result from incorrect jetting, air leaks or over-advanced ignition timing.

   c. Preignition will cause a sand-blasted appearance on the piston crown. This condition is discussed in Chapter Two.

   d. If the piston damage is confined to the exhaust port area on the front of the piston, look for incorrect jetting (too lean) or over-advanced ignition timing.

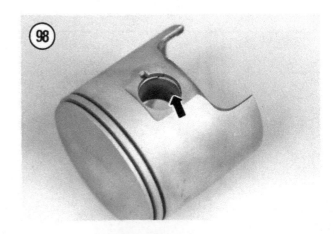

   e. If the piston has a melted pocket starting in the crown or if there is a hole in the piston crown, the engine is running too lean. This can be due to incorrect jetting, an air leak or over-advanced ignition timing. A spark plug that is too hot will also cause this type of piston damage.

f. If the piston is seized around the skirt, but the dome color indicates proper lubrication (no signs of dryness or excessive heat), the damage may result from a condition referred to as cold seizure. This condition typically results from running a water-cooled engine too hard without first properly warming it up. A lean fuel mixture can also cause skirt seizure.

4. Check the piston ring locating pins in the piston (**Figure 97**). The pins should be tight and the piston should show no signs of cracking around the pins. If a locating pin is loose, replace the piston. A loose pin will fall out and cause severe engine damage.

5. Check the wrist pin clip grooves (**Figure 98**) in the piston for cracks or other damage that could allow a clip to fall out. This would cause severe engine damage. Replace the piston if any one groove shows signs of wear or damage.

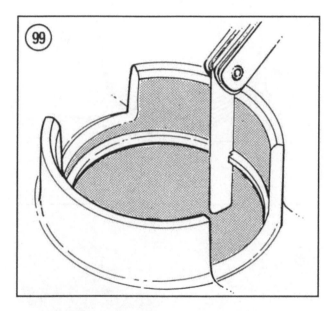

> *NOTE*
> *Maintaining proper piston ring end gap helps to ensure peak engine performance. Excessive ring end gap reduces engine performance and can cause overheating. Insufficient ring end gap will cause the ring ends to butt together and cause the ring to break, resulting in severe engine damage. So that you don't have to wait for parts, always order extra cylinder head and base gaskets to have on hand for routine top end inspection and maintenance.*

6. Measure piston ring end gap. Place a ring into the bottom of the cylinder and push it in with the crown of the piston until it is just below the transfer ports. This ensures that the ring is square in the cylinder bore. Measure the gap with a flat feeler gauge (**Figure 99**) and compare to the wear limit in **Table 3**, **Table 4** or **Table 5**. If the gap is greater than specified, the rings should be replaced as a set.

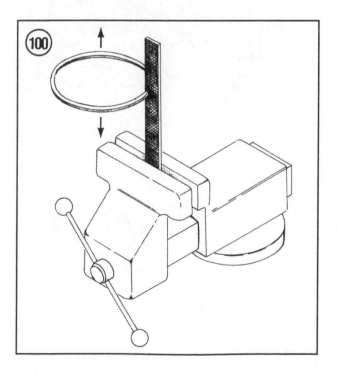

> *NOTE*
> *When installing new rings, measure the end gap in the same manner as for old ones. If the gap is less than specified, make sure you have the correct piston rings. If the replacement rings are correct, but the end gap is too small, carefully file the ends with a fine cut file until the gap is correct (**Figure 100**).*

Insufficient ring end gap will allow the rings to butt together. This can result in ring breakage, ring seizure or rapid cylinder wear. These conditions will reduce engine performance and may cause seizure or severe damage to the cylinder wall.

7. Carefully remove all carbon buildup from the ring grooves with a broken ring (**Figure 101**). Inspect the grooves carefully for burrs, nicks, or broken and cracked lands. Recondition or replace the piston if necessary.

8A. *467, 537 and 536 engines:* Install the bottom ring in its groove and measure the groove clearance with a flat feeler guage (**Figure 102**). Compare to dimensions in **Table 3** or **Table 4**. If the clearance is greater than specified, the rings must be replaced as a set. If the clearance is still excessive with new rings, replace the piston.

NOTE
The top piston ring is a Dykes ring and its side clearance cannot be checked as described in Step 8A.

8B. *583 engine:* Install the piston ring in its groove and measure the groove clearance with a flat feeler gauge (**Figure 102**). Compare to dimensions in **Table 5**. If the clearance is greater than specified, the ring must be replaced. If the clearance is still excessive with a new ring, replace the piston.

9. Observe the condition of the piston crown (**Figure 103**). Normal carbon buildup can be removed with a wire wheel mounted on a drill press. If the piston shows signs of overheating, pitting or other abnormal conditions, the engine may be experiencing preignition or detonation; both conditions are discussed in Chapter Two.

CAUTION
Do not wire brush piston skirts or ring lands. The wire brush removes aluminum and increases piston clearance. The brush also rounds the corners of the ring lands which results in decreased support for the piston rings.

10. If the piston checked out okay after performing these inspection procedures, measure

the piston outside diameter as described under *Piston/Cylinder Clearance* in this chapter.

11. If new piston rings are required, the cylinders should be honed before assembling the engine. Refer to *Cylinder Honing* in this chapter.

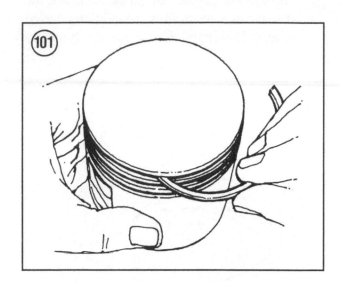

Piston/Cylinder Clearance

The following procedure requires the use of highly specialized and expensive measuring tools. If such equipment is not readily available, have the measurements performed by a dealer or machine shop. Always replace both pistons as a set.

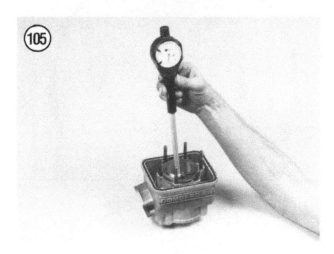

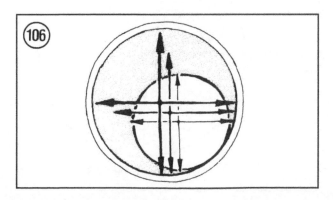

1. Measure the outside diameter of the piston with a micrometer approximately 16 mm (5/8 in.) above the bottom of the piston skirt, at a 90° angle to the piston pin (**Figure 104**). If the diameter exceeds the wear limit in **Table 3**, **Table 4** or **Table 5**, install new pistons.

NOTE
Always install new rings when installing a new piston.

2. Wash the cylinder block in solvent to remove any oil and carbon particles. The cylinder bore must be cleaned thoroughly before attempting any measurement as incorrect readings may be obtained.

3. Check cylinder taper and out-of-round with bore gauge or telescoping gauge (**Figure 105**). Measure the bore gauge or telescoping gauge with a micrometer to determine the dimension. Measure the cylinder bore at the points shown in **Figure 106**. Measure in 3 axes—in line with the piston pin and at 90° to the pin. If cylinder taper or out-of-round is greater than specification (**Tables 3-5**), the cylinders must be rebored to the next oversize and new pistons and rings installed.

NOTE
*The new pistons should be obtained first before the cylinders are bored so that the pistons can be measured. The cylinders must be bored to match the pistons. Piston-to-cylinder clearance is specified in **Tables 3-5**.*

4. Piston clearance is the difference between the maximum piston diameter and the minimum cylinder diameter. For a run-in (used) piston and cylinder, subtract the dimension of the piston from the cylinder dimension. If the clearance exceeds the dimension in **Tables 3-5**, the cylinders should be rebored and new pistons and rings installed.

Cylinder Honing

The surface condition of a worn cylinder bore is normally very shiny and smooth. New piston rings installed in a cylinder with even minimum wear, will not seat properly and engine performance will suffer from compression losses. Cylinder honing, often referred to as bead breaking or deglazing, is required whenever new piston rings are installed. When a cylinder bore is honed, the surface is slightly roughed up to provide a textured or crosshatched surface. This surface finish controls wear of the new rings and helps them to seat and seal properly. *Whenever new rings are installed, the cylinder surface should be honed.* This service can be performed by a Ski-Doo dealer or independent repair shop. The cost of having the cylinder honed by a dealer is usually minimal compared to the cost of purchasing a hone and doing the job yourself. If you choose to hone the cylinder yourself, follow the hone manufacturer's directions closely.

> *CAUTION*
> *After a cylinder has been reconditioned by boring or honing, the bore should be properly cleaned to remove all material left from the machining operation. Refer to the inspection procedure under **Cylinder** in this chapter. Improper cleaning will not remove all of the machining residue resulting in rapid wear of the new piston and rings.*

Piston Installation

1. Prior to assembly, perform the inspection procedures to make sure all worn or defective

parts have been cleaned or replaced. All parts should be thoroughly cleaned before installation or assembly.

2. Apply assembly oil to the needle bearing and install it in the connecting rod (**Figure 107**).

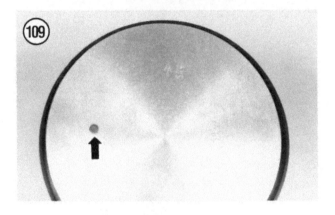

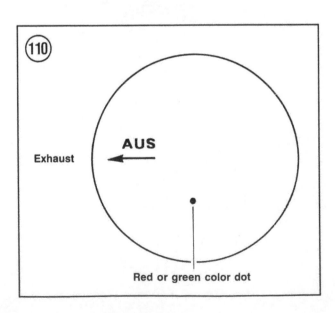

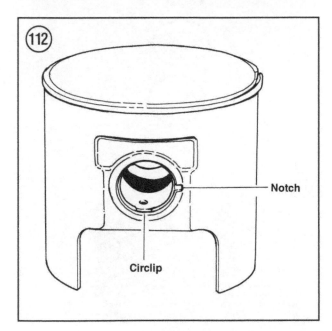

3. Oil the wrist pin and install it in the piston until the end of it extends slightly beyond the inside of the boss (**Figure 108**).

NOTE
*Piston crowns are marked with a green or red dot (**Figure 109**) that must match the same color ID mark placed on the cylinders. If your are installing new pistons, match the piston dot color with the cylinder dot color. If you are installing used pistons, install pistons according to marks made during disassembly.*

4. Place the piston over the connecting rod with the AUS over the arrow on the piston crown pointing toward the front (exhaust side) of the engine (**Figure 110**). This is essential so the piston ring ends will be correctly positioned and will not snag in the ports. Line up the pin with the bearing and push the pin into the piston until it is even with the wrist pin clip grooves.

CAUTION
*If the wrist pin will not slide in the piston smoothly, use the homemade tool described during **Piston and Piston Ring Removal** to install the wrist pin (**Figure 85**). When using the homemade tool, the pipe and pad is not required. Instead, run the threaded rod through the wrist pin. Secure the end of the wrist pin next to the piston with the small washer and nut. Slide the large washer onto the threaded rod so that it is next to the wrist pin. Install the nut next to the large washer and tighten it to push the wrist pin into the piston. Do not drive the wrist pin into the piston or you may damage the needle bearing and connecting rod.*

5. Install *new* wrist pin clips (**Figure 111**), making sure they are completely seated in their grooves with the open ends of the clips facing down. See **Figure 112**.

NOTE
***Figure 113** shows wrist pin clips being installed with the Ski-Doo piston circlip*

*installer (part No. 529 0086 00). This
tool ensures positive installation without
the possibility of distorting or damaging
the clip by conventional installation
methods.*

CAUTION
*New wrist pin clips should always be
installed. If old clips are installed, they
must snap securely into piston grooves.
A weak clip could disengage during
engine operation and cause severe
engine damage.*

6. Check the installation by rocking the piston
back and forth around the pin axis and from side-
to-side along the axis. It should rotate freely back
and forth but not from side-to-side.

7A. *467, 537 and 536 engines:* See **Figure 114**.
Install the piston rings—first the bottom one, then
the top—by carefully spreading the ends of the
ring with your thumbs and slipping the ring over
the top of the piston. Make sure manufacturer's
mark on the piston rings are toward the top of
the piston. If you are installing used rings, install
them by referring to the identification marks
made during removal.

7B. *583 engine:* Install the piston ring by
spreading the ends of the ring with your thumbs
and slipping the ring over the top of the piston.
See **Figure 115**.

8. Make sure the ring(s) are seated completely
in the grooves, all the way around the
circumference, and that the ends are aligned with
the locating pin(s). See **Figure 115** or **Figure 116**.

9. If new components were installed, the engine
must be broken in as if it were new. Refer to
Break-in Procedure in Chapter Three.

CRANKCASE, CRANKSHAFT AND ROTARY VALVE

Disassembly of the crankcase—splitting the
cases—and removal of the crankshaft assembly
requires engine removal from the frame.
However, the cylinder head, cylinder and all
other attached assemblies should be removed
with the engine in the frame.

The crankcase is made in 2 halves of precision
diecast aluminum alloy and is of the "thin-
walled" type (**Figure 117**). To avoid damage to
them, do not hammer or pry on any of the
interior or exterior projected walls. These areas
are easily damaged if stressed beyond what they
are designed for. They are assembled without a
gasket; only gasket sealer is used while dowel

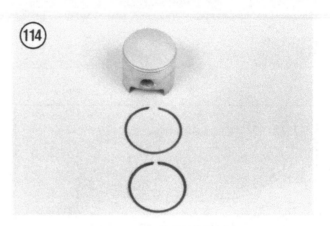

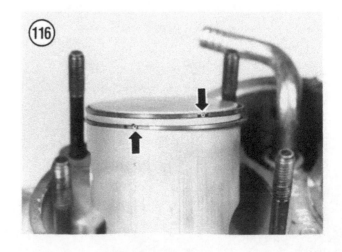

pins align the crankcase halves when they are bolted together. The crankcase halves are sold as a matched set only. If one crankcase half is severely damaged, both must be replaced.

Crankshaft components are available as individual parts. However, crankshaft service—replacement of unsatisfactory parts or crankshaft alignment—should be entrusted to a dealer or engine specialist. Special measuring and alignment tools, a hydraulic press and experience are necessary to disassemble, assemble and accurately align the crankshaft assembly. Which, in the case of the average twin cylinder engine, is made up of a number of pressed-together pieces, not counting the bearings, seals and connecting rods. However, you can save considerable expense by disassembling the engine and taking the crankshaft in for service.

The procedure which follows is presented as a complete, step-by-step major lower end overhaul that should be followed if the engine is to be completely reconditioned.

5

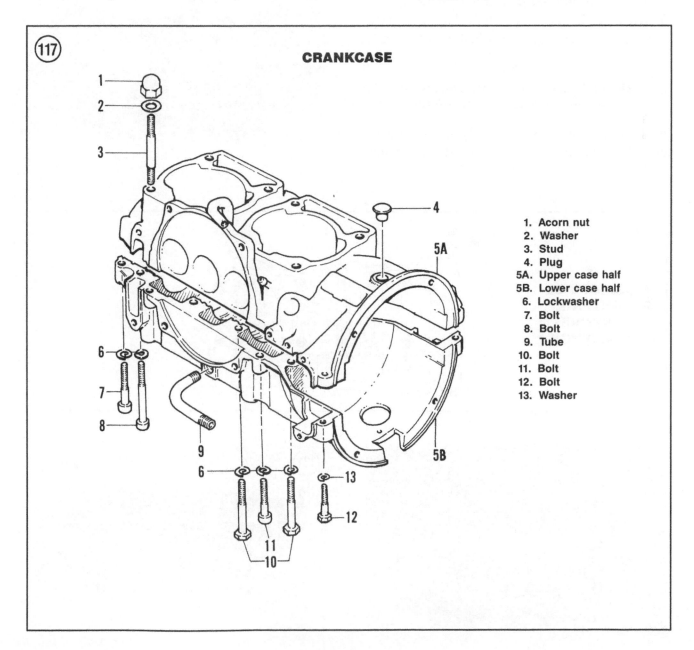

(117)   CRANKCASE

1. Acorn nut
2. Washer
3. Stud
4. Plug
5A. Upper case half
5B. Lower case half
6. Lockwasher
7. Bolt
8. Bolt
9. Tube
10. Bolt
11. Bolt
12. Bolt
13. Washer

Crankcase Disassembly

This procedure describes disassembly of the crankcase halves and removal of the crankshaft.

1. Remove the engine from the frame as described in this chapter.

2. Remove all of the engine mounts from the crankcase as described in this chapter.

3. Remove the flywheel and stator plate as described in Chapter Seven.

4. Remove the rotary valve as described under *Rotary Valve and Shaft* in this chapter.

5. Remove the pistons if they have not been previously removed, as described in this chapter.

CAUTION
Do not damage the crankcase studs when performing the following procedures.

6. Turn the crankcase assembly so that it rests up-side-down on its crankcase studs (**Figure 118**).

7. Loosen the crankcase bolts in 2 stages in a crisscross pattern by reversing the torque pattern in **Figure 119**. Remove all of the crankcase bolts.

8. Tap on the large bolt bosses with a soft mallet to break the crankcase halves apart and then lift the bottom case half (**Figure 120**) off of the top half.

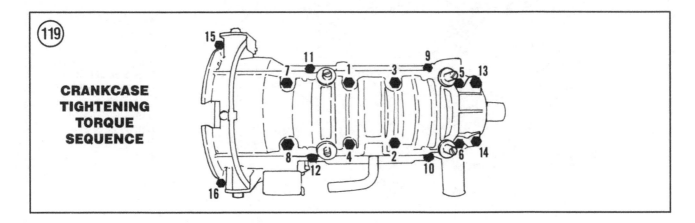

CRANKCASE
TIGHTENING
TORQUE
SEQUENCE

5

CAUTION
Make sure that you have removed all the fasteners. If the cases are hard to separate, check for any fasteners you may have missed.

CAUTION
Do not pry the cases apart with a screwdriver or any other sharp tool, otherwise the sealing surface will be damaged.

9. Remove the crankshaft (**Figure 121**) from the upper crankcase half. Support the crankshaft on the workbench so that it cannot roll off.

10. If necessary, remove the rotary valve shaft as described under *Rotary Valve and Shaft* in this chapter.

Cleaning

Refer to **Figure 117** for this procedure.

1. Clean both crankcase halves with cleaning solvent. Thoroughly dry with compressed air and wipe off with a clean shop cloth. Be sure to remove all traces of old gasket sealer from all mating surfaces.

2. Clean the oil passages in the upper crankcase half. See **Figure 122**, typical. Use compressed air to ensure that they are clean.

3. Clean the crankshaft assembly in solvent and dry with compressed air. Oil the bearings with injection oil to prevent rusting.

Crankcase Inspection

Refer to **Figure 117** for this procedure.

1. Carefully inspect the case halves (**Figure 123**) for cracks and fractures. Also check the areas around the stiffening ribs, around bearing bosses and threaded holes. If any are found, have them repaired by a shop specializing in the repair of precision aluminum castings or replace them.

2. Check the bearing surface areas in the upper (**Figure 124**) and lower (**Figure 125**) halves.

3. Check the threaded holes in both crankcase halves for thread damage or dirt or oil buildup. If necessary, clean or repair the threads with a suitable size metric tap. Coat the tap threads with kerosene or an aluminum tap fluid before use.

4. Check the upper crankcase studs (**Figure 126**) for thread damage. If necessary, replace damaged studs as described under *Crankcase Stud Replacement* in this chapter.

5. Check the oil seal grooves (**Figure 127**) in the upper and lower crankcase halves for cracks or damage.

6. Inspect the rotary valve oil seal and bearing as described under *Rotary Valve and Shaft* in this chapter.

7. Check the rotary valve machined surface on the upper and lower crankcase halves (**Figure 128**). Check the surface for gouges or other damage that would indicate rotary valve or rotary valve shaft damage.

8. Inspect the rotary valve as described in this chapter.

9. If there is any doubt as to the condition of the crankcase halves, and they cannot be repaired, replace the crankcase halves as a set.

Crankcase Stud Replacement

Damaged crankcase studs will prevent the cylinders from being properly torqued. This will cause engine damage.

Because Acorn nuts are used to tighten the cylinder, the crankcase studs must be installed to a specified length. See **Figure 129**.

A tube of Loctite 242 (blue), 2 nuts, 2 wrenches and a new stud will be required. Replace damaged studs as follows:

1. Measure the stud height with a vernier caliper prior to removal (**Figure 129**).

2. Thread two nuts onto the damaged stud. Tighten the 2 nuts against each other so that they are locked.

NOTE
If the threads on the damaged stud do not allow installation of the 2 nuts, you will have to remove the stud with a pair of Vise Grips.

3. Turn the bottom nut counterclockwise and unscrew the stud.

4. Clean the threads with solvent or electrical contact cleaner and allow to dry thoroughly.

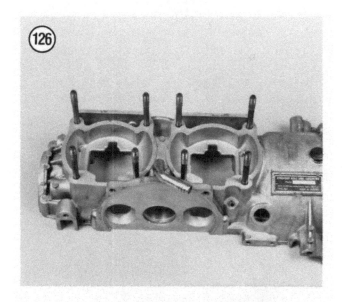

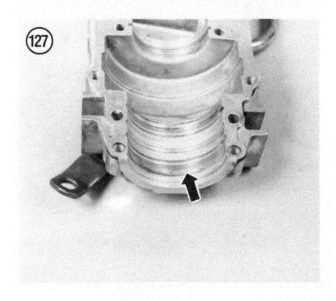

5. Install 2 nuts on the top half of the new stud as in Step 1. Make sure they are locked securely.

6. Coat the bottom half of a new stud with Loctite 242 (blue).

7. Turn the top nut clockwise and thread the new stud securely. Measure the portion of the installed stud from the cylinder surface to the top of the stud (**Figure 129**). The stud should be installed

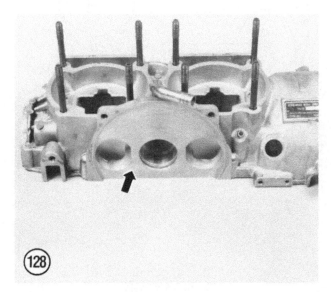

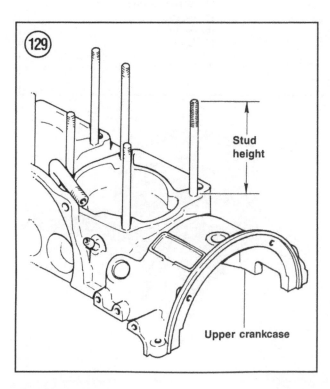

to the height recorded prior to removal. If not, reposition the stud as necessary.

8. Remove the nuts and repeat for each stud as required.

9. Follow Loctite's directions on cure time before assembling the component.

Crankshaft Inspection

Refer to **Figure 130** for this procedure.

1. Check the left- (**Figure 131**) and right-hand (**Figure 132**) crankshaft oil seals. Replace the seals if damaged.

2. Check the connecting rod small-end (**Figure 133**) and big-end (**Figure 134**) for excessive heat discoloration or other damage. Place the crankshaft on V-blocks and spin the connecting rod by hand; check for excessive noise or roughness.

NOTE
A set of V-blocks can be made out of hardwood to perform the check described in Step 2. However, only machined V-blocks should be used to check crankshaft runout described later in this procedure.

3. Check connecting rod big-end axial play with a feeler gauge. Insert the gauge between the crankshaft and the connecting rod as shown in **Figure 135**. If the play meets or exceeds the service limit, the big-end bearings and connecting rod must be replaced. See **Tables 3-5** for service limit.

4. Repeat Steps 2 and 3 for both connecting rods.

5. Carefully examine the condition of the crankshaft ball bearings (A, **Figure 136**). Clean the bearings in solvent and allow to dry thoroughly. Oil each bearing before checking it. Roll each bearing around by hand, checking that it turns quietly and smoothly and that there are no rough spots. There should be no apparent radial play. Defective bearings should be replaced.

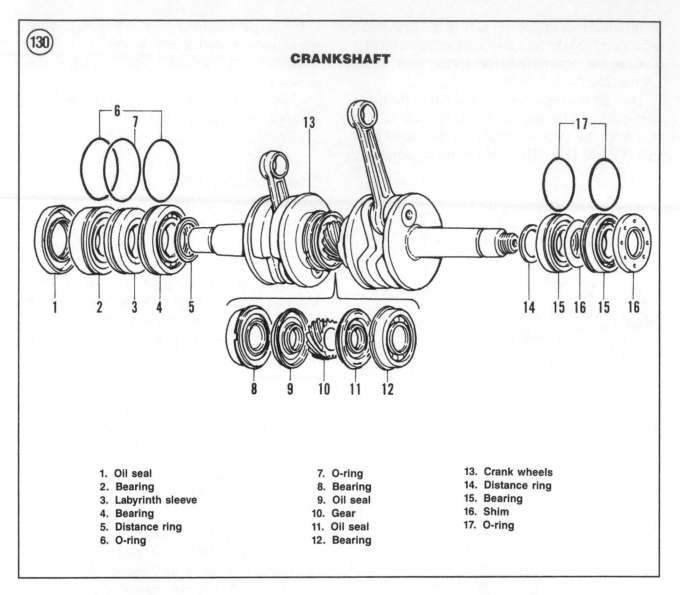

CRANKSHAFT

1. Oil seal
2. Bearing
3. Labyrinth sleeve
4. Bearing
5. Distance ring
6. O-ring

7. O-ring
8. Bearing
9. Oil seal
10. Gear
11. Oil seal
12. Bearing

13. Crank wheels
14. Distance ring
15. Bearing
16. Shim
17. O-ring

6. Check the central gear (**Figure 137**) for cracks, deep scoring or excessive wear. If the central gear is damaged, check the gear on the rotary valve shaft for damage. If one gear appears damaged while the other appears okay, replace *both* gears.

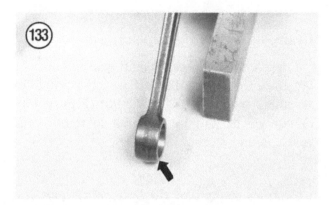

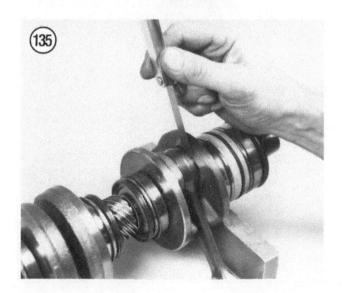

5

NOTE
*The bearings installed on the outside of the crank wheels can be replaced as described under **Crankshaft Bearing Replacement** in this chapter. To replace the bearings, seals and central gear installed between the crank wheels, the crankshaft will have to be disassembled; refer service to a qualified dealer or crankshaft specialist.*

7. Support the crankshaft by placing it onto 2 V-blocks. Check runout with a dial indicator at the points indicated in **Figure 138**. Turn the crankshaft slowly and note the gauge reading. The maximum difference recorded is crankshaft runout. If the runout at any position exceeds the

service limit (**Tables 3-5**), the crankshaft should be serviced by a dealer or crankshaft specialist.

NOTE
Do not check crankshaft runout with the crankshaft placed between centers. V-blocks must be used as described in Step 7.

8. Check the crankshaft threads (**Figure 139**) for stripping, cross-threading or other damage. Have threads repaired by a dealer or machine shop.

9. Check the key seat (C, **Figure 136**) for cracks or other damage. If the key seat is damaged, refer service to a dealer or machine shop.

10. If the crankshaft exceeded any of the service limits in Step 3 or Step 4, or if one or more bearings are worn or damaged, have the crankshaft rebuilt by a dealer or crankshaft specialist.

11. Replace all of the outer bearing and labyrinth sleeve O-rings (B, **Figure 136**) before reinstalling the crankshaft. When purchasing new O-rings, note that 3 different size O-rings are used (**Figure 130**). When purchasing new O-rings, have the parts manager identify O-ring position.

12. Check the rotary valve clearance as described in the inspection procedure under *Rotary Valve and Shaft* in this chapter.

Crankshaft Bearing Replacement

Replace the outer crankshaft bearings and the labyrinth sleeve as follows. The Ski-Doo puller assembly (part No. 420 8762 96) (**Figure 140**) or equivalent will be required to remove the bearings and sleeve.

CAUTION
*When using the Ski-Doo puller (or equivalent puller) to remove the bearings in Step 1, place a protective cap over the end of the crankshaft (**Figure 140**) to prevent the puller screw from damaging the crankshaft.*

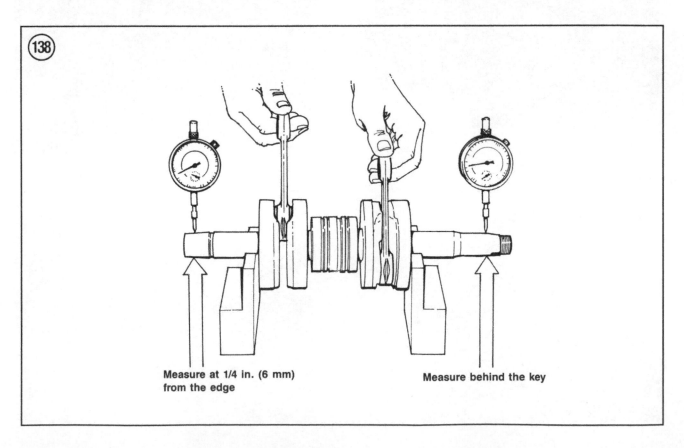

(138)

Measure at 1/4 in. (6 mm) from the edge

Measure behind the key

1. Using the Ski-Doo puller assembly, remove the left- (**Figure 141**) and right-hand (**Figure 142**) bearing assembly.

2. Clean the crankshaft bearing area with solvent or electrical contact cleaner and dry thoroughly.

3. Coat both crankshaft bearing areas with anti-seize lubricant (part No. 413 7010 00) or equivalent.

5

NOTE
*Three different bearings are used on the outside of the crankshaft. See **Figure 130**. When you purchase the replacement bearings, have the parts manager identify each bearing and its position on the crankshaft. It is critical that the bearings are properly installed.*

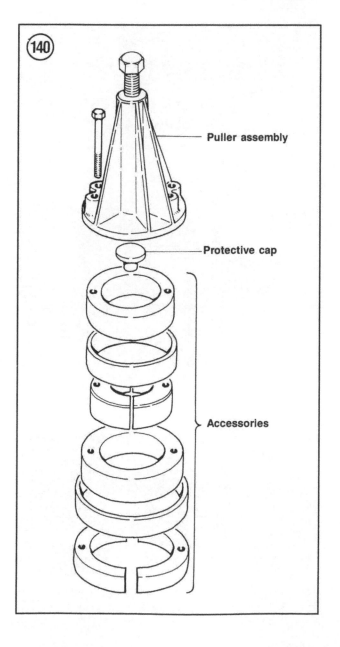

Puller assembly

Protective cap

Accessories

NOTE
Ski-Doo recommends that the bearings are heated in oil before installation. Completely read Step 4 through before heating and installing bearings. During bearing installation, the crankshaft should be supported securely so that the bearings can be installed quickly. If a bearing cools and tightens on the crankshaft before it is completely installed, remove the bearing with the puller and reheat.

4. Install the bearings and labyrinth sleeve as follows:

   a. Lay the bearings on a clean, lint-free surface in the order of assembly.

   b. When installing the bearings and labyrinth sleeve, make sure that the bearing and sleeve O-ring grooves face in the direction shown in **Figure 141** and **Figure 142**.

NOTE
If the bearings are installed incorrectly with regards to O-ring alignment, the O-rings will not line up with the O-ring grooves in the upper and lower crankcase halves.

   c. Refer to *Shrink Fit* under *Ball Bearing Replacement* in Chapter One.

   d. Heat and install the bearings and labyrinth sleeve. Refer to **Figure 130**, **Figure 141** and **Figure 142** during installation.

   e. After the bearings have cooled, install the bearing and labyrinth sleeve O-rings.

Crankshaft Installation

Refer to **Figure 130** for this procedure.

1. Install the rotary valve shaft as described in this chapter.

2. Fill both of the outer crankshaft oil seal lip cavities with a low temperature lithium base grease. Then, install the left- (**Figure 131**) and right-hand (**Figure 132**) oil seals.

3. Support the upper crankcase by its crankcase studs as shown in **Figure 143**.

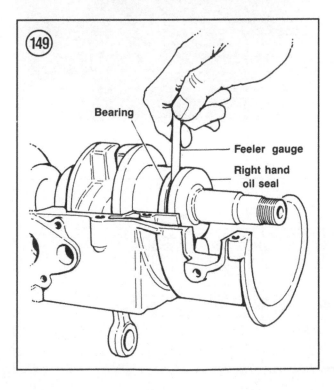

Bearing

Feeler gauge

Right hand
oil seal

NOTE
*Step 4 describes crankshaft (**Figure 144**) installation. However, because of the number of separate procedures required during installation, read Step 4 through first before actually installing the crankshaft.*

4. Align the crankshaft with the upper crankcase half and install the crankshaft (**Figure 144**). Note the following:

  a. Make sure that the left-hand outer crankshaft oil seal ring and the 3 O-rings fit into the crankcase grooves as shown in **Figure 145**.

  b. Make sure that the crankshaft gear (**Figure 146**) meshes properly with the rotary valve shaft gear.

  c. Check that the 2 center bearing clips and oil seal ring fit into the crankcase grooves as shown in **Figure 147**.

  d. Make sure that the 2 right-hand bearing O-rings fit into the crankcase grooves as shown in A, **Figure 148**.

  e. Position the right-hand oil seal (B, **Figure 148**) so that there is a 1 mm (0.040 in.) gap between the oil seal and bearing. Check gap with a feeler gauge as shown in **Figure 149**.

CAUTION
The bearing-to-oil seal gap set in sub-step e provides room for bearing lubrication. If the gap is not set, bearing failure may result.

5. Recheck Step 4.

Crankcase Assembly

1. Install the crankshaft in the upper crankcase half as described in this chapter.

2. Oil the crankshaft gear and the bottom end bearings with injection oil.

NOTE
Follow the manufacturer's directions when applying primer and crankcase sealer in Step 3.

5

3. Apply crankcase sealer as follows:
   a. Make sure the crankcase mating surfaces are completely clean.
   b. Apply Loctite Primer N or NF to both crankcase surfaces.
   c. Apply Loctite Gasket Eliminator 515 to both crankcase surfaces.
4. Put the lower case half (**Figure 150**) onto the upper half. Check the mating surfaces all the way around the case halves to make sure they are even (**Figure 151**).
5. Apply a light coat of oil to the crankcase bolt threads before installing them.
6. Install the crankcase bolts and run them down hand-tight.
7. Torque the crankcase bolts in 2-steps in the sequence shown in **Figure 152**. Refer to **Table 7** for torque specifications.

CAUTION
While tightening the crankcase fasteners, make frequent checks to ensure that the crankshaft turns freely and that the crankshaft locating rings and O-rings properly fit in the case halves.

8. Check again that the crankshaft turns freely. If it is binding, separate the crankcase halves and determine the cause of the problem.
9. Turn the engine right side up.
10. Liberally coat the crank pins, bearings with 2-stroke injection oil. Apply the same oil though the oil delivery holes in the transfer ports.
11. Install the engine top end as described in this chapter.
12. Install the rotary valve and impeller as described in this chapter.
13. Install the stator plate and flywheel as described in Chapter Seven.
14. Install the engine mounts as described in this chapter.

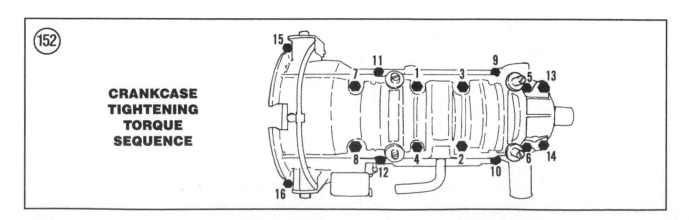

CRANKCASE TIGHTENING TORQUE SEQUENCE

15. Install the engine in the frame as described in this chapter.

16. If new components were installed, the engine must be broken in as if it were new. Refer to *Break-in Procedure* in Chapter Three.

ROTARY VALVE AND SHAFT

Refer to **Figure 153** and **Figure 154** when performing procedures in this section.

Rotary Valve Removal

The rotary valve can be removed with the engine mounted in the frame. This procedure is shown with the engine removed for clarity.

1. If the engine is mounted in the frame, perform the following:

   a. Remove the carburetors as described in Chapter Six.

   b. Disconnect the oil injection hose at the oil pump. Plug the hose to prevent oil leakage and contamination.

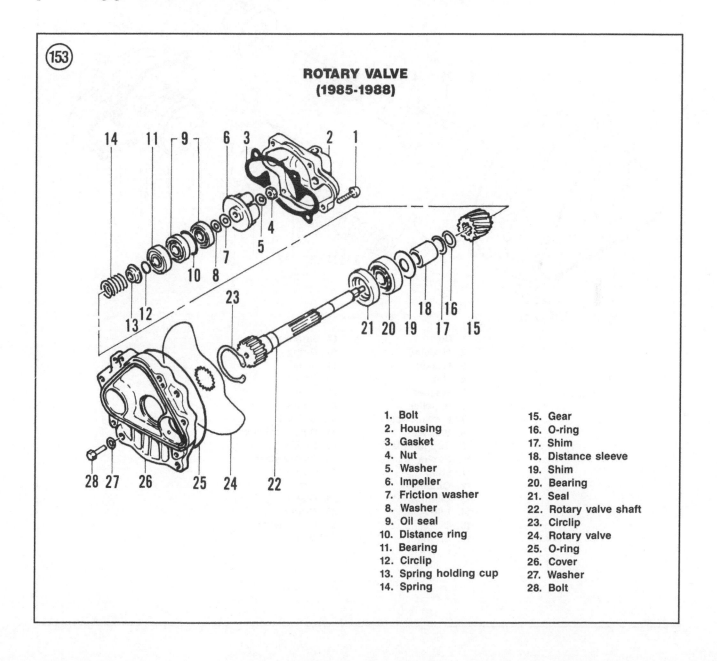

ROTARY VALVE (1985-1988)

1. Bolt
2. Housing
3. Gasket
4. Nut
5. Washer
6. Impeller
7. Friction washer
8. Washer
9. Oil seal
10. Distance ring
11. Bearing
12. Circlip
13. Spring holding cup
14. Spring
15. Gear
16. O-ring
17. Shim
18. Distance sleeve
19. Shim
20. Bearing
21. Seal
22. Rotary valve shaft
23. Circlip
24. Rotary valve
25. O-ring
26. Cover
27. Washer
28. Bolt

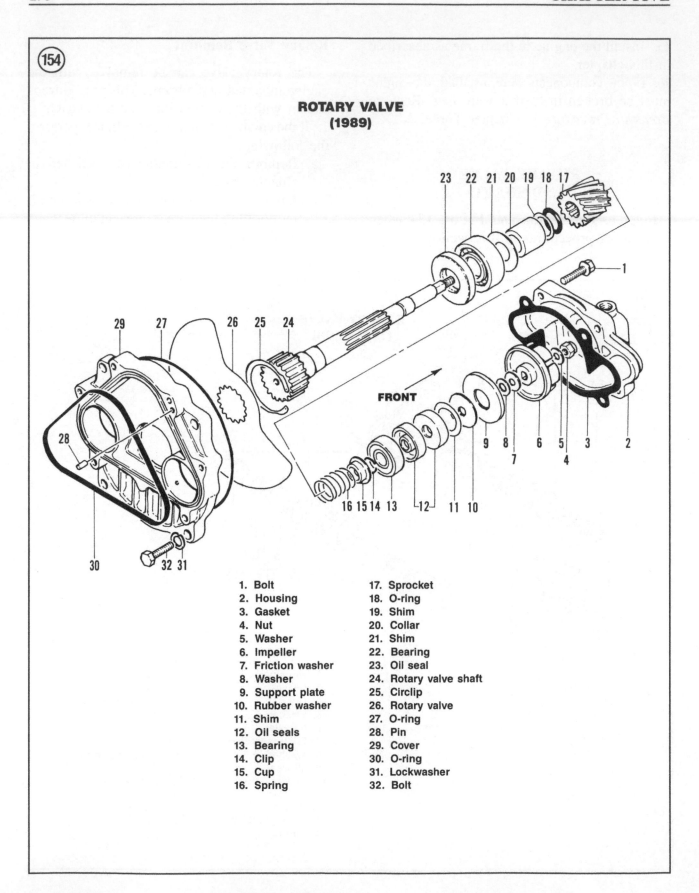

ROTARY VALVE
(1989)

FRONT

1. Bolt
2. Housing
3. Gasket
4. Nut
5. Washer
6. Impeller
7. Friction washer
8. Washer
9. Support plate
10. Rubber washer
11. Shim
12. Oil seals
13. Bearing
14. Clip
15. Cup
16. Spring
17. Sprocket
18. O-ring
19. Shim
20. Collar
21. Shim
22. Bearing
23. Oil seal
24. Rotary valve shaft
25. Circlip
26. Rotary valve
27. O-ring
28. Pin
29. Cover
30. O-ring
31. Lockwasher
32. Bolt

c. Disconnect the oil pump cable at the oil pump.

NOTE
It is not necessary to remove the oil pump from the rotary valve cover in Step 2. If service to the oil pump is required, refer to Chapter Eight.

2. Remove the bolts holding the rotary valve cover to the crankcase. Remove the rotary valve cover (**Figure 155**) and its O-ring.

NOTE
The rotary valve is asymmetrical. Mark the valve with a felt disc pen to assist valve installation.

3. Remove the rotary valve (**Figure 156**).
4. If necessary, remove the rotary valve shaft as described in this chapter.

Inspection

1. Inspect the rotary valve (**Figure 157**) for cracks, splitting or other damage. Replace the rotary valve if necessary.
2. Whenever the crankcase is disassembled, check the rotary valve clearance as follows:
   a. With the crankcase disassembled, install the rotary valve onto the end of the rotary valve shaft (**Figure 158**).
   b. Install the rotary valve cover without its O-ring. Install the cover bolts and tighten the cover securely.
   c. Insert a feeler gauge between the rotary valve and the upper crankcase (**Figure 158**). If the clearance is not within 0.27-0.48 mm (0.011-0.019 in.), check the rotary valve, crankcase and rotary valve cover for damage.

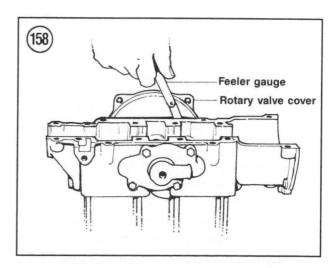

Feeler gauge
Rotary valve cover

5

Rotary Valve Timing

The upper crankcase on most models has a ridge mark (**Figure 159**) that can be used for rotary valve installation. If your crankcase is equipped with a ridge mark, install rotary valve as described under *Rotary Valve Installation*. However, if your crankcase does not have a ridge mark or if you are installing a modified rotary valve, perform the following. A dial indicator and degree wheel will be required to perform the following procedure.

1. Install the rotary valve shaft, if previously removed, as described in this chapter.

2. Install the engine top end assembly, if previously removed.

3. Find top dead center (TDC) as follows:
   a. Remove both spark plugs as described in Chapter Three.

   NOTE
   Removal of both spark plugs will ease crankshaft rotation.

   b. Screw the extension onto a dial indicator and insert the dial indicator into the adapter.
   c. Screw the dial indicator adapter into the cylinder head (**Figure 160**) on the MAG side. Do not lock the dial indicator in the adapter at this time.

   NOTE
   Remove the drive belt, if necessary, before performing sub-step d.

   d. Rotate the flywheel (by turning the primary sheave) until the dial indicator rises all the way up in its holder (piston is approaching top dead center). Then, slide the indicator far enough into the holder to obtain a reading.
   e. Lightly tighten the set screw on the dial indicator adapter to secure the dial gauge.
   f. Rotate the flywheel until the dial on the gauge stops and reverses direction. This is top dead center. Zero the dial gauge by aligning the zero with the dial (**Figure 161**).

g. Tighten the set screw on the dial indicator adapter securely.

NOTE
*When performing Step 4, refer to **Table 6** for rotary valve opening and closing degrees for your model.*

4. Set the engine at TDC as described in Step 3. Install a degree wheel onto the rotary valve

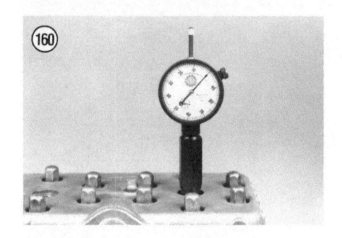

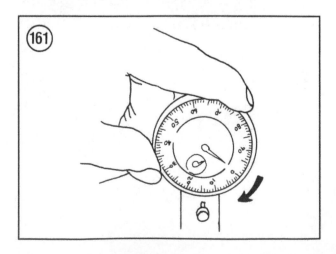

shaft. Mark the "opening" point (on crankcase) from the bottom edge of the magneto side intake port as shown in **Figure 162**. Mark the "closing" point (on crankcase) from the top edge of the intake port as shown in **Figure 163**.

5. Position rotary valve on gear (**Figure 164**) so edges align with timing marks made in Step 4.

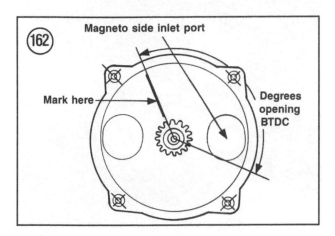

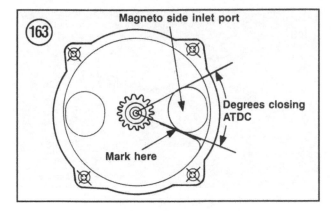

NOTE
The rotary valve is asymmetrical. Position each side of valve disc on gear to determine position in which greater alignment accuracy can be achieved.

Rotary Valve Installation

Coat the rotary valve with injection oil before final assembly.

1. Install the rotary valve shaft, if previously removed, as described in this chapter.

2. Install the engine top end assembly, if previously removed.

3. Find top dead center as described under *Rotary Valve Timing* in this section.

4. Locate the ridge mark on the upper crankcase. See **Figure 159**.

5. Position rotary valve on gear (**Figure 164**) so valve edge aligns with the ridge mark as shown in **Figure 165**.

NOTE
The rotary valve is asymmetrical. Position each side of valve disc on gear to determine position in which greater alignment accuracy can be achieved.

NOTE
*If your crankcase does not have a ridge mark, install rotary valve as described under **Rotary Valve Timing** in this section.*

6. Install the O-ring into the groove in the rotary valve cover (**Figure 166**).

7. Install the rotary valve cover (**Figure 155**). Install the cover bolts and tighten to the torque specification in **Table 7**.

8. Reverse Steps 1 and 2 under *Removal* to complete installation.

Rotary Valve Shaft Removal

Refer to **Figure 153** or **Figure 154** for this procedure.

1. Remove the engine from the frame as described in this chapter.

2. Remove the rotary valve as described in this chapter.

3. Remove the bolts holding the water pump cover (**Figure 167**, typical) to the crankcase. Remove the water pump cover and gasket (**Figure 168**).

4. Remove the impeller assembly as follows:
   a. Loosen and remove the impeller nut (**Figure 169**).
   b. Remove the washer (**Figure 170**).
   c. Carefully pull the impeller (**Figure 171**) off of the rotary valve shaft.
   d. Remove the serrated washer (**Figure 172**).

NOTE
The washer removed in sub-step e (Figure 173) is different from the one shown in (Figure 170). Do not intermix them.

   e. Remove the friction washer (**Figure 173**).

5. Split the crankcase and remove the crankshaft as described in this chapter.

6. Remove the circlip from the rotary valve side with circlip pliers (**Figure 174**).

7. Carefully tap the rotary valve shaft out of the crankcase from the exhaust side. Remove the rotary valve shaft (**Figure 175**).

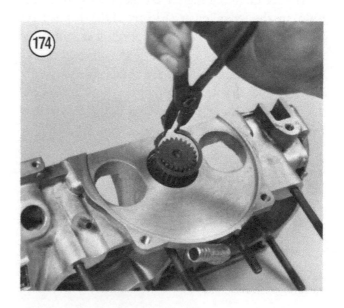

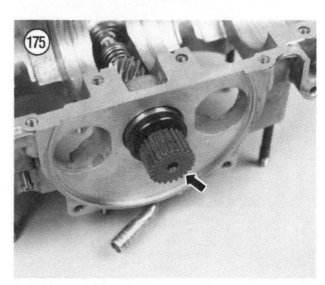

Rotary Valve Shaft Crankcase
Bearing and Seal Replacement

Refer to **Figure 153** or **Figure 154** for this procedure.

1A. *1985-1988:* Using the seal pusher (part No. 420 876 512) (**Figure 176**), drive the bearing, oil seals and spacer ring out of the upper crankcase. See **Figure 177**.

1B. *1989:* Using the seal pusher (part No. 420 876 512) (**Figure 176**), drive the bearing, oil seals, shim and rubber washer out of the upper crankcase. See **Figure 177**.

2. Before cleaning the crankcase, check the drain hole in the crankcase (**Figure 178**) for oil or coolant leakage. If oil or coolant residue is present, the oil or coolant seal was leaking.

3. Clean the bearing bore in the crankcase with solvent and dry thoroughly.

4. Check the bearing/seal area in the crankcase for cracks or other damage.

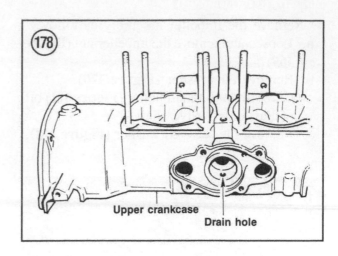

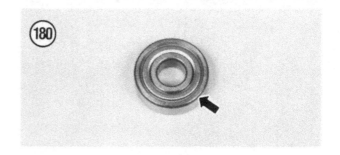

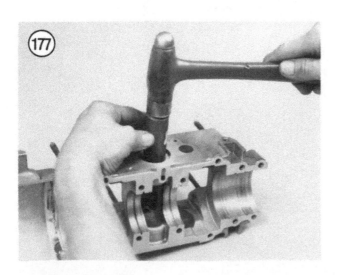

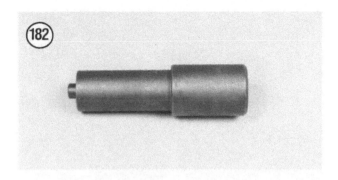

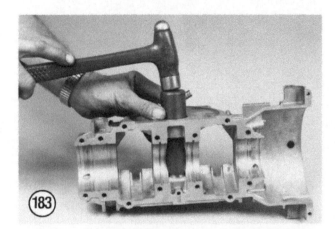

5. Hold the bearing outer race and turn the inner race by hand (**Figure 179**). Check for roughness or excessive noise. Replace the bearing if necessary.

NOTE
*Install the bearing in Step 6 so that the shielded side (**Figure 180**) faces toward the rotary valve.*

6. Place the bearing into the bearing bore as shown in **Figure 181**. Then, drive the bearing into the crankcase with the ball bearing pusher (part No. 420 8765 00) (**Figure 182**) until the bearing seats against the crankcase shoulder (**Figure 183**). See **Figure 184**.

7. Apply a low temperature lithium base grease to both oil seals.

NOTE
*Use the seal pusher (part No. 420 876 512) (**Figure 176**) when installing the oil seals in Step 8.*

8A. *1985-1988:* Install the seal assembly as follows:
  a. Align the first oil seal with the crankcase so that the closed side faces out. Drive the oil seal into the crankcase until it seats against the bearing (**Figure 177**).
  b. Insert the distance ring into the crankcase (**Figure 185**) so that it seats against the oil seal. See **Figure 186**.

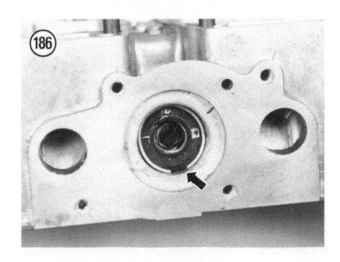

5

c. Fill the area on the outside of the seal with the same lithium base grease used in Step 7.

d. Align the second oil seal with the crankcase so that the flat side faces *out* (**Figure 187**). Install the second oil seal so that there is a gap between the first and second seal (**Figure 188**) and the drain hole (**Figure 178**) is not obstructed.

8B. *1989*: Install the seal assembly as follows:

a. Align the first oil seal with the crankcase so that the closed side faces out. Drive the oil seal into the crankcase until it seats against the bearing (**Figure 177**).

b. Fill the area on the outside of the seal with the same lithium base grease used in Step 7.

c. Align the second oil seal with the crankcase so that the flat side faces *in*. Install the second oil seal so that there is a gap between the first and second seal.

d. Install the shim so that it rest against the second oil seal.

e. Install the rubber washer.

Rotary Valve Shaft Inspection

Refer to **Figure 189** for this procedure.

1. Turn the bearings (A) and check for roughness or excessive noise.

2. Visually check the gears (B) for cracks, deep scoring or excessive wear.

3. Check the oil seal lip (C) for cuts or any sign of damage.

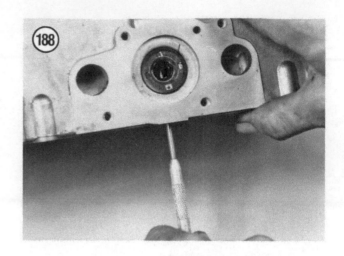

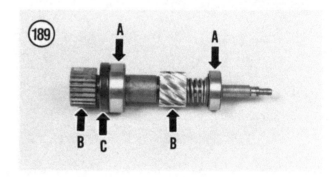

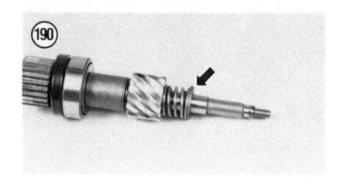

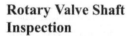

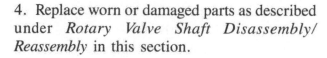

4. Replace worn or damaged parts as described under *Rotary Valve Shaft Disassembly/ Reassembly* in this section.

Rotary Valve Shaft Disassembly/Reassembly

A press will be required to disassemble and reassemble the rotary valve shaft. Read the complete procedure through before beginning disassembly. If you are not equipped with special tools, refer service to a Ski-Doo dealer.

1. Compress the cup (**Figure 190**) and remove the circlip (**Figure 191**) from the end of the rotary valve shaft.
2. Remove the spring retainer cup (**Figure 192**).
3. Remove the spring (**Figure 193**).
4. Remove the gear (**Figure 194**).
5. Remove the O-ring (**Figure 195**).
6. Remove the shim (**Figure 196**).
7. To remove the distance sleeve, shim, bearing and oil seal (**Figure 197**), perform the following:

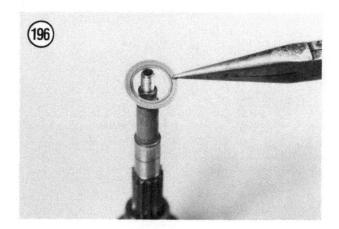

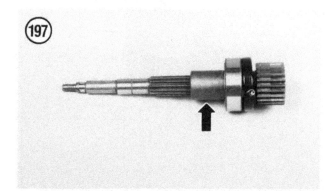

5

a. Heat the distance sleeve with a torch to break the Loctite bond.

b. Using a press and bearing puller as shown in **Figure 198**, remove the distance sleeve, shim and bearing.

c. Remove the oil seal.

8. Clean all parts in solvent and dry thoroughly.

9. Install oil seal, bearing, shim and distance sleeve as follows:

a. Fill the oil seal lip with a low temperature lithium base grease.

b. Install the oil seal (**Figure 199**) so that the shield side faces toward the rotary valve.

c. Apply Loctite RC 609 to the inside of the distance sleeve.

d. Install the bearing, shim and distance sleeve with a press as shown in **Figure 200**.

10. Install the shim (**Figure 196**).

11. Install a new O-ring (**Figure 195**).

12. Install the gear (**Figure 194**).

13. Install the spring (**Figure 193**).

14. Install the spring retainer cup (**Figure 192**).

15. Compress the cup and install the circlip (**Figure 191**). Make sure the circlip seats in the rotary valve shaft completely (**Figure 190**).

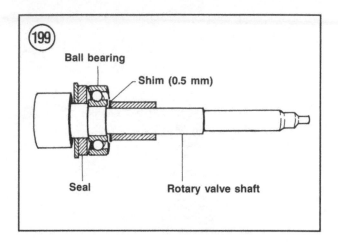

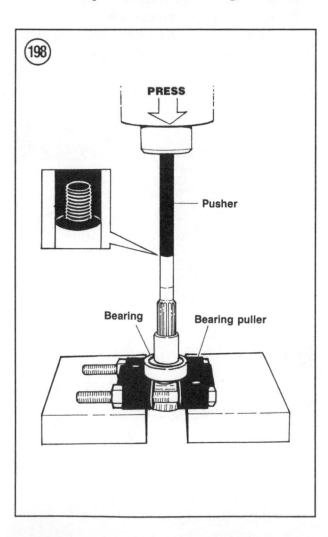

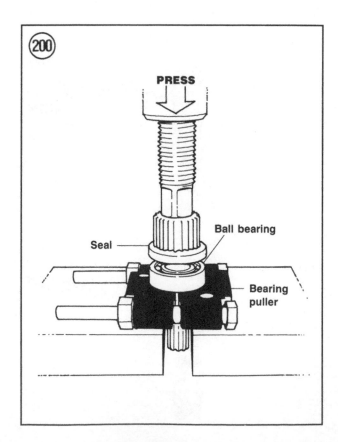

(201)

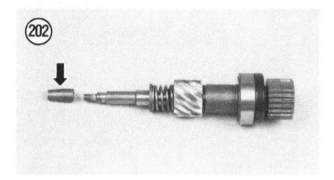

(202)

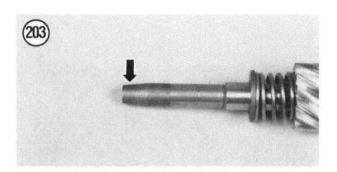

(203)

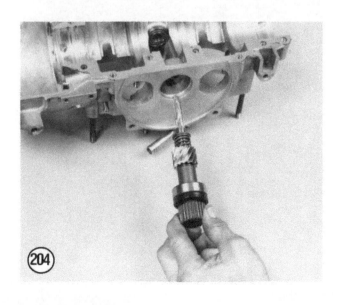

(204)

Impeller Inspection

Inspect the impeller (**Figure 201**) for cracks or other damage. Replace if necessary.

Rotary Valve Shaft Installation

To prevent oil seal damage, the seal sleeve (part No. 420 8769 80) (**Figure 202**) must be installed onto the end of the rotary valve shaft. See **Figure 203**.

1. Install the seal sleeve onto the end of the rotary valve shaft (**Figure 203**).

2. Coat the seal sleeve with a low temperature lithium base grease and insert the rotary valve shaft into the upper crankcase (**Figure 204**). Push the shaft into the crankcase until it stops (**Figure 205**). Then, use the pusher (part No. 520 8766 05) (**Figure 206**) and drive the rotary valve shaft

5

(205)

(206)

all the way into the crankcase. See **Figure 207**. Remove the seal sleeve from the end of the rotary valve shaft (**Figure 208**).

3. Secure the rotary valve shaft with the circlip (**Figure 209**). Make sure the circlip seats in the crankcase groove completely.

4. Install the crankshaft and assemble the crankcases as described in this chapter.

5. Install the impeller assembly (**Figure 210**) as follows:

   a. Install the friction washer (**Figure 211**).

   b. Install the serrated washer (**Figure 212**).

   c. Align the flat on the impeller with the flat on the shaft (**Figure 213**) and install the impeller. See **Figure 214**.

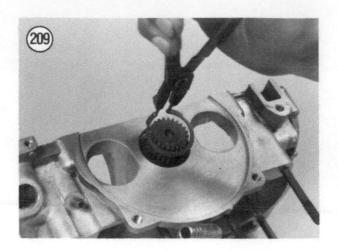

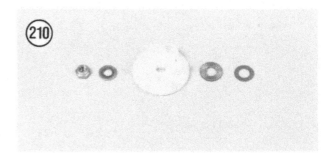

d. Install the washer (**Figure 215**).
e. Install a new impeller nut (**Figure 216**). Tighten the nut securely.
f. Install the water pump cover with a new gasket (**Figure 217**). Tighten the cover bolts securely.

5

Tables are on the following pages.

Table 1 ENGINE NUMBER IDENTIFICATION

Year	Model	Engine number
1985	Formula MX	467
	Formula Plus	537
1986	Formula MX	467
	Formula Plus	537
	Formula MX (H/A*)	467
1987	Formula MX	467
	Formula Plus	537
	Formula MX LT	467
1988	Formula MX	467
	Formula Plus	537
	Formula MX LT	467
1989	Formula MX	467
	Formula MX LT	467
	Formula Plus	536
	Formula Plus LT	536
	Formula Mach I	583

*High altitude

Table 2 GENERAL ENGINE SPECIFICATIONS

Bore	
MX, MX (H/A) and MX LT	69.5 mm (2.736 in.)
Plus and Plus LT	72.0 mm (2.835 in.)
Mach I	76.0 mm (2.992 in.)
Stroke	
MX, MX (H/A) and MX LT	61.0 mm (2.402 in.)
Plus and Plus LT	64.0 mm (2.520 in.)
Mach I	64.0 mm (2.520 in.)
Displacement	
MX, MX (H/A) and MX LT	462.8 cc (28.2 cu. in.)
Plus and Plus LT	521.2 cc (31.8 cu. in.)
Mach I	580.8 cc (35.4 cu. in.)
Compression ratio	
MX, MX (H/A) and MX LT	7.5:1
Plus (1985-1988)	6.5:1
Plus and Plus LT (1989)	6.1:1
Mach I	5.9:1

Table 3 467 ENGINE SPECIFICATIONS

	Specification mm (in.)	Wear limit mm (in.)
Cylinder head		
Warp limit	N.A.	—
Cylinder		
Taper	—	0.08
	—	(0.0031)
Out-of-round	—	0.05
	—	(0.0020)
Piston-to-cylinder clearance		
1985-1986	0.08-0.10	0.20
	(0.0031-0.0039)	(0.008)
1987-on	0.10-0.12	0.20
	(0.0039-0.0047)	(0.008)
Piston ring end gap		
1985-1987	0.20	1.00
	(0.008)	(0.039)
1988-on	0.20-0.35	1.00
	(0.008-0.014)	(0.0039)
Piston ring groove		
clearance	0.04-0.11	0.20
	(0.002-0.004)	(0.008)
Connecting rod big end axial play		
1985-1988	0.40-0.73	1.20
	(0.016-0.029)	(0.047)
1989	0.40	1.20
	(0.016)	(0.047)
Crankshaft end play		
1985-1987	N.A.	—
1988	0.10-1.00	—
	(0.004-0.039)	—
1989	1.00	—
	(0.039)	—
Crankshaft deflection		
	—	0.08
	—	(0.0031)

Table 4 536 AND 537 ENGINE SPECIFICATIONS

	Specification mm (in.)	Wear limit mm (in.)
Cylinder head		
Warp limit	N.A.	—
Cylinder		
Taper	—	0.08
	—	(0.0031)
Out-of-round	—	0.05
	—	(0.0020)
Piston-to-cylinder clearance		
1985	0.09-0.11	0.20
	(0.0035-0.0043)	(0.008)
	(continued)	

Table 4 536 AND 537 ENGINE SPECIFICATIONS (continued)

	Specification mm (in.)	Wear limit mm (in.)
Piston-to-cylinder clearance		
1986-1988	0.11-0.13 (0.0043-0.0051)	0.20 (0.008)
1989	0.09-0.10 (0.0035-0.0039)	0.20 (0.008)
Piston ring end gap		
1985-1987	0.20 (0.008)	1.00 (0.039)
1988-on	0.20-0.35 (0.008-0.014)	1.00 (0.039)
Piston ring groove clearance	0.04-0.11 (0.002-0.0043)	0.20 (0.008)
Connecting rod big end axial play		
1985-1988	0.40-0.73 (0.016-0.029)	1.20 (0.047)
1989	0.40 (0.016)	1.20 (0.047)
Crankshaft end play		
1985-1987	N.A.	—
1988	0.10-1.00 (0.004-0.039)	— —
1989	1.00 (0.039)	— —
Crankshaft deflection	— —	0.08 (0.0031)

Table 5 583 ENGINE SPECIFICATIONS

	Specification mm (in.)	Wear limit mm (in.)
Cylinder head		
Warp limit	N.A.	—
Cylinder		
Taper	— —	0.08 (0.0031)
Out-of-round	— —	0.05 (0.0020)
Piston-to-cylinder clearance	0.11-0.13 (0.0043-0.0051)	0.20 (0.008)
Piston ring end gap	0.20-0.35 (0.008-0.014)	1.00 (0.039)
Piston ring groove clearance	0.04-0.10 (0.002-0.0039)	0.20 (0.008)
Connecting rod big end axial play	0.40 (0.016)	1.20 (0.047)
Crankshaft end play	0.10-0.30 (0.004-0.012)	—
Crankshaft deflection	— —	0.08 (0.0031)

Table 6 ROTARY VALVE TIMING

	Opening BTDC	Closing ATDC
1985-1988		
All models	132°	52°
1989		
Formula MX	132°	52°
Formula MX LT	132°	52°
Formula Plus	117°	52°
Formula Plus LT	117°	52°
Formula Mach I	140°	61°

Table 7 ENGINE TIGHTENING TORQUES

	N·m	ft.-lb.
Cylinder head nuts		
1985-1988		
All models	20	15
1989		
All models	22	16
Cylinder base nuts		
1985-1988	21	15
1989	22	16
Crankcase nut or screws		
1985-1988		
M6	9	7
M8	21	15
1989		
M6	9	7
M8	22	16
Exhaust manifold bolts or nuts		
1985-1987	N.A.	—
1988-on	25	18
Flywheel nut		
1985-1988		
All models	100	74
1989		
All models	105	77
Engine mounts		
1985-1987	11	8
1988-on	38	28
Rotary valve cover	20	15

5

Chapter Six

Fuel And Exhaust Systems

The fuel system consists of the fuel tank, fuel pump, carburetors and air silencer. There are slight differences among the various models and they are noted in the various procedures.

The exhaust system consists of an exhaust pipe assembly and a muffler.

This chapter includes service procedures for all parts of the fuel and exhaust systems.

Carburetor specifications are listed in **Tables 1-5** at the end of the chapter.

AIR SILENCER

The air silencer (**Figure 1**), sometimes referred to as the air box, should be periodically inspected for cleanliness, cracks or debris buildup.

> *CAUTION*
> *Never run the engine without the air silencer installed. Running the engine without the air silencer will cause a lean fuel mixture and engine seizure. The air silencer must be installed during carburetor adjustment.*

Removal/Installation

Refer to **Figure 1** for this procedure.
1. Open the shroud.
2. Loosen the intake boot clamps at the carburetor.
3. Disconnect the intake hoses at the air silencer.
4. Remove the air silencer (**Figure 2**).
5. Cover the carburetors to prevent dirt or moisture from entering.
6. Installation is the reverse of these steps.

Inspection

Check the air silencer box for cracks or other damage. Repair cracks before reinstalling the silencer box onto the engine.

CARBURETOR

All models are equipped with 2 slide type Mikuni carburetors. Refer to **Tables 1-4** for carburetor identification and specifications.

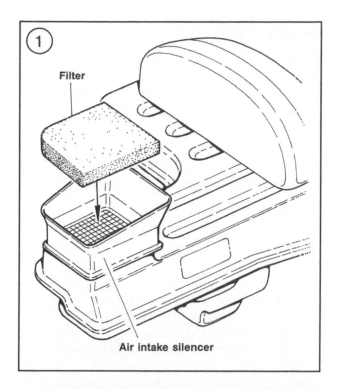

Filter

Air intake silencer

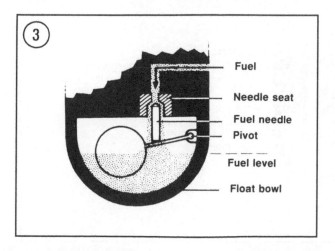

Fuel

Needle seat

Fuel needle

Pivot

Fuel level

Float bowl

A hand operated primer located on the right-hand side of the cowl is used for cold starting. Some models are also equipped with a choke lever. Fuel is supplied by a remote pulse-type fuel pump.

CARBURETOR OPERATION

For proper operation, a gasoline engine must be supplied with fuel and air mixed in proper proportions by weight. A mixture in which there is an excess of fuel is said to be rich. A lean mixture is one which contains insufficient fuel. A properly adjusted carburetor supplies the proper mixture to the engine under all operating conditions.

The carburetors installed on all models consist of several major systems. A float and float valve mechanism maintain a constant fuel level in the float bowl. The pilot system supplies fuel at low speeds. The main fuel system supplies fuel at medium and high speeds. Finally, a primer system supplies a rich mixture needed to start a cold engine.

Float Mechanism

To assure a steady supply of fuel, the carburetor is equipped with a float valve through which fuel flows by the pulse operated fuel pump into the float bowl (**Figure 3**). Inside the bowl is a combined float assembly that moves up and down with the fuel level. Resting on the float arm is a float needle, which rides inside the float valve. The float valve regulates fuel flow into the float bowl. The float needle and float valve contact surfaces which are accurately machined to ensure correct fuel flow calibration. As the float rises, the float needle rises inside the float valve and blocks it, so that when the fuel has reached the required level in the float bowl, no more fuel can enter.

Pilot and Main Fuel Systems

The carburetor's purpose is to supply and atomize fuel and mix it in correct proportions with air that is drawn in through the air intake. At primary throttle openings (from idle to 1/8 throttle) a small amount of fuel is siphoned through the pilot jet by suction from the incoming air (**Figure 4**). As the throttle is opened further, the air stream begins to siphon fuel through the main jet and needle jet. The tapered needle increases the effective flow capacity of the needle jet as it rises with the throttle slide, in that it occupies decreasingly less of the area of the needle jet (**Figure 5**). In addition, the amount of cutaway in the leading edge of the throttle slide aids in controlling the fuel/air mixture during partial throttle openings.

At full throttle, the carburetor venturi is fully open and the needle is lifted far enough to permit the main jet to flow at full capacity. See **Figure 6** and **Figure 7**.

Starting System

The starting system consists of a push-pull primer button mounted on the outside of the snowmobile below the handlebar. When the engine is cold, the primer should be pushed 2-3 times.

Removal/Installation

1. Open the shroud.
2. Remove the air silencer (**Figure 2**) as described in this chapter.

NOTE
Before removing the carburetors, note that a red dot is painted on one carburetor and on the mating oil pump mounting flange. These marks are used

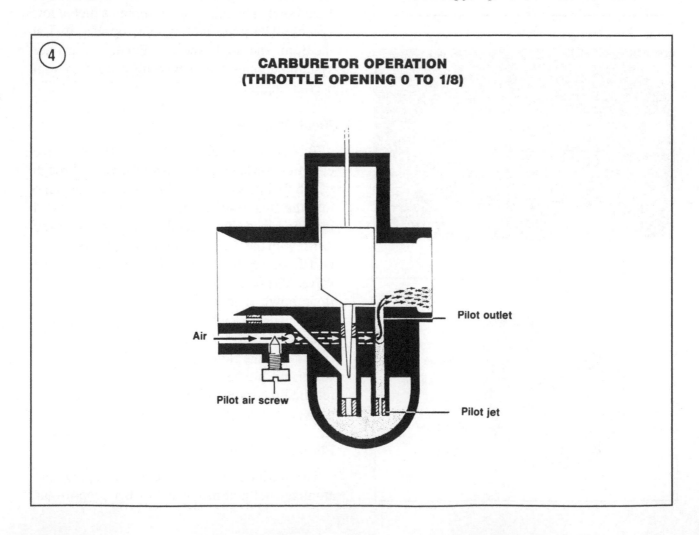

④

**CARBURETOR OPERATION
(THROTTLE OPENING 0 TO 1/8)**

Pilot outlet

Air

Pilot air screw

Pilot jet

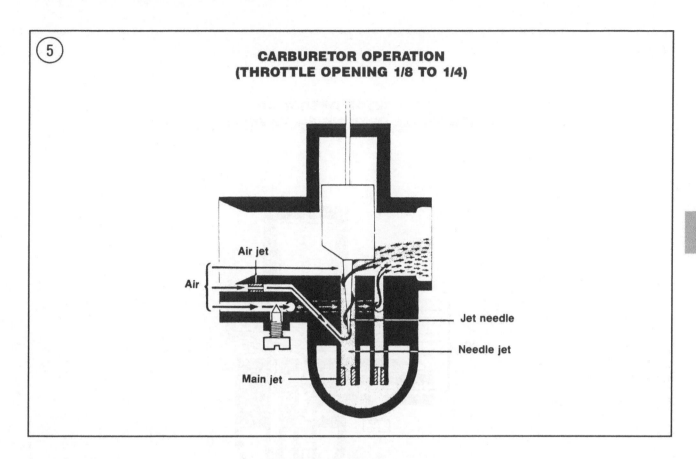

**⑤ CARBURETOR OPERATION
(THROTTLE OPENING 1/8 TO 1/4)**

Air jet

Air

Jet needle

Needle jet

Main jet

6

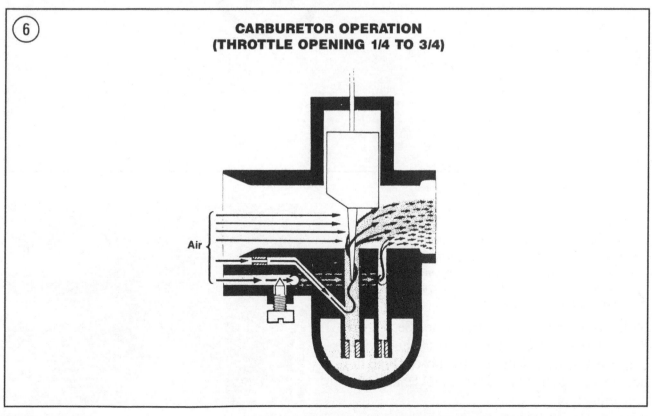

**⑥ CARBURETOR OPERATION
(THROTTLE OPENING 1/4 TO 3/4)**

Air

**CARBURETOR OPERATION
(THROTTLE OPENING 3/4 TO FULL)**

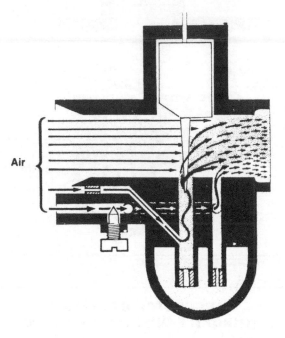

Air

to identify carburetor mounting positions. This is due to the different jetting found in the 2 carburetor bodies. *Figure 8* shows a typical mark found on the MAG side of the oil pump mounting flange. Locate the marks on your snowmobile before removal.

3. Label the hoses at the carburetors. See **Figure 9** or **Figure 10**. Loosen the metal hose clips (**Figure 11**) and disconnect the hoses. Plug the hoses with a golf tee to prevent fuel leakage and contamination.

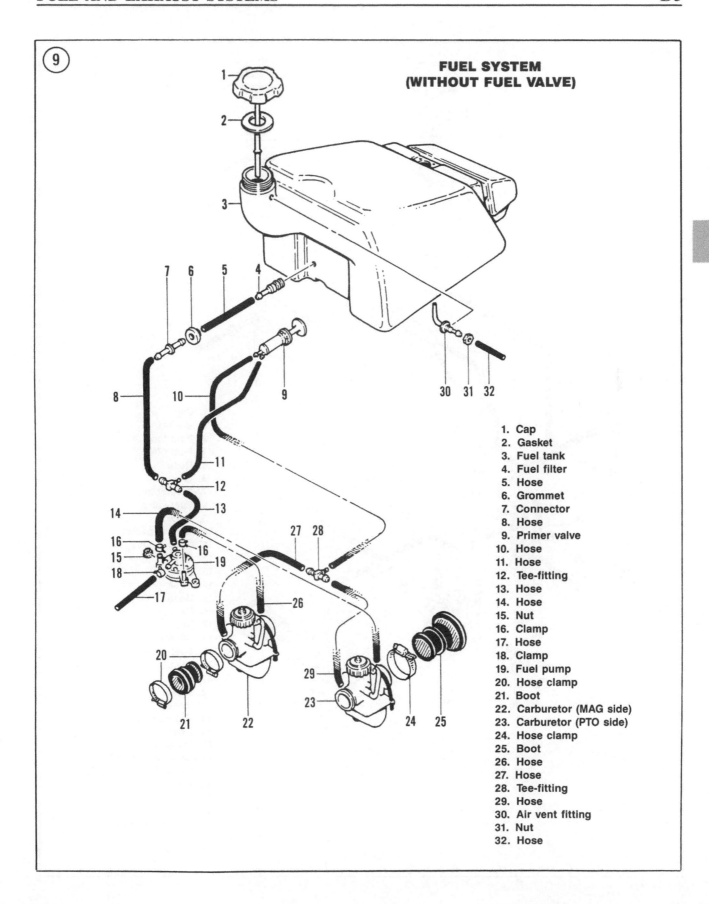

**FUEL SYSTEM
(WITHOUT FUEL VALVE)**

1. Cap
2. Gasket
3. Fuel tank
4. Fuel filter
5. Hose
6. Grommet
7. Connector
8. Hose
9. Primer valve
10. Hose
11. Hose
12. Tee-fitting
13. Hose
14. Hose
15. Nut
16. Clamp
17. Hose
18. Clamp
19. Fuel pump
20. Hose clamp
21. Boot
22. Carburetor (MAG side)
23. Carburetor (PTO side)
24. Hose clamp
25. Boot
26. Hose
27. Hose
28. Tee-fitting
29. Hose
30. Air vent fitting
31. Nut
32. Hose

6

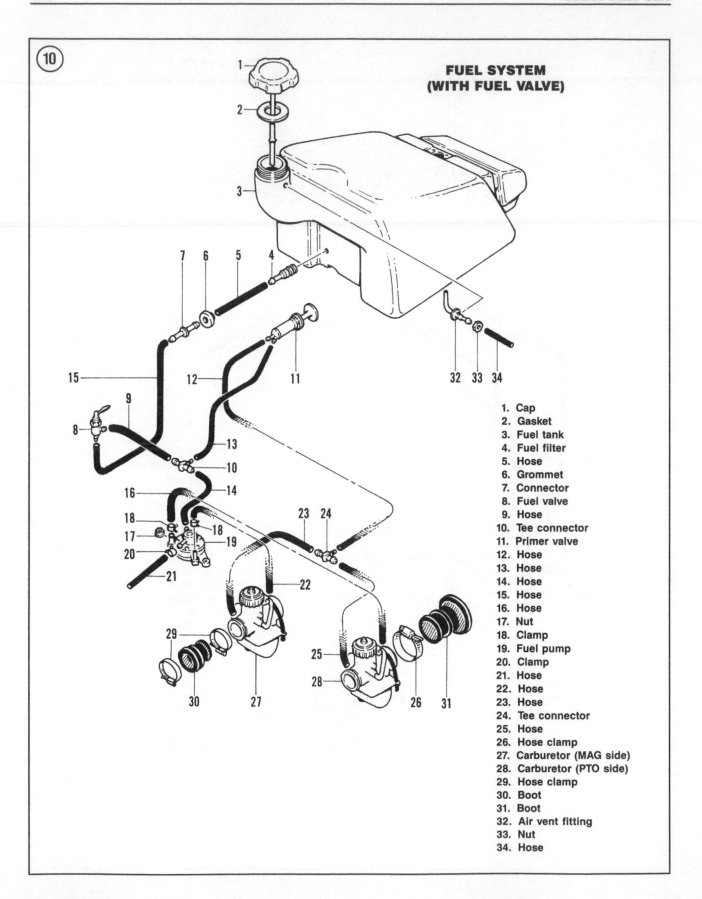

10

**FUEL SYSTEM
(WITH FUEL VALVE)**

1. Cap
2. Gasket
3. Fuel tank
4. Fuel filter
5. Hose
6. Grommet
7. Connector
8. Fuel valve
9. Hose
10. Tee connector
11. Primer valve
12. Hose
13. Hose
14. Hose
15. Hose
16. Hose
17. Nut
18. Clamp
19. Fuel pump
20. Clamp
21. Hose
22. Hose
23. Hose
24. Tee connector
25. Hose
26. Hose clamp
27. Carburetor (MAG side)
28. Carburetor (PTO side)
29. Hose clamp
30. Boot
31. Boot
32. Air vent fitting
33. Nut
34. Hose

4. Loosen the carburetor caps (**Figure 12**) and remove the throttle slide assembly from the carburetor body.

> *CAUTION*
> *Handle the slide carefully to prevent scratching or damaging the slide and needle jet in any way.*

5. Loosen the hose clamps at the intake manifolds.

6. Remove the carburetors.

7. Installation is the reverse of these steps. Note the following.

8. Install the carburetors in their original mounting positions. Refer to the *NOTE* in this procedure for identification.

9. Adjust the carburetors as described in Chapter Three.

10. Adjust the oil pump cable as described in Chapter Three.

11. Make sure the fuel hoses are properly connected to prevent a fuel leak. Secure the fuel hoses with new clips.

> *WARNING*
> *Do not start the engine if the fuel hoses are leaking.*

12. Make sure the air hoses are properly positioned and the hose clamps tightened securely to prevent an air leak.

Intake Manifolds

The intake manifolds (**Figure 8**) should be inspected frequently for tears or other damage that could cause a lean fuel mixture. If the intake manifolds are removed, apply a thin coat of silicone sealant onto the manifold nozzles (**Figure 13**).

Carburetor Identification

Due to the number of models covered in this manual and the slight differences in the specific carburetors the following will help identify your specific carburetor.

6

The Mikuni carburetor used on all of these models is basically the same and to help clarify this the carburetors are split into 3 types. Slight differences do occur between models, so it is important to pay particular attention to the location and order of parts during disassembly.

The Type I carburetor is found on the following models:

a. 1985-1988 467 engine.
b. 1985-1988 537 engine.

The Type II carburetor is found on the following model:

1989 536 engine.

The Type III carburetor is found on the following model:

1989 583 engine.

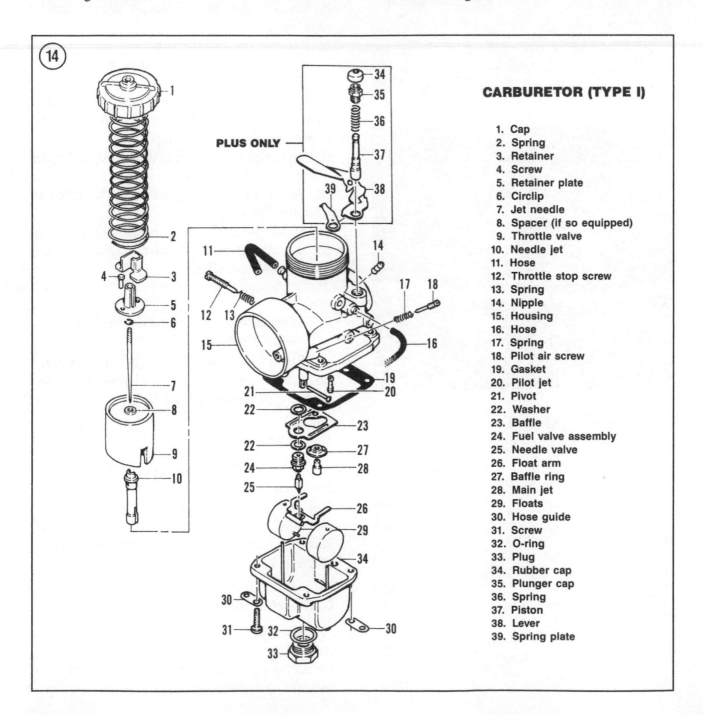

CARBURETOR (TYPE I)

1. Cap
2. Spring
3. Retainer
4. Screw
5. Retainer plate
6. Circlip
7. Jet needle
8. Spacer (if so equipped)
9. Throttle valve
10. Needle jet
11. Hose
12. Throttle stop screw
13. Spring
14. Nipple
15. Housing
16. Hose
17. Spring
18. Pilot air screw
19. Gasket
20. Pilot jet
21. Pivot
22. Washer
23. Baffle
24. Fuel valve assembly
25. Needle valve
26. Float arm
27. Baffle ring
28. Main jet
29. Floats
30. Hose guide
31. Screw
32. O-ring
33. Plug
34. Rubber cap
35. Plunger cap
36. Spring
37. Piston
38. Lever
39. Spring plate

PLUS ONLY

Disassembly
(Type I and Type II)

Refer to **Figure 14** (Type I) or **Figure 15** (Type II) for this procedure.

NOTE
Because the jetting between the PTO and MAG side carburetors are different, store components in separate boxes.

1. Lightly seat the pilot air screw, counting number of turns required for reassembly reference, then back screw out and remove from carburetor with spring.

2. Remove the throttle stop screw and spring.

3. Remove the screws holding the float bowl to the carburetor housing. Remove the float bowl (**Figure 16**).

4. Remove the pilot jet (**Figure 17**) with a flat-tipped screwdriver.

5. Remove the main jet (**Figure 18**).

6. Remove the baffle ring. See **Figure 19** or **Figure 20**.

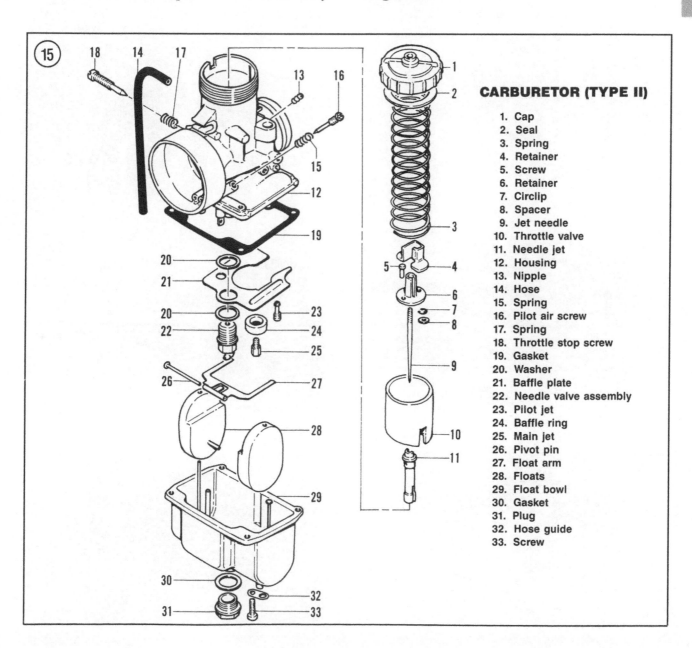

CARBURETOR (TYPE II)

1. Cap
2. Seal
3. Spring
4. Retainer
5. Screw
6. Retainer
7. Circlip
8. Spacer
9. Jet needle
10. Throttle valve
11. Needle jet
12. Housing
13. Nipple
14. Hose
15. Spring
16. Pilot air screw
17. Spring
18. Throttle stop screw
19. Gasket
20. Washer
21. Baffle plate
22. Needle valve assembly
23. Pilot jet
24. Baffle ring
25. Main jet
26. Pivot pin
27. Float arm
28. Floats
29. Float bowl
30. Gasket
31. Plug
32. Hose guide
33. Screw

7. Remove the needle jet through the top of the carburetor (**Figure 21**).

8. Remove the pivot pin (A, **Figure 22**) and pivot arm (B, **Figure 22**).

9. Remove the needle valve assembly as follows:

   a. Remove the clip (**Figure 23**), if so equipped.

   b. Remove the needle (**Figure 24**).

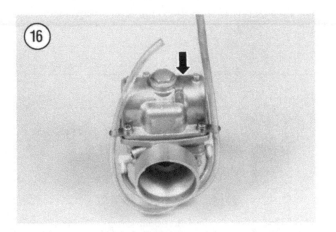

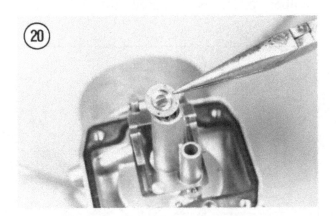

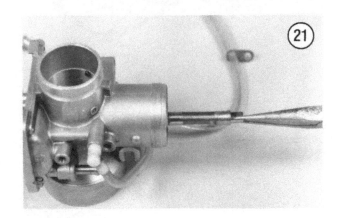

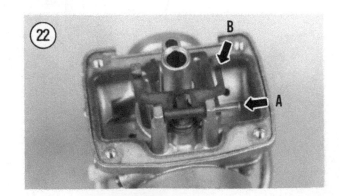

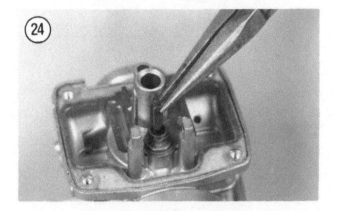

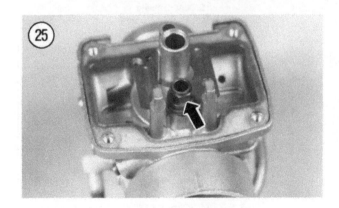

c. Remove the seat (**Figure 25**).

d. Remove the washer (**Figure 26**).

e. Remove the baffle plate (**Figure 27**).

f. Remove the washer (**Figure 28**).

10. Remove the float bowl gasket (**Figure 29**).

11. Clean and inspect the carburetor assembly as described in this chapter.

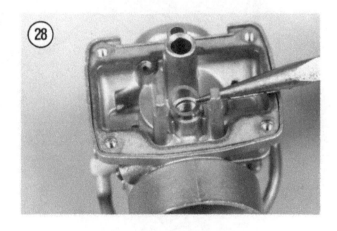

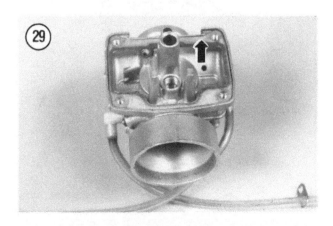

Assembly
(Type I and Type II)

Refer to **Figure 14** (Type I) or **Figure 15** (Type II) for this procedure.

1. Install a new float bowl gasket (**Figure 29**), if the other gasket is torn or otherwise damaged.

2. Install the needle valve assembly as follows:

NOTE
Replace the washers during assembly.

a. Install the first washer (**Figure 28**).

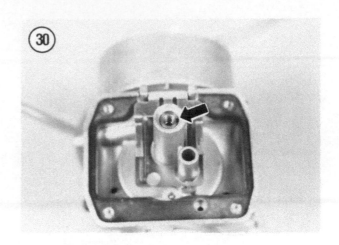

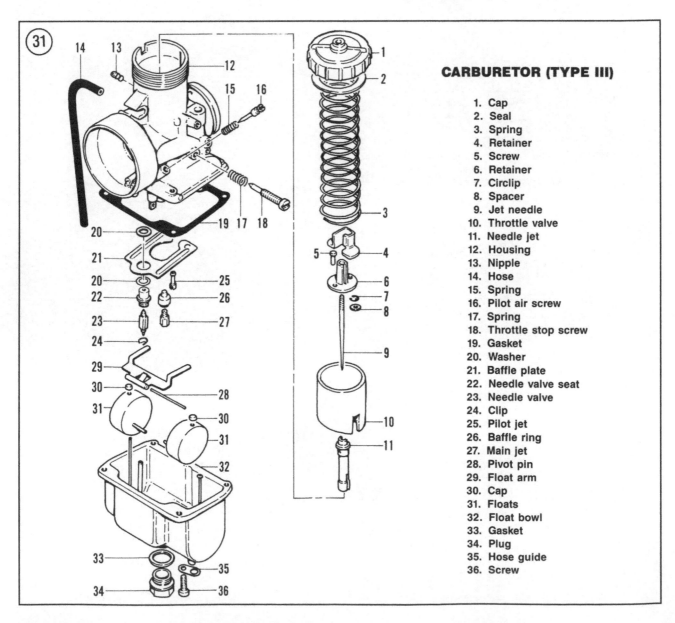

CARBURETOR (TYPE III)

1. Cap
2. Seal
3. Spring
4. Retainer
5. Screw
6. Retainer
7. Circlip
8. Spacer
9. Jet needle
10. Throttle valve
11. Needle jet
12. Housing
13. Nipple
14. Hose
15. Spring
16. Pilot air screw
17. Spring
18. Throttle stop screw
19. Gasket
20. Washer
21. Baffle plate
22. Needle valve seat
23. Needle valve
24. Clip
25. Pilot jet
26. Baffle ring
27. Main jet
28. Pivot pin
29. Float arm
30. Cap
31. Floats
32. Float bowl
33. Gasket
34. Plug
35. Hose guide
36. Screw

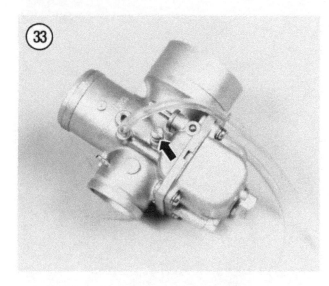

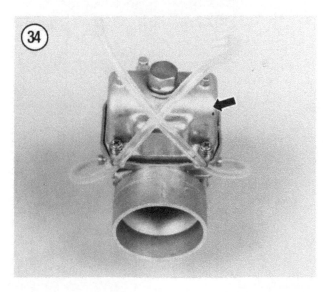

b. Install the baffle plate (**Figure 27**).

c. Install the second washer (**Figure 26**).

d. Install and tighten the seat (**Figure 25**).

e. Install the fuel valve so that the tapered portion faces down (**Figure 24**).

f. Secure the valve with the clip (**Figure 23**), if so equipped.

3. Install the float arm (B, **Figure 22**) and float pin (A, **Figure 22**). Push the pin in all the way.

4. Install the needle jet (**Figure 21**) so that the notch in the bottom of the needle jet aligns with the pin in the needle jet bore. See **Figure 30**.

5. Install the baffle ring. See **Figure 19** or **Figure 20**.

6. Install the main jet and tighten it securely (**Figure 18**).

7. Install the pilot jet (**Figure 17**) with a suitable screwdriver. Tighten the pilot jet securely.

8. Install the float bowl (**Figure 16**) and screws. Tighten the screws securely.

9. Install the throttle stop screw and spring.

10. Slide the spring onto the pilot air screw. Install the screw into the carburetor body until it lightly seats, then back out the number of turns noted during disassembly.

11. After installing the carburetor, perform the carburetor adjustments as described in Chapter Three.

Disassembly
(Type III)

Refer to **Figure 31** for this procedure.

NOTE
Because the jetting between the PTO and MAG side carburetors are different, store components in separate boxes.

1. Lightly seat the pilot air screw (**Figure 32**), counting number of turns required for reassembly reference, then back screw out and remove from carburetor with spring.

2. Remove the throttle stop screw and spring (**Figure 33**).

3. Remove the screws holding the float bowl to the carburetor housing. Remove the float bowl (**Figure 34**).

4. Remove the pilot jet (**Figure 35**) with a flat-tipped screwdriver.

5. Remove the main jet (**Figure 36**).

6. Remove the needle jet extension (**Figure 37**).

7. Remove the needle jet through the top of the carburetor (**Figure 38**).

8. Remove the pivot pin (A, **Figure 39**) and pivot arm (B, **Figure 39**).

9. Remove the needle valve assembly as follows:

  a. Remove the clip (**Figure 40**).

  b. Remove the needle (**Figure 41**).

  c. Remove the seat (**Figure 42**).

  d. Remove the washer (**Figure 43**).

  e. Remove the baffle plate (**Figure 44**).

  f. Remove the washer (**Figure 45**).

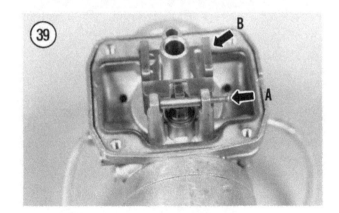

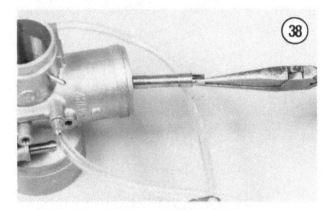

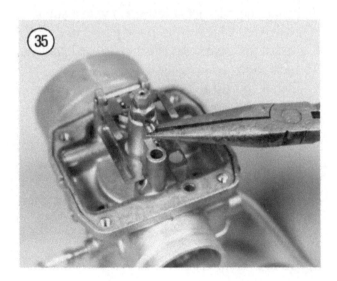

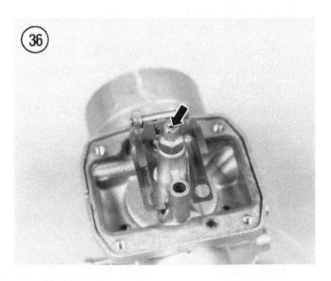

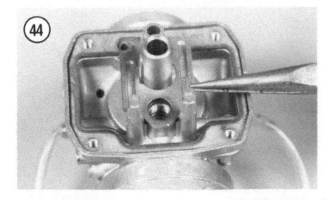

10. Remove the float bowl gasket (**Figure 46**).
11. Clean and inspect the carburetor assembly as described in this chapter.

Assembly
(Type III)

Refer to **Figure 31** for this procedure.
1. Install a new float bowl gasket (**Figure 46**), if the other gasket is torn or otherwise damaged.
2. Install the needle valve assembly as follows:

NOTE
Replace the washers during installation.

a. Install the first washer (**Figure 45**).
b. Install the baffle plate (**Figure 44**).

6

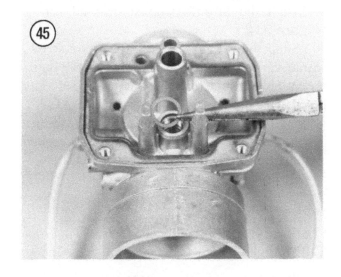

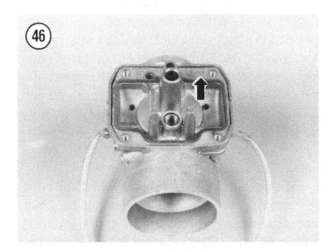

c. Install the second washer (**Figure 43**).

d. Install and tighten the seat (**Figure 42**).

e. Install the fuel valve so that the tapered portion faces down (**Figure 41**).

f. Secure the valve with the clip (**Figure 40**).

3. Install the float arm (B, **Figure 39**) and float pin (A, **Figure 39**). Push the pin in all the way.

4. Install the needle jet (**Figure 38**) so that the notch in the bottom of the needle jet aligns with the pin in the needle jet bore. See **Figure 47**.

5. Install the needle jet extension (**Figure 37**).

6. Install the main jet and tighten securely (**Figure 36**).

7. Install the pilot jet (**Figure 35**) with a suitable screwdriver. Tighten the pilot jet securely.

8. Install the float bowl (**Figure 34**) and screws. Tighten the screws securely.

9. Install the throttle stop screw and spring (**Figure 33**).

10. Slide the spring onto the pilot air screw. Install the screw into the carburetor body until it lightly seats, then back out the number of turns noted during disassembly. See **Figure 32**.

11 After installing the carburetor, perform the carburetor adjustments as described in Chapter Three.

Cleaning/Inspection (All Models)

Submersion of the carburetor housing assembly in a carburetor cleaner will damage rubber O-rings, seals and plastic parts.

1. Clean the carburetor castings and metal parts with aerosol solvent and a brush. Spray the aerosol solvent on the casting and scrub off any gum or varnish with a small bristle brush.

2. After cleaning the castings and metal parts, wash them thoroughly in hot soapy water. Rinse with clean water and dry thoroughly.

3. Inspect the carburetor body and float bowl for fine cracks or evidence of fuel leaks. Minor damage can be repaired with an epoxy or liquid aluminum type filler.

4. Blow out the jets with compressed air.

CAUTION
Do not use wire or a drill bit to clean carburetor passages or jets. This can enlarge the passages and change the carburetor calibration. If a passage or jet is severely clogged, use a piece of broom straw to clean it.

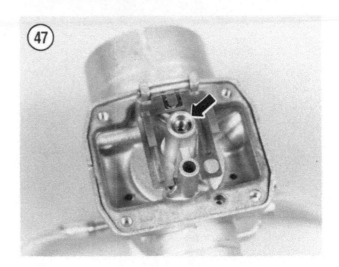

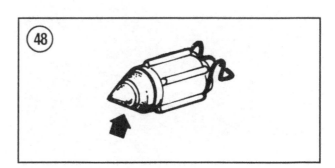

5. Inspect the tip of the float valve for wear or damage (**Figure 48**). Replace the valve and seat as a set if they are less than perfect.

> *NOTE*
> *A damaged float valve will result in flooding of the carburetor float chamber and impair performance. In addition, accumulation of raw gasoline in the engine compartment presents a severe fire hazard.*

6. Inspect the pilot air screw taper for scoring and replace it if less than perfect.
7. Inspect the jets for internal damage and damaged threads. Replace any jet that is less than

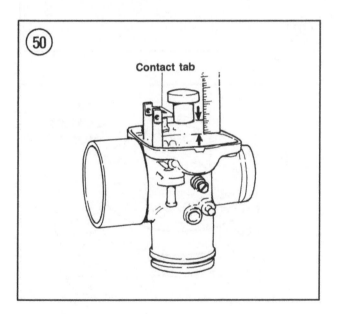

Contact tab

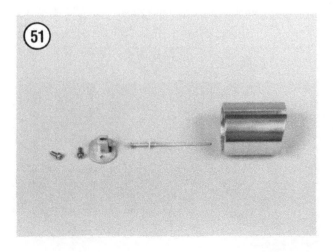

perfect. Make certain replacement jets are the same size as the originals.

> *CAUTION*
> *The jets must be scrupulously clean and shiny. Any burring, roughness or abrasion could cause a lean mixture that could result in major engine damage.*

8. Check the movement of the float arm on the pivot pin. It must move freely without binding.
9. O-ring seals tend to become hardened after prolonged use and heat and therefore lose their ability to seal properly. Inspect all O-rings and replace if necessary.
10. Check the floats (**Figure 49**) for fuel saturation, deterioration or excessive wear where it contacts the float arm. Replace as required. If the float is in good condition, check it for leakage as follows. Fill the float bowl with water and push the floats down. There should be no signs of bubbles. Replace the floats if necessary.

Float Height
Check and Adjustment

1. Remove the float bowl as described in this chapter.
2. Remove the float bowl gasket.
3. Invert the carburetor. Allow the float arm to contact the fuel valve, but don't compress the spring-loaded plunger on the needle.
4. Measure the distance from the float arm to the float bowl gasket surface (**Figure 50**). Refer to **Table 5** for float height.
5. If the float height is incorrect, remove the float pin (A, **Figure 39**) and float arm (B, **Figure 39**). Bend tang in center of float arm to adjust.
6. Reinstall float arm and pin. Recheck adjustment.
7. Install float bowl as described in this chapter.

Jet Needle/Throttle Valve
Removal/Installation

A typical jet needle/throttle valve assembly is shown in **Figure 51**.

1. Unscrew the carburetor cap and pull the throttle valve out of the carburetor.

2. At the end of the throttle cable, push up on the throttle spring. Then, disconnect the cable from the retainer.

3. Remove the screws holding the retainer plate and remove the plate.

4. Remove the jet needle.

5. Installation is the reverse of these steps. Note the following:

    a. Carburetor tuning is described in Chapter Four.

    b. Some models have a nylon washer installed between the jet needle E-clip and throttle valve (**Figure 52**). During reassembly, make sure the nylon washer is installed.

> *CAUTION*
> *Failure to install the nylon washer can cause engine seizure.*

    c. When installing the throttle valve into the carburetor, align the groove in the throttle valve with the pin in the carburetor bore while at the same time aligning the jet needle with the needle jet opening.

FUEL PUMP

Removal/Installation

1. Open the shroud.

2. Remove the air silencer as described in this chapter.

> *NOTE*
> *To remove the wire hose clamps in Step 3, squeeze the ends together with pliers and slide the clamp off the fuel fitting. Do not pull the hoses off without first removing the clamps or you may damage the ends of the hose.*

3. Label the hoses at the fuel pump (**Figure 53**). Plug the hoses with golf tees to prevent fuel leakage or contamination.

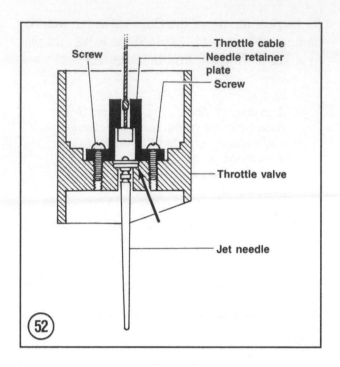

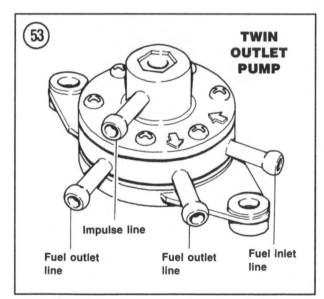

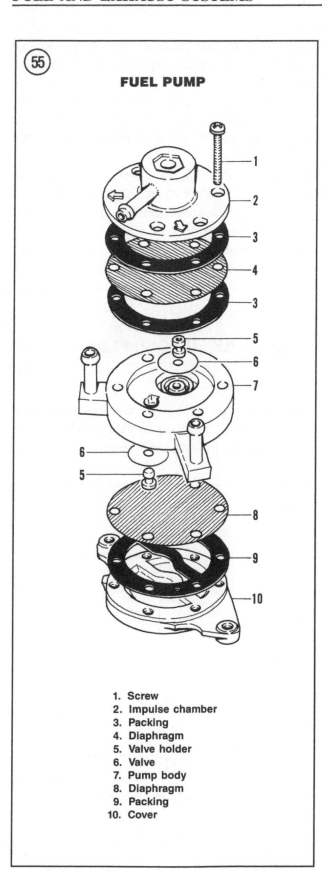

FUEL PUMP

1. Screw
2. Impulse chamber
3. Packing
4. Diaphragm
5. Valve holder
6. Valve
7. Pump body
8. Diaphragm
9. Packing
10. Cover

4. Remove the bolts holding the fuel pump to the bulkhead and remove the fuel pump (**Figure 54**).

5. Replace the wire hose clamps when they have lost tension.

6. Installation is the reverse of these steps. Reconnect the hoses according to the ID marks made before disassembly.

Disassembly/Assembly

Refer to **Figure 55** for this procedure.

1. Plug the 4 pump body fuel fittings and clean the pump in solvent; dry thoroughly.

2. Test the pump outlet check valve by sucking and blowing through one of the outlet openings (block other outlet opening). See **Figure 53** for outlet identification. You should be able to draw air through the valve, but not blow air through it.

3. Test the pump inlet check valve by sucking and blowing through the inlet opening. You should be able to blow through the valve, but not draw air through it.

4. If the check valves do not operate as specified in Step 2 or Step 3, replace them as follows.

5. Remove the 6 Phillips screws holding the covers on each side of the pump body. Separate the covers from the body.

6. Remove the diaphragms and gaskets. Separate the diaphragms from the gaskets. Discard the gaskets.

NOTE
Do not remove the valve holders and valves unless replacement is required. Do not reinstall valves which have been previously used.

7. Remove the valve holders and valves. Discard the valves.

8. Clean and inspect the pump components as described in this chapter.

9. Install new valves as follows:

   a. Align the new valve with its seat. Make sure the valve is flat.

   b. Lubricate the tip of the valve holder with a drop of oil. Then, push the holder through

6

the valve with a pin punch or rod with a diameter of 3/32 in. (2.5 mm) as shown in **Figure 56**.

c. Repeat for both valves.

10. Install the diaphragms and gaskets in the order shown in **Figure 55**. Secure the pump with the 6 Phillips screws.

11. Repeat Step 2 and Step 3 to test the check valves after installation.

Cleaning and Inspection

1. Clean pump body and covers in solvent. Dry housing and covers with compressed air.

2. Check body condition. Make sure the valve seats provide a flat contact area for the valve disc. Replace the body if cracks or rough gasket mating surfaces are found.

3. Check diaphragms for holes or tearing. Replace diaphragms as required.

FUEL TANK

Removal/Installation

Refer to **Figure 57** or **Figure 58** for this procedure.

1. Remove the seat and center cover.

2. Label and disconnect all hoses at the fuel tank. Plug the hoses to prevent leakage.

3. Remove the fuel tank mounting bolts and remove the fuel tank.

4. Installation is the reverse of these steps. Check all hose connections for leaks. When mounting fuel tank, position fuel tank mounting bracket so that tank has 1.6 mm (1/16 in.) of play for expansion.

Cleaning/Inspection

> *WARNING*
> *The fuel tank should be cleaned in an open area well away from all sources of flames or sparks.*

1. Pour old gasoline from the tank into a sealable container manufactured specifically for gasoline storage.

2. Pour about 1 quart of fresh gasoline into the tank and slosh it around for several minutes to loosen sediment. Pour the contents into a sealable container.

3. Examine the tank for cracks and abrasions, particularly at points where the tank contacts the body. Abraded areas can be protected and cushioned by coating them with a non-hardening silicone sealer and allowing it to dry before installing the tank. However, if abrading is extensive, or if the tank is leaking, replace it.

FUEL VALVE (IF SO EQUIPPED)

Removal/Installation

To remove the fuel valve, disconnect the hoses at the fuel valve as shown in **Figure 58**. Remove the mounting bracket and remove the valve.

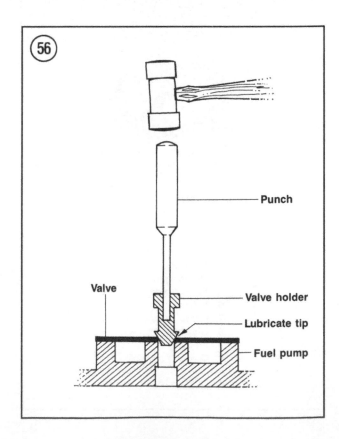

(56)

Punch

Valve

Valve holder

Lubricate tip

Fuel pump

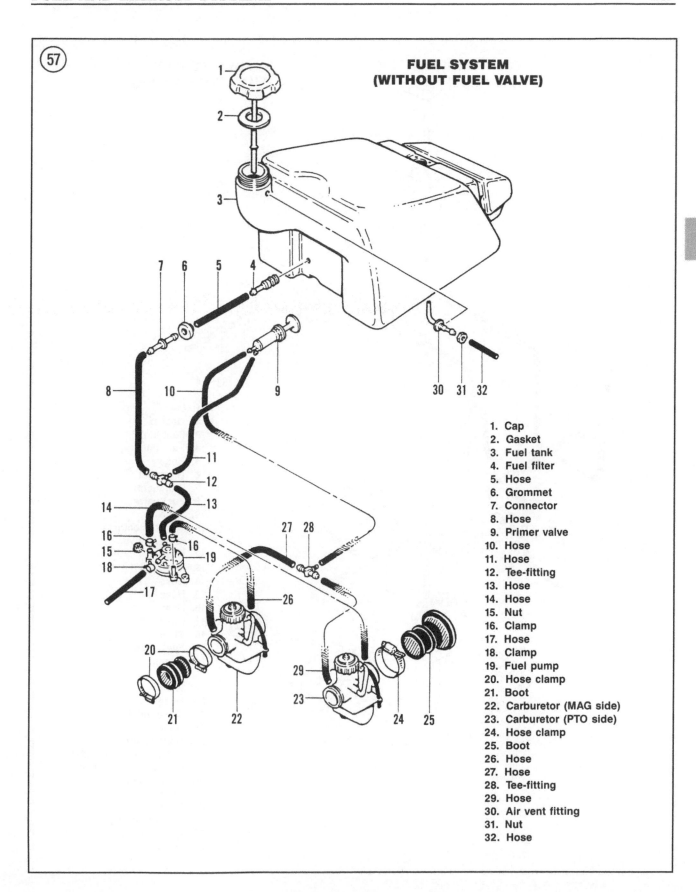

57

**FUEL SYSTEM
(WITHOUT FUEL VALVE)**

1. Cap
2. Gasket
3. Fuel tank
4. Fuel filter
5. Hose
6. Grommet
7. Connector
8. Hose
9. Primer valve
10. Hose
11. Hose
12. Tee-fitting
13. Hose
14. Hose
15. Nut
16. Clamp
17. Hose
18. Clamp
19. Fuel pump
20. Hose clamp
21. Boot
22. Carburetor (MAG side)
23. Carburetor (PTO side)
24. Hose clamp
25. Boot
26. Hose
27. Hose
28. Tee-fitting
29. Hose
30. Air vent fitting
31. Nut
32. Hose

6

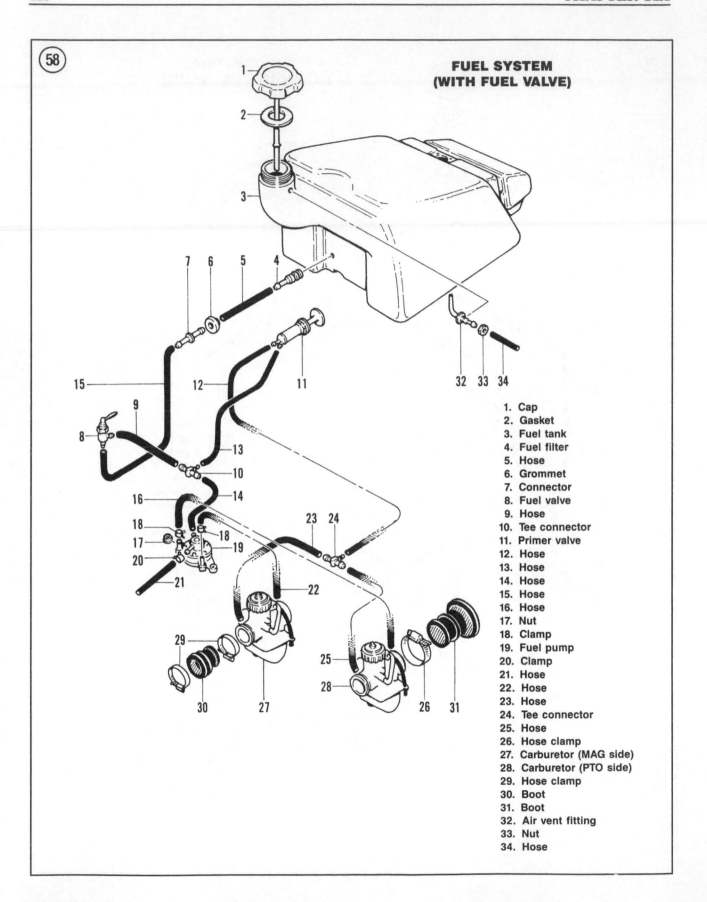

58

**FUEL SYSTEM
(WITH FUEL VALVE)**

1. Cap
2. Gasket
3. Fuel tank
4. Fuel filter
5. Hose
6. Grommet
7. Connector
8. Fuel valve
9. Hose
10. Tee connector
11. Primer valve
12. Hose
13. Hose
14. Hose
15. Hose
16. Hose
17. Nut
18. Clamp
19. Fuel pump
20. Clamp
21. Hose
22. Hose
23. Hose
24. Tee connector
25. Hose
26. Hose clamp
27. Carburetor (MAG side)
28. Carburetor (PTO side)
29. Hose clamp
30. Boot
31. Boot
32. Air vent fitting
33. Nut
34. Hose

During installation, make sure the lever stoppers are positioned so that the inner tip of the fuel valve lever contacts the stoppers while in the closed or open position. Because vibration could accidentally close the fuel valve during engine operation, the stoppers hold the valve ON. **Figure 59** shows proper adjustment of the stoppers with the lever in the closed and open position. If necessary, bend the stoppers by hand until their position is correct.

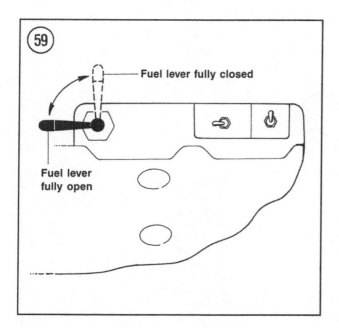

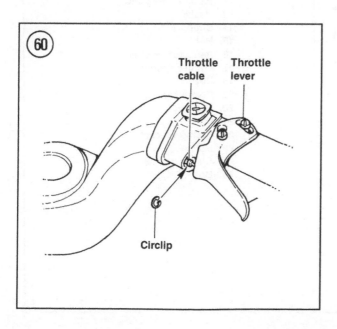

THROTTLE CABLE REPLACEMENT

When replacing the throttle cable, note the following:
  a. Make sure the circlip is installed at the upper cable end as shown in **Figure 60**.
  b. Adjust the carburetors as described in the carburetor synchronization section under *Engine Tune-up* in Chapter Three.
  c. Adjust the oil injection cable as described in Chapter Three.
  d. Operate the throttle lever to make sure the carburetor throttle slides operate correctly.
  e. Check cable routing.

EXHAUST SYSTEM

Removal/Installation

Refer to **Figure 61** or **Figure 62** for this procedure.
1. Open the shroud.

> *WARNING*
> *If the exhaust system is hot, wait until it cools down before removing it.*

2. Disconnect the tailpipe springs (**Figure 63**) at the muffler and exhaust pipe. Disconnect the muffler springs.
3. Disconnect the long exhaust pipe spring (**Figure 64**).
4. Disconnect the 2 exhaust pipe springs at the manifold (**Figure 65**).
5. Remove the exhaust pipe.
6. Remove the muffler and tailpipe (**Figure 66**).
7. Remove the bolts and washers holding the exhaust manifold onto the cylinders. Remove the exhaust manifold (**Figure 67**).
8. Examine the exhaust pipe and muffler (**Figure 68**) for cracks or other damage. Repair as described in this chapter.
9. Installation is the reverse of these steps.
10. Install new exhaust manifold gaskets.
11. Check the exhaust system for leaks after installation.

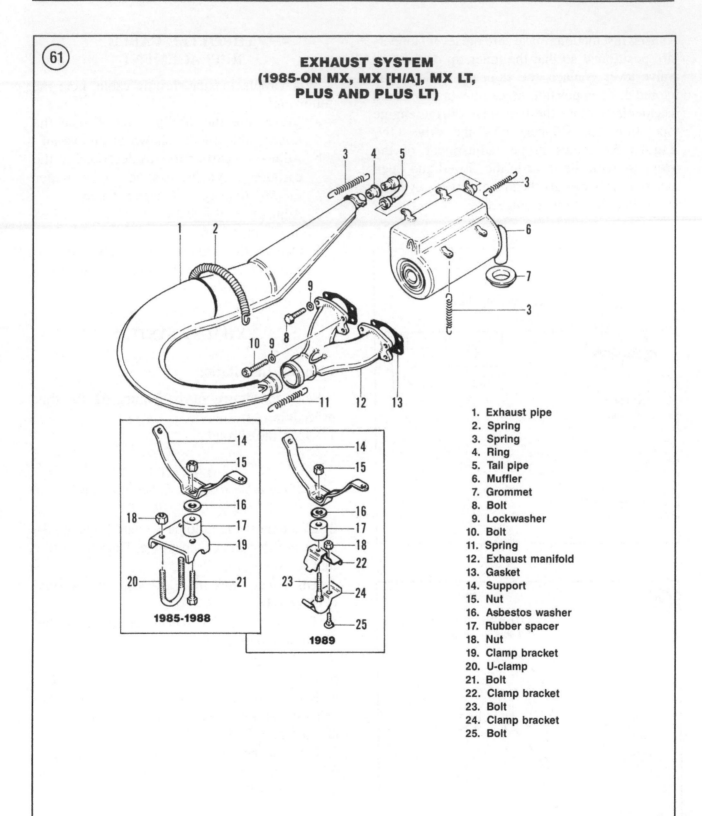

⑥⑴

EXHAUST SYSTEM
(1985-ON MX, MX [H/A], MX LT,
PLUS AND PLUS LT)

1985-1988

1989

1. Exhaust pipe
2. Spring
3. Spring
4. Ring
5. Tail pipe
6. Muffler
7. Grommet
8. Bolt
9. Lockwasher
10. Bolt
11. Spring
12. Exhaust manifold
13. Gasket
14. Support
15. Nut
16. Asbestos washer
17. Rubber spacer
18. Nut
19. Clamp bracket
20. U-clamp
21. Bolt
22. Clamp bracket
23. Bolt
24. Clamp bracket
25. Bolt

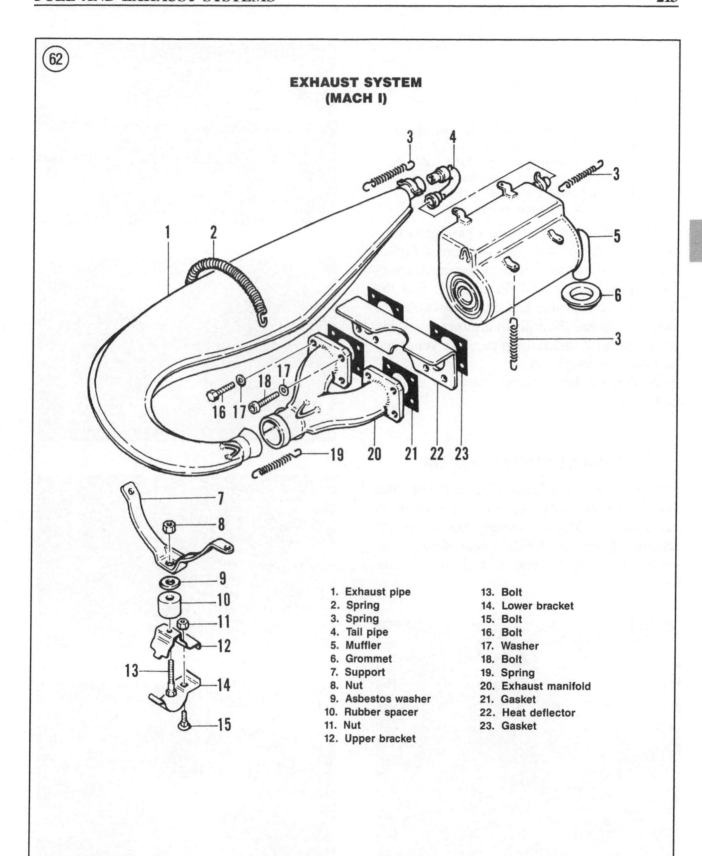

**EXHAUST SYSTEM
(MACH I)**

1. Exhaust pipe
2. Spring
3. Spring
4. Tail pipe
5. Muffler
6. Grommet
7. Support
8. Nut
9. Asbestos washer
10. Rubber spacer
11. Nut
12. Upper bracket
13. Bolt
14. Lower bracket
15. Bolt
16. Bolt
17. Washer
18. Bolt
19. Spring
20. Exhaust manifold
21. Gasket
22. Heat deflector
23. Gasket

6

Cleaning

> *WARNING*
> *When performing Step 1, do not start the drill until the cable is inserted into the pipe. At no time should the drill be running when the cable is free of the pipe. The whipping action of the cable could cause serious personal injury. Wear heavy shop gloves and a face shield when using this equipment.*

1. Clean all accessible exhaust passages with a blunt, roundnose tool. To clean areas further down the pipe, chuck a piece of discarded control cable in an electric drill (**Figure 69**). Fray the loose end of the cable and insert the cable into the pipe. Operate the drill while moving the cable back and forth inside the pipe. Take your time and do a thorough job.

2. Shake the large pieces out into a trash container.

EXHAUST SYSTEM REPAIR

A dent in the exhaust pipe will alter the system's flow characteristics and degrade performance. Minor damage can be easily repaired if you have welding equipment, some simple body tools, and a bodyman's slide hammer.

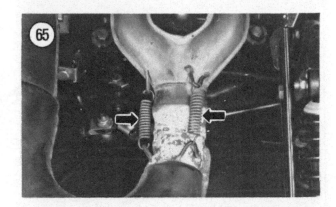

Small Dents

1. Drill a small hole in the center of the dent. Screw the end of the slide hammer into the hole.

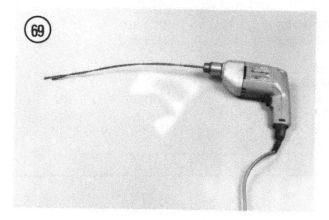

2. Heat the area around the dent evenly with a torch.

3. When the dent is heated to a uniform orange-red color, operate the slide hammer to raise the dent.

4. When the dent is removed, unscrew the slide hammer and weld the drilled hole closed.

Large Dents

Large dents that are not crimped can be removed with heat and a slide hammer as previously described. However, several holes must be drilled along the center of the dent so that it can be pulled out evenly.

If the dent is sharply crimped along the edges, the affected section should be cut out with a hacksaw, straightened with a body dolly and hammer and welded back into place.

Before cutting the exhaust pipe apart, scribe alignment marks over the area where the cuts will be made to aid correct alignment when the pipe is rewelded.

After the welding is completed, wire brush and clean up all welds. Paint the entire pipe with a high-temperature paint to prevent rusting.

6

Tables are on the following pages.

Table 1 CARBURETOR SPECIFICATIONS (467 ENGINE)

Year	1985	1986	1986 H/A*
Type	Mikuni	Mikuni	Mikuni
Size			
PTO	VM34-320	VM34-352	VM34-355
MAG	VM34-321	VM34-353	VM34-356
Main jet			
PTO	240	220	160
MAG	250	240	170
Needle jet	P-4 (159)	P-4 (159)	P-4 (159)
Pilot jet	40	40	40
Jet needle/clip pos.	6DH4/3	6DH7/3	6DH7/3
Throttle valve	2.5	2.5	2.5
Year	**1987**	**1988**	**1989**
Type	Mikuni	Mikuni	Mikuni
Size			
PTO	VM34-372	VM34-352	VM34-352
MAG	VM34-373	VM34-353	VM34-353
Main jet			
PTO	200	220	220
MAG	220	240	240
Needle jet	P-4 (159)	P-4 (159)	P-4 (159)
Pilot jet	50	40	40
Jet needle/clip pos.	6DH4/3	6DH7/3	6DH7/3
Throttle valve	2.5	2.5	2.5
*High altitude			

Table 2 CARBURETOR SPECIFICATIONS (537 ENGINE)

Year	1985	1986	1987	1988
Type	Mikuni	Mikuni	Mikuni	Mikuni
Size				
PTO	VM40-29	VM40-29	VM40-20	VM40-29
MAG	VM40-30	VM40-30	VM40-30	VM40-30
Main jet				
PTO	330	330	330	330
MAG	380	350	350	350
Needle jet	AA5 (224)	AA5 (224)	AA5 (224)	AA5 (224)
Pilot jet	40	40	40	40
Jet needle/clip pos.	7DH2/3	7DH2/2	7DH2/2	6DH2/2
Throttle valve	2.5	2.5	2.5	2.5

Table 3 CARBURETOR SPECIFICATIONS (536 ENGINE)

Year	1989
Type	Mikuni
Size	
PTO	VM34-381
MAG	VM34-381
Main jet	
PTO	250
MAG	250
Needle jet	Q-4 (159)
Pilot jet	30
Jet needle/clip pos.	6FJ6/3
Throttle valve	2.0

6

Table 4 CARBURETOR SPECIFICATIONS (583 ENGINE)

Year	1989
Type	Mikuni
Size	
PTO	VM38-171
MAG	VM38-172
Main jet	
PTO	260
MAG	290
Needle jet	P-2 (480)
Pilot jet	45
Jet needle/clip pos.	6DH8/2
Throttle valve	2.5

Table 5 FLOAT HEIGHT

Carburetor size	mm	in.
VM34	22-24	0.86-0.94
VM38	16-18	0.63-0.71
VM40	17-19	0.67-0.75

Chapter Seven

Electrical System

This chapter provides service procedures for the ignition system and lights. Electrical troubleshooting procedures are described in Chapter Two. Wiring diagrams are at the end of the book. **Table 1** and **Table 2** are at the end of the chapter.

CAPACITOR DISCHARGE IGNITION (CDI)

All models use a capacitor discharge ignition system.

> *NOTE*
> *Refer to Chapter Two for troubleshooting and test procedures.*

Flywheel
Removal/Installation

The flywheel must be removed to service the stator coils. Flywheel replacement is usually necessary only if the magnets have been damaged by mechanical heat or shock. Refer to **Figure 1** when performing this procedure.

1. Remove the muffler as described in Chapter Six.
2. Remove the MAG side spark plug. Then, reconnect the spark plug to the plug cap and ground the plug. Turn the recoil starter to bring the MAG side piston to TDC. Reinstall the spark plug and reconnect the cap.
3. Remove the recoil starter housing as described in Chapter Ten.

> *NOTE*
> *The flywheel can be removed with the engine installed in the frame. The following photographs show the engine removed for clarity.*

4. If the engine is installed in the frame, disconnect the fuel pump pulse hose at the crankcase.
5. Insert the crankshaft locking tool (part No. 420 8766 40) through the pulse hose nozzle and

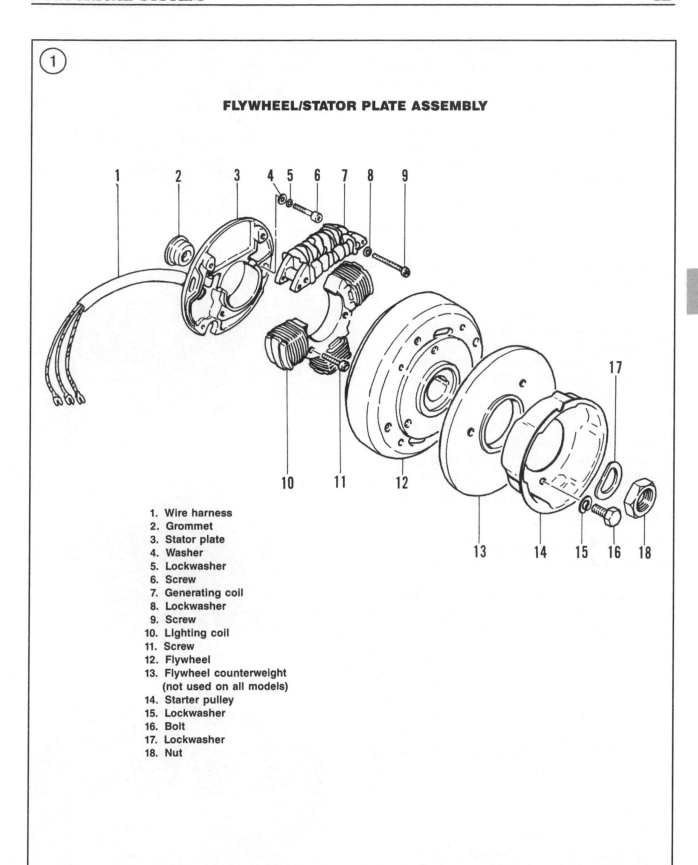

FLYWHEEL/STATOR PLATE ASSEMBLY

1. Wire harness
2. Grommet
3. Stator plate
4. Washer
5. Lockwasher
6. Screw
7. Generating coil
8. Lockwasher
9. Screw
10. Lighting coil
11. Screw
12. Flywheel
13. Flywheel counterweight
    (not used on all models)
14. Starter pulley
15. Lockwasher
16. Bolt
17. Lockwasher
18. Nut

engage the tool with the crankshaft (**Figure 2**).
After installing the tool, rotate the crankshaft
slightly until the tool "locks" the crankshaft.

CAUTION
*Do not substitute the crankshaft locking
tool described in Step 5. Substituting the
tool with a softer piece of metal may
cause crankshaft damage.*

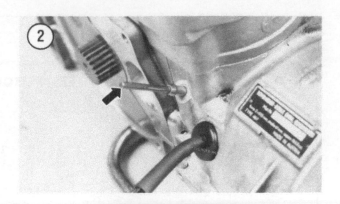

6. Remove the bolts and washers holding the
starter pulley (A, **Figure 3**) to the flywheel.
Remove the starter pulley.
7. Remove the flywheel counterweight (**Figure
1**), if so equipped.

NOTE
*The flywheel nut has been secured with
Loctite. To prevent thread damage, tap
on the flywheel nut before attempting to
loosen it in Step 8. **Do not** strike the
flywheel.*

8. Loosen and remove the flywheel nut (B,
Figure 3) and lockwasher.
9. Bolt the flywheel ring (A, **Figure 4**) (part No.
420 8766 55) onto the face of the flywheel.
Thread the puller (B, **Figure 4**) (part No. 420
8760 65) into the ring.

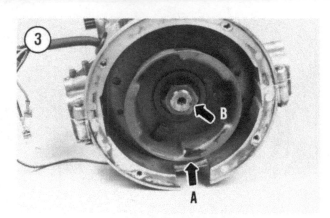

CAUTION
*Do not use heat or hammer on the
flywheel to remove it in Step 10. Heat
may cause the flywheel to seize on the
crankshaft, while hammering can
damage the flywheel or bearings.*

10. Tighten the puller bolt (B, **Figure 4**) and
break the flywheel free of the crankshaft taper.
You may have to alternate tapping on the center
puller bolt sharply with a hammer and tightening
the bolt some more, but don't hit the flywheel.
11. Remove the flywheel and the puller
assembly. Remove the puller from the flywheel.
12. Remove the Woodruff key from the flywheel
(**Figure 5**).

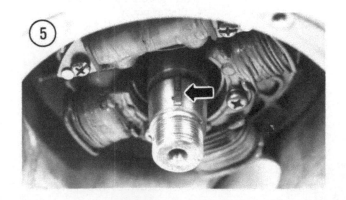

NOTE
*Before installing the flywheel, check the
magnets (**Figure 6**) for metal trash they*

may have picked up. Debris stuck to the magnets can cause coil damage.

13. Spray the flywheel and crankshaft tapers with a rust inhibitor, such as WD-40.

14. Place the Woodruff key in the crankshaft key slot (**Figure 5**). Position flywheel (**Figure 7**) over crankshaft with key slot in flywheel hub aligned with key in crankshaft.

15. Apply Loctite 242 (blue) to the flywheel threads and install the lockwasher and flywheel nut (B, **Figure 3**). Lock the crankshaft with same tool used during removal and tighten the flywheel nut (A, **Figure 8**) to the torque specification in **Table 1**.

16. Install the flywheel counterweight (if so equipped) by aligning the mark on the counterweight with the mark on the flywheel.

17. Install the starter pulley (A, **Figure 3**), washers and bolts. Tighten the bolts securely.

18. Remove the crankshaft locking tool from the pulse hose nozzle (**Figure 2**).

19. Reinstall the recoil starter housing as described in Chapter Ten.

20. If the engine is installed in the frame, note the following:
   a. Reconnect the pulse hose at the pulse hose nozzle on the crankcase. Secure the hose with the wire clip.
   b. Reinstall the muffler as described in Chapter Six.

Inspection

1. Check the flywheel carefully for cracks or breaks.

> *WARNING*
> *A cracked or chipped flywheel must be ·replaced. A damaged flywheel may fly apart at high rpm causing severe engine damage. Do not attempt to repair a damaged flywheel.*

2. Check tapered bore of flywheel and crankshaft taper for signs of fretting or working.

3. Check key slot in flywheel (**Figure 9**) for cracks or other damage. Check keyseat in crankshaft (**Figure 5**) for cracks or other damage.

4. Check the Woodruff key for cracks or damage.

5. Check crankshaft and flywheel nut threads for wear or damage.

6. Replace flywheel, flywheel nut, crankshaft half or Woodruff key as required.

Stator Plate
Removal/Installation

> *NOTE*
> *Refer to Chapter Two for troubleshooting and test procedures.*

1. Remove the flywheel as described in this chapter.

2. If the engine is installed in the frame, disconnect the stator plate electrical connectors.

3. Make a mark on the stator plate and crankcase for alignment during reassembly.

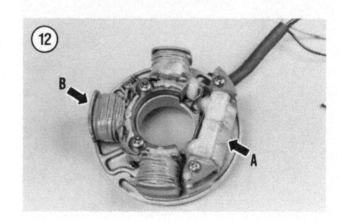

4. Remove the screws securing the stator plate to the crankcase. Remove the stator plate (**Figure 10**) by removing the grommet (B, **Figure 8**) and pulling the wires through the case opening.

5. Installation is the reverse of these steps. Note the following.

   a. Check the coil wires (**Figure 11**) for chafing or other damage. Replace coil harness, if necessary.

   b. Check and adjust ignition timing as described in Chapter Three.

   c. Make sure all electrical connections are tight and free from corrosion. This is absolutely necessary with electronic ignition systems.

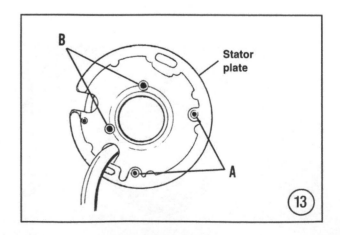

Generating Coil Replacement

If necessary, replace the generating coil (A, **Figure 12**) as follows:

1. Remove the stator plate as described in this chapter.

2. Heat the back of the stator plate at the points shown (A, **Figure 13**) to loosen the Loctite used

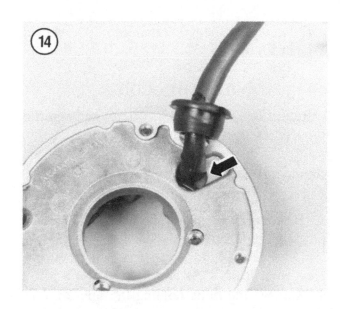

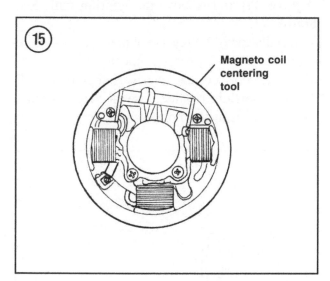

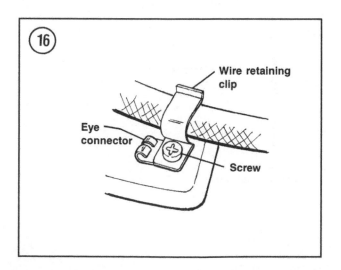

to secure the coil screws. Do not exceed 200° F (93° C) when heating the screws.

CAUTION
*Protect the wire harness at the stator plate (**Figure 14**) when heating the screws in Step 2.*

3. Remove the Phillips screws holding the generating coil (A, **Figure 12**) to the stator plate.
4. Cut the 4 wires at the coil as close to the coil as possible.
5. Reverse to install the new coil. Note the following:
   a. Resolder the wires to the new coil terminals with rosin core solder.
   b. Remove the shipping nuts from the new coil screws.
   c. Position the new generating coil onto the stator plate.
   d. Coat the new screws with Loctite 242 (blue) and install the screws. Place the centering tool (part No. 420 8769 22) over the coils (**Figure 15**) and tighten the generating coil screws.

Lighting Coil Replacement

If necessary, replace the lighting coil (B, **Figure 12**) as follows:
1. Remove the stator plate as described in this chapter.
2. Heat the back of the stator plate at the points shown (B, **Figure 13**) to loosen the Loctite used to secure the coil screws. Do not exceed 200° F (93° C) when heating the screws.

CAUTION
*Protect the wire harness at the stator plate (**Figure 14**) when heating the screws in Step 2.*

3. Remove the Phillips screws holding the lighting coils (B, **Figure 12**) to the stator plate.
4. Remove the wire retaining clip from the stator plate (**Figure 16**).

5. Each of the lighting coil wires are protected by a harness tube (**Figure 17**). To disconnect the coil wires from the main wire harness, first pull the harness tubes toward the lighting coil to uncover the soldered joint. Then unsolder both coil wires.

6. Reverse to install the new coil. Note the following:

   a. See **Figure 17**. Solder the lighting coil's white harness tube wire to the yellow wire.

   b. See **Figure 17**. Solder the lighting coil's black harness tube wire to the yellow/black wire.

   c. Use rosin core solder when soldering wires in sub-step a and b.

   d. Position the new lighting coil onto the stator plate.

   e. Coat the new screws with Loctite 242 (blue) and install the screws. Place the centering tool (part No. 420 8769 22) over the coils (**Figure 15**) and tighten the lighting coil screws.

   f. Reconnect the wire retaining clip as shown in **Figure 16**.

IGNITION COIL

Refer to **Figure 18** when performing procedures in this section.

Removal/Installation

1. Open the shroud.

2. Disconnect the spark plug caps at the spark plugs (**Figure 19**).

3. Disconnect the ignition coil 2-prong connector at the CDI box.

4. Remove the bolts holding the ignition coil (**Figure 20**) and remove the ignition coil. See **Figure 21**.

5. Installation is the reverse of these steps. Before connecting the 2-prong connector at the coil, make sure both connector halves are clean of all dirt and moisture residue. Use electrical contact cleaner to clean the connectors.

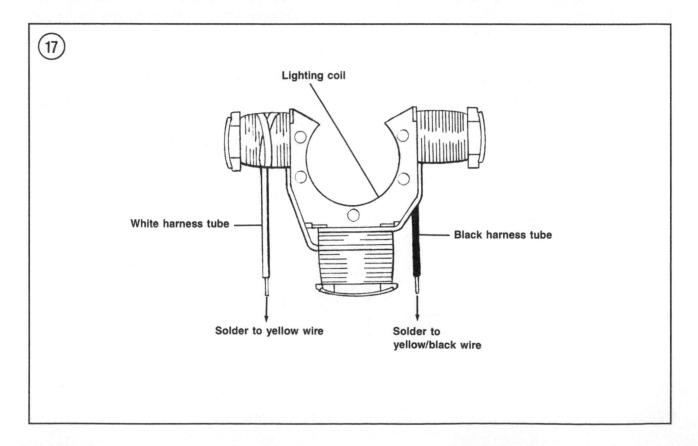

(17)

Lighting coil

White harness tube

Black harness tube

Solder to yellow wire

Solder to yellow/black wire

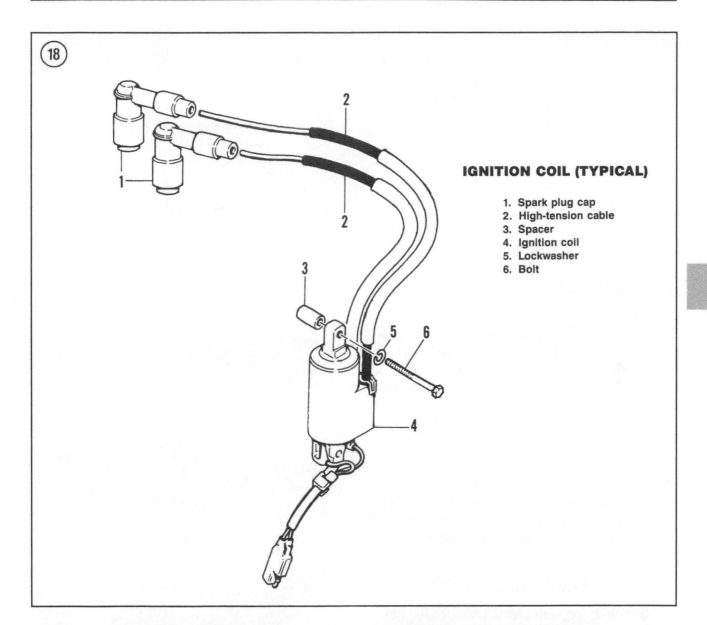

IGNITION COIL (TYPICAL)

1. Spark plug cap
2. High-tension cable
3. Spacer
4. Ignition coil
5. Lockwasher
6. Bolt

7

Spark Plug Caps

The spark plug caps (**Figure 22**) can be replaced by pulling the old caps off of the coil's high tension wire. Reverse to install. Make sure the cap is pushed all the way on the high tension wire.

Testing

Refer to Chapter Two.

CDI BOX

Removal/Installation

1. Open the shroud.
2. Disconnect the 2 connectors at the CDI box.
3. Locate the CDI box in the engine compartment. On some models, it will be necessary to remove the coolant tank. Remove the nuts holding the CDI box to its mounting bracket. Remove the CDI box. See **Figure 23**, typical.
4. Install by reversing these removal steps. Before connecting the electrical wire connectors at the unit, make sure the connectors are clean of all dirt and moisture residue. Use electrical contact cleaner to clean the connectors.

VOLTAGE REGULATOR

The voltage regulator is mounted on the right-hand side of the vehicle.

1. Disconnect voltage regulator wires.
2. Remove the nut or the bolt and screw and remove the voltage regulator. See **Figure 24** (early model) or **Figure 25** (late model).
3. Installation is the reverse of these steps. Before connecting the electrical wire connectors at the unit, make sure the connectors are clean of all dirt and moisture residue. Use electrical contact cleaner to clean the connectors.

Testing

Refer to Chapter Two.

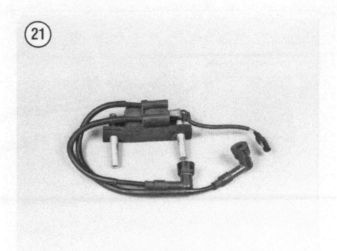

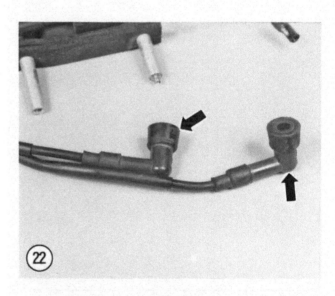

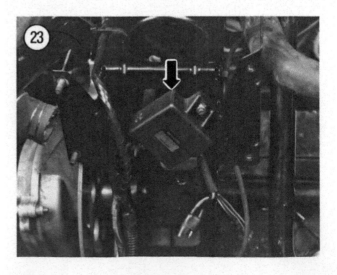

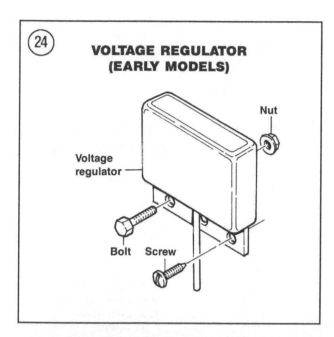

VOLTAGE REGULATOR (EARLY MODELS)

Nut

Voltage regulator

Bolt   Screw

LIGHTING SYSTEM

The lighting system consists of the headlight, taillight/brake light combination, meter illumination lights and pilot lamps. In the event of trouble with any light, the first thing to check is the affected bulb itself. If the bulb is good, check all wiring and connections with a test light or ohmmeter. Replacement bulbs are listed in **Table 2**.

Headlight Replacement

CAUTION
*Most models are equipped with quartz-halogen bulbs (**Figure 26**). Do not touch the bulb glass with your fingers because traces of oil on the bulb will drastically reduce the life of the bulb. Clean any traces of oil from the bulb with a cloth moistened in alcohol or lacquer thinner.*

WARNING
*If the headlight has just burned out or turned off, it will be **hot**. Don't touch the bulb until it cools off.*

Refer to **Figure 27** or **Figure 28**.
1. Open the shroud.
2. Disconnect the electrical connector at the bulb.
3. Remove the rubber boot (**Figure 29**) and disconnect the bulb retainer clip. Remove the bulb.
4. Installation is the reverse of these steps. Make sure the bulb engages the bulb housing correctly.

Headlight Adjustment

1. Park the snowmobile on a level surface 12 1/2 ft. (381 cm) from a vertical wall (**Figure 30**).

NOTE
A rider should be seated on the snowmobile when performing the following.

2. Measure the distance from the floor to the center of the headlight lens (**Figure 30**). Make

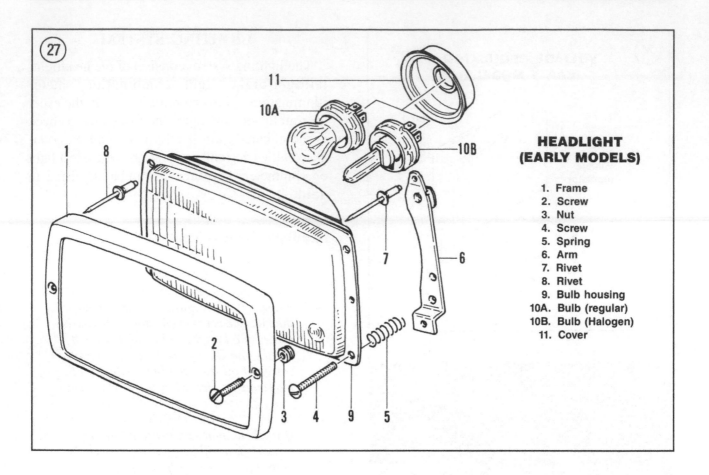

**HEADLIGHT
(EARLY MODELS)**

1. Frame
2. Screw
3. Nut
4. Screw
5. Spring
6. Arm
7. Rivet
8. Rivet
9. Bulb housing
10A. Bulb (regular)
10B. Bulb (Halogen)
11. Cover

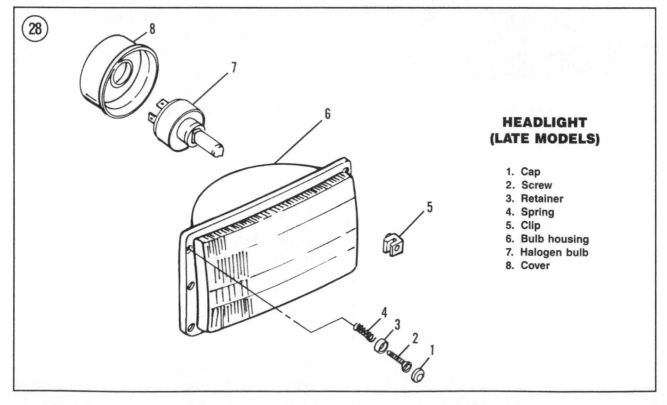

**HEADLIGHT
(LATE MODELS)**

1. Cap
2. Screw
3. Retainer
4. Spring
5. Clip
6. Bulb housing
7. Halogen bulb
8. Cover

a mark on the wall the same distance from the floor. For instance, if the center of the headlight lens is 2 feet above the floor, mark "A" in **Figure 30** should also be 2 feet above the floor.

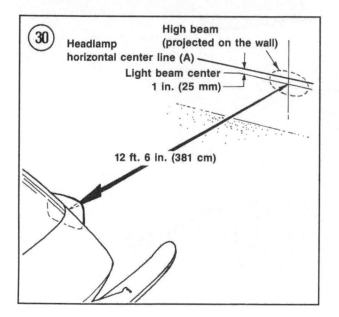

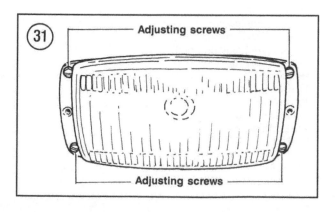

3. Start the engine, turn on the headlight and set the beam selector to HIGH. Do not adjust the headlight beam with the selector set at LOW.

4. The most intense area of the beam on the wall should be 1 in. (25 mm) below the "A" mark (**Figure 30**) and in line with the imaginary vertical centerlines from the headlight to the wall.

5. To raise the beam, tighten the top adjuster screws and loosen the bottom screws (**Figure 31**). To lower the beam, tighten the bottom screws and loosen the top screws. To move the beam to the right, tighten the right screws and loosen the left. To move the beam to the left, tighten the left screws and loosen the right.

Taillight Bulb Replacement

1. Remove the taillight lens mounting screws and remove the lens (**Figure 32**).

2. Turn the bulb counterclockwise and remove it (**Figure 33**).

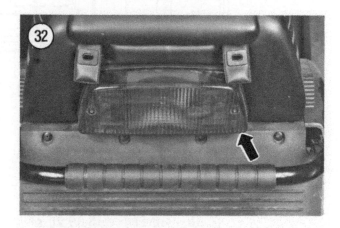

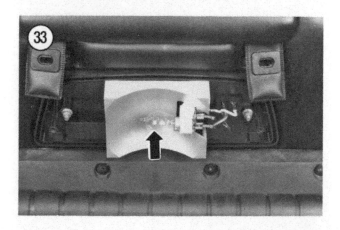

METER ASSEMBLY

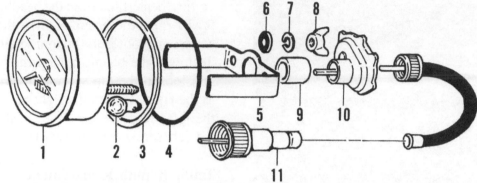

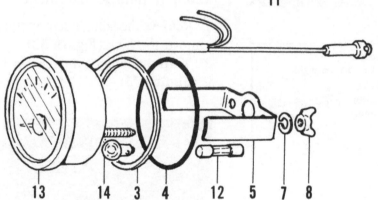

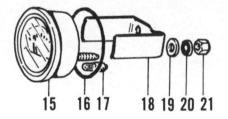

1. Speedometer	13. Tachometer
2. Bulb	(if so equipped)
3. O-ring	14. Bulb (tachometer
4. Retainer ring	equipped models only)
5. Holder	15. Fuel level gauge
6. Flat washer	16. O-ring
7. Lockwasher	17. Bulb
8. Wing nut	18. Holder
9. Spacer	19. Flat washer
10. Speedometer corrector	20. Lockwasher
11. Speedometer cable	21. Nut
12. Fuse (tachometer	22. Temperature gauge
equipped models only)	23. Bulb

3. Clean the lens in a mild detergent and check for cracks.

4. Installation is the reverse of these steps.

METER ASSEMBLY

The meter assembly consists of a speedometer, tachometer, fuel level gauge and temperature gauge (**Figure 34**, typical). The meters are installed in the shroud (**Figure 35**).

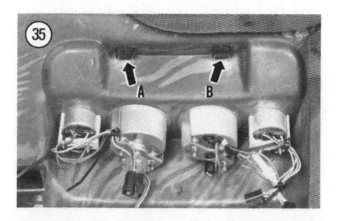

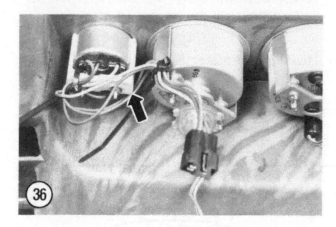

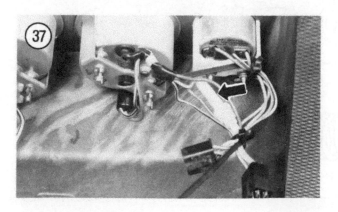

Bulb Replacement

1. Open the shroud.

2. Locate the bulb socket at the affected meter assembly and pull the bulb socket out of the meter assembly. If necessary, remove the meter mounting bracket to gain access to the bulb socket (**Figure 36**, typical). Replace the bulb.

3. Installation is the reverse of these steps.

Meter Removal/Installation

1. Open the shroud.

2. Disconnect the electrical connectors at the affected meter.

3. Pull the bulb socket out of the meter assembly.

4. *Speedometer:* Disconnect the speedometer cable at the meter.

5. Remove the nuts securing the meter mounting bracket and remove the meter.

6. Installation is the reverse of these steps.

Tachometer Fuse

The tachometer is protected by a 0.1 amp fuse (**Figure 37**). If the tachometer stops during engine operation, remove the fuse and check its condition. Replace the fuse if necessary.

CAUTION
Never substitute tinfoil or wire for a fuse. Never use a higher amperage fuse than specified. An overload could result in tachometer damage.

PILOT LAMPS

High Beam Pilot Lamp

The high beam pilot lamp (A, **Figure 35**) lights up whenever the headlight is on HIGH beam. If the lamp does not light, replace it.

1. Open the shroud.

2. Disconnect the electrical connector at the high beam pilot lamp and remove the pilot lamp (A, **Figure 35**).

3. Reverse to install. Check pilot lamp operation by turning the headlight to HIGH; the lamp should come on.

Oil Level Pilot Lamp

The oil tank is equipped with an oil level sensor that is wired to the injection oil level pilot lamp on the instrument panel (B, **Figure 35**). When the oil level in the tank reaches a specified low point, the pilot lamp will light.

The oil injection level pilot lamp lights up whenever the brake lever is operated. If the lamp does not light up during brake operation, replace the lamp as follows:

1. Open the shroud.
2. Disconnect the electrical connector at the oil level pilot lamp and remove the pilot lamp (B, **Figure 35**).
3. Reverse to install. Check pilot lamp operation by applying the brake lever with the engine running; the lamp should come on.

SWITCHES

Switches can be tested for continuity with an ohmmeter (see Chapter One) or a test light at the switch connector plug by operating the switch in each of its operating positions and comparing results with the switch operation.

When testing switches, note the following:
a. When separating 2 connectors, pull on the connector housings and not the wires.
b. After locating a defective circuit, check the connectors to make sure they are clean and properly connected. Check all wires going into a connector housing to make sure each wire is properly positioned and that the wire end is not loose.
c. When joining connectors, push them together until they click into place.
d. When replacing handlebar switch assemblies, make sure the cables are routed correctly so that they are not crimped when the handlebar is turned from side-to-side.

Tether Switch
Removal/Installation

Refer to **Figure 38**.
1. Disconnect the tether switch electrical connector.
2. Unscrew the tether switch and remove it.
3. Installation is the reverse of these steps.

Headlight Dimmer Switch

The headlight dimmer switch (**Figure 39**) is mounted in the brake lever housing. To replace the switch, disassemble the brake housing.

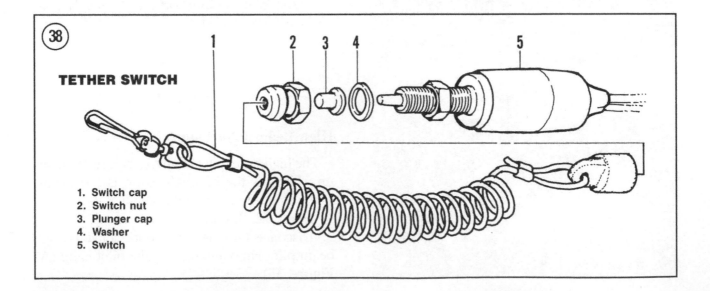

38

TETHER SWITCH

1. Switch cap
2. Switch nut
3. Plunger cap
4. Washer
5. Switch

Disconnect the switch connectors and remove the switch. Reverse to install.

Brake Light Switch

The brake light switch (**Figure 40**) is mounted in the brake lever housing. To replace the switch,

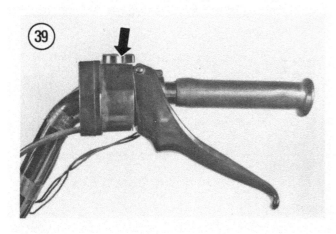

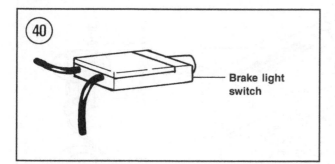

Brake light switch

disassemble the brake housing. Disconnect the switch connectors and remove the switch. Reverse to install.

Emergency Cut-out Switch

The emergency cut-out switch (**Figure 41**) is mounted in the throttle housing. To replace the switch, disassemble the throttle housing. Disconnect the switch connectors and remove the switch. Reverse to install.

Ignition Switch

The ignition switch (**Figure 42**) is mounted on the console. Refer to **Figure 42** when replacing the switch.

7

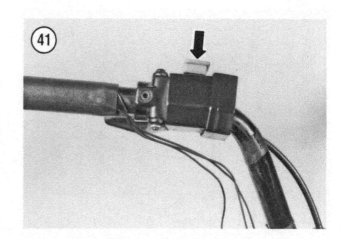

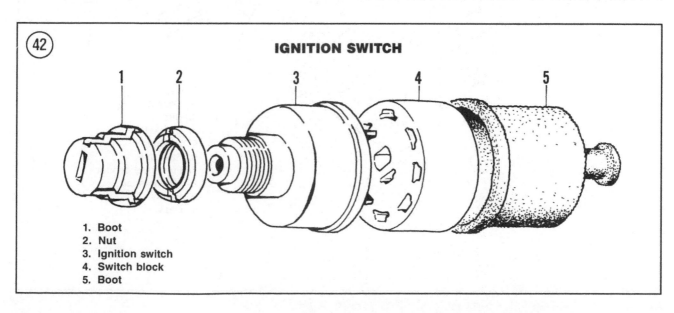

IGNITION SWITCH

1. Boot
2. Nut
3. Ignition switch
4. Switch block
5. Boot

Thermoswitch

The thermoswitch (**Figure 43**) is mounted in the cylinder head. To replace the switch, disconnect the connector at the switch and unscrew the switch from the cylinder head. Reverse to install.

FUEL LEVEL SENSOR

Some models are equipped with a fuel level sensor that is mounted in the fuel tank (**Figure 44**). The fuel level sensor works together with the fuel gauge on the instrument panel. To replace the fuel level sensor, note the following:

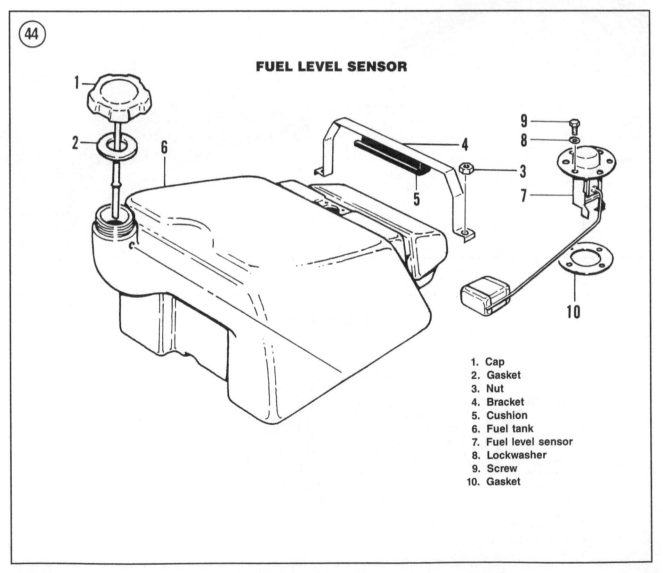

FUEL LEVEL SENSOR

1. Cap
2. Gasket
3. Nut
4. Bracket
5. Cushion
6. Fuel tank
7. Fuel level sensor
8. Lockwasher
9. Screw
10. Gasket

1. Remove and drain the fuel tank as described in Chapter Six.

2. Remove the screws and washers securing the sensor to the fuel tank.

3. Remove the sensor and gasket. Discard the gasket.

4. Reverse to install. Install a new fuel level sensor gasket.

OIL LEVEL GAUGE

All models are equipped with an oil level gauge mounted in the oil tank (**Figure 45**). To replace the gauge, open the shroud and disconnect the oil gauge electrical connector. Pull the oil gauge out of the oil tank (**Figure 46**) and remove it. Reverse to install.

WIRING DIAGRAMS

Wiring diagrams are located at the end of this book.

7

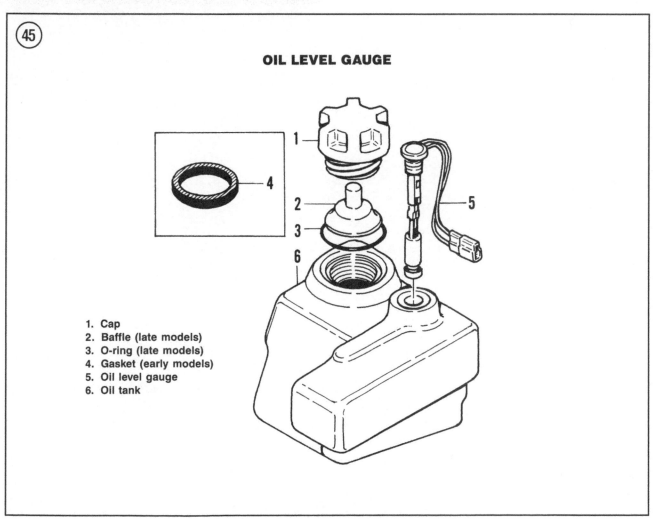

OIL LEVEL GAUGE

1. Cap
2. Baffle (late models)
3. O-ring (late models)
4. Gasket (early models)
5. Oil level gauge
6. Oil tank

Table 1 FLYWHEEL NUT TIGHTENING TORQUE

	N·m	ft.-lb.
1985-1988	100	74
1989	105	77

Table 2 REPLACEMENT BULBS

	Watt
Headlight	
MX, MX (H/A) and MX LT	60/60
Plus and Plus LT	60/55*
Mach I	60/55*
Taillight	5/21
Tachometer and speedometer	5
Fuel and temperature gauge	2
* Halogen bulb.	

Chapter Eight

Oil Injection System

The fuel:oil ratio required by snowmobile engines depends upon engine demand. Without oil injection, oil must be hand-mixed with gasoline to assure that sufficient lubrication is provided at all operating speeds and engine load conditions. This ratio is adequate for high-speed

operation, but contains more oil than is required to lubricate the engine properly during idle.

With oil injection, the amount of oil provided with the fuel sent to the engine cylinders can be varied instantly and accurately to provide the optimum ratio for proper engine lubrication at any operating speed or engine load condition.

All models are equipped with an oil injection system. The system consists of a mechanical gear-driven pump, external oil tank, oil injection hoses and throttle/oil pump cable assembly. A junction box connects the throttle and pump cable so that they operate simultaneously.

This chapter covers complete oil injection system service.

SYSTEM COMPONENTS

The oil injection pump (**Figure 1**) is mounted on the rotary valve cover. The oil pump gear is mounted onto the oil pump. The pump gear engages the rotary valve shaft gear. The oil pump is connected to the throttle by a cable. An oil reservoir tank is mounted in the engine compartment (**Figure 2**). Oil injection hoses connect the oil tank to the pump and connect the pump to the engine. A warning light on the dash comes on when a low oil level is reached in the reservoir tank.

OIL PUMP SERVICE

Oil Pump Bleeding

The oil pump must be bled whenever one of the following conditions has been met:

   a. The oil tank ran empty.
   b. Any one of the oil injection hoses were disconnected.
   c. The machine was turned on its side.
   d. Pre-delivery service.

1. Check that the oil tank is full. See Chapter Three.

> *NOTE*
> *It is not necessary to disconnect the throttle cable, starter cables or switch leads from the carburetors when performing Step 2.*

2. Check that all hoses are connected to the oil pump and to the oil reservoir tank.
3. *Main oil line:* To bleed the main oil line (line between the oil tank and pump), loosen the bleeder screw (**Figure 3**) and allow oil to bleed until there are no air bubbles in the main oil line. Tighten the bleeder screw.

> *WARNING*
> *Never lean into the snowmobile's engine compartment while wearing a scarf or other loose clothing when the engine is running or when the driver is attempting to start the engine. If the scarf or clothing should catch in the drive belt or clutch, severe injury or death could occur. Make sure the belt guard is in place.*

4. *Small oil lines:* To bleed the small oil lines (lines between oil pump and intake manifold) (A, **Figure 4**), start the engine and allow to idle. Hold the pump lever (B, **Figure 4**) in the fully open position. When there are no air bubbles in the hoses, release the pump lever and turn the engine off.

> *NOTE*
> *Figure 4 shows the engine and oil pump cable removed for clarity.*

COMPONENT REPLACEMENT

Oil Reservoir Tank Removal/Installation

Refer to **Figure 5** for this procedure.
1. Open the shroud.

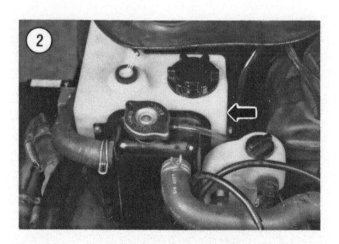

2. Label the hoses at the tank before removal.

NOTE
If the tank has oil in it, steps should be taken to prevent oil leakage during removal. If the carburetors are installed and access to the outlet hoses at the oil pump are difficult, purchase a length of hose with the same ID. Cut the hoses into 2 separate lengths. Block one end of each hose with a bolt. These hoses can then be used to plug the tank outlet ports when the original hoses are disconnected.

3. Remove the oil level gauge from the oil tank (**Figure 2**).
4. Remove the bolts holding the oil tank to the frame.
5. Lift the oil tank up slightly and disconnect the hoses at the tank. Plug the outlet ports to prevent oil leakage.

6. Installation is the reverse of these steps.
7. Bleed the oil pump as described in this chapter.

Oil Level Gauge

Refer to Chapter Seven.

Oil Hoses

Fresh oil hoses should be installed whenever the old hoses become hard and brittle. When replacing damaged or worn oil hoses, make sure to install transparent hoses with the correct ID. Non-transparent hoses will not allow you to visually inspect the hoses for air pockets or other blockage that could cause engine seizure. When reconnecting hoses, secure each hose end with a clamp.

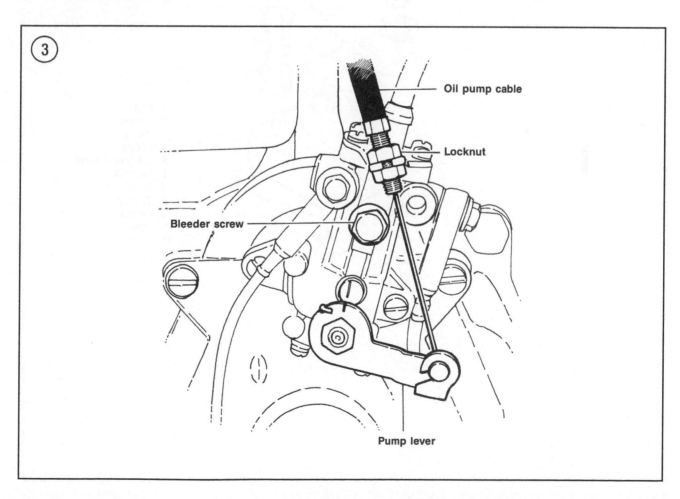

OIL PUMP

Removal/Disassembly

The following describes how to remove the oil pump and gear assembly. Refer to **Figure 6** when performing this procedure.

1. If the engine is installed in the frame, note the following:
   a. Remove the carburetors as described in Chapter Six.
   b. Disconnect the oil pump cable at the oil pump.
   c. Disconnect the main oil hose at the oil pump.

2. Remove the screws holding the 2 rotary valve cover halves together and separate the covers.
3. Remove the bolts holding the rotary valve cover to the crankcase. Remove the rotary valve cover assembly together with the oil pump (**Figure 7**).
4. Remove the oil pump gear (A, **Figure 8**) as follows:

NOTE
To prevent oil pump gear damage when loosening the gear nut, the gear must be held with a special tool. You can use the Ski-Doo gear holder (part No. 420 2779

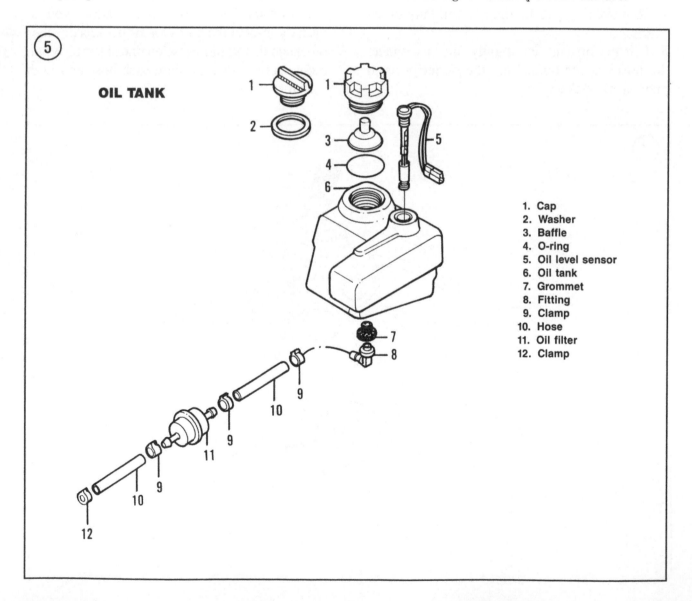

OIL TANK

1. Cap
2. Washer
3. Baffle
4. O-ring
5. Oil level sensor
6. Oil tank
7. Grommet
8. Fitting
9. Clamp
10. Hose
11. Oil filter
12. Clamp

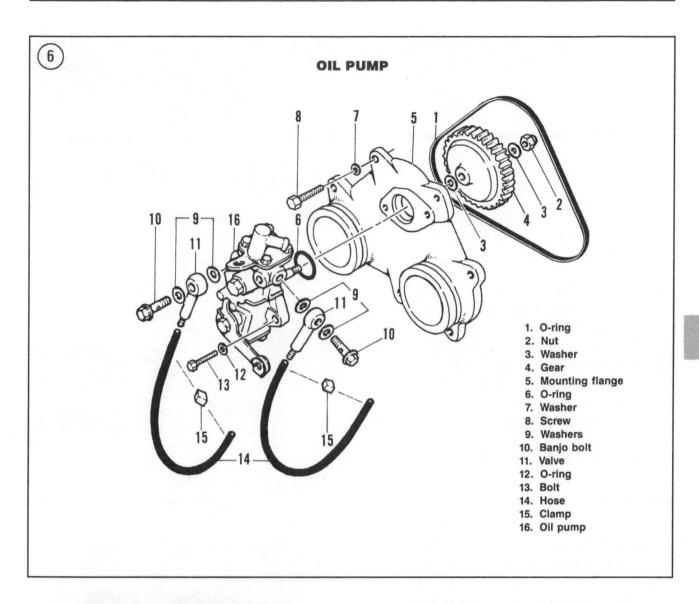

⑥ **OIL PUMP**

1. O-ring
2. Nut
3. Washer
4. Gear
5. Mounting flange
6. O-ring
7. Washer
8. Screw
9. Washers
10. Banjo bolt
11. Valve
12. O-ring
13. Bolt
14. Hose
15. Clamp
16. Oil pump

8

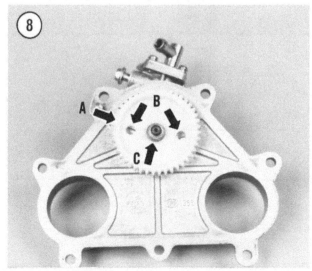

*00) or you can fabricate the tool shown in **Figure 9**. The pins used in the tool will fit into the 2 gear holes (B, **Figure 8**). The diameter of the 2 pins should fit the holes in the gear snugly (B, **Figure 8**). The pins should be placed in the tool according to the center-to-center distance (**Figure 10**) of the 2 gear holes.*

> *CAUTION*
> *Do not use a pair of pliers or wedge the gear in any way when loosening the gear nut. The gear is made of nylon and will break if mishandled.*

a. Hold the oil pump gear with the special tool and loosen the gear nut (C, **Figure 8**).
b. Remove the nut.
c. Remove the washer (**Figure 11**).
d. Remove the gear (**Figure 12**).
e. Remove the washer (**Figure 13**).

5. Remove the screws holding the oil pump to the outer cover and remove the oil pump (**Figure 14**).

6. Remove the pin (**Figure 15**).

Inspection

1. While it is generally not recommended to disassemble the oil pump, **Figure 16** identifies those parts which are replaceable. If a non-replaceable part is damaged, the entire oil pump assembly will have to be replaced.

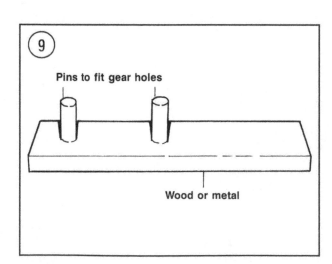

Pins to fit gear holes

Wood or metal

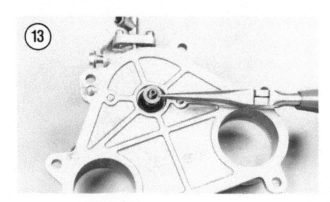

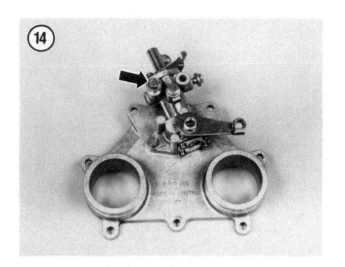

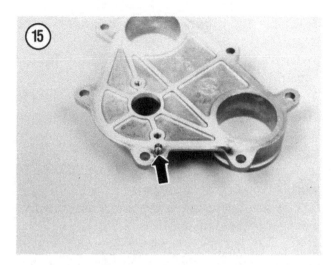

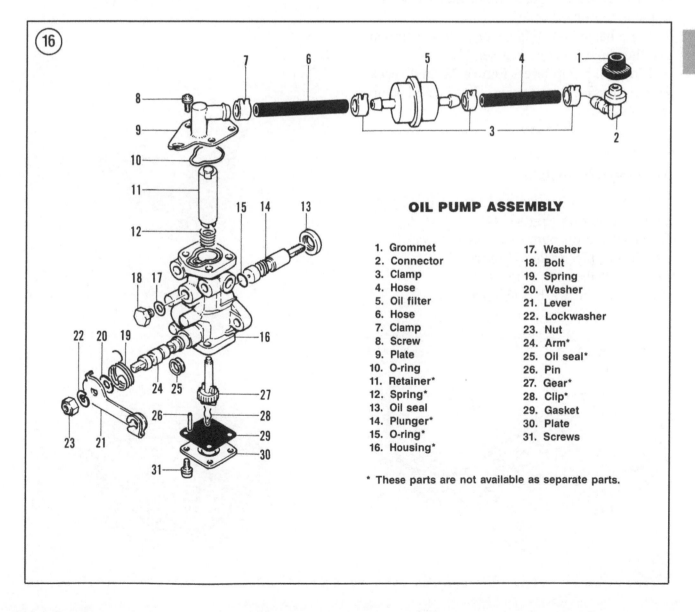

8

OIL PUMP ASSEMBLY

1. Grommet
2. Connector
3. Clamp
4. Hose
5. Oil filter
6. Hose
7. Clamp
8. Screw
9. Plate
10. O-ring
11. Retainer*
12. Spring*
13. Oil seal
14. Plunger*
15. O-ring*
16. Housing*

17. Washer
18. Bolt
19. Spring
20. Washer
21. Lever
22. Lockwasher
23. Nut
24. Arm*
25. Oil seal*
26. Pin
27. Gear*
28. Clip*
29. Gasket
30. Plate
31. Screws

*These parts are not available as separate parts.

2. Check the oil hose fittings (**Figure 17**) for tightness.

3. Check the oil hose valves (**Figure 18**) for contamination or damage.

4. Check the oil pump oil seal (**Figure 19**) for cuts, damage or any signs of damage. Replace the oil seal if it appears damaged or if there are signs of oil leakage.

5. Replace the cover O-ring (**Figure 20**) if it is worn or otherwise damaged.

6. Check the oil pump gear for cracks, excessive wear or other damage.

7. Inspect machined surfaces for burrs, cracks or other damage. Repair minor damage with a fine-cut file or oilstone.

8. Flush banjo bolts (**Figure 21**) with solvent and dry thoroughly before reassembly.

9. Move oil pump lever (**Figure 22**) and check for tightness or other damage. Replace oil pump if necessary.

Assembly/Installation

1. Install the oil pump and secure it with its mounting screws (**Figure 14**).

2. Install the oil pump gear as follows:
   a. Install the washer (**Figure 13**).
   b. Install the gear (**Figure 12**).
   c. Install the washer (**Figure 11**).

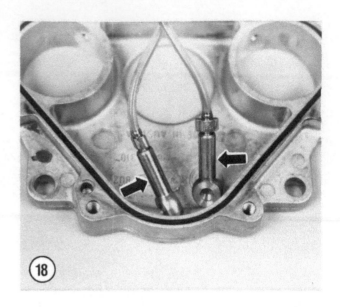

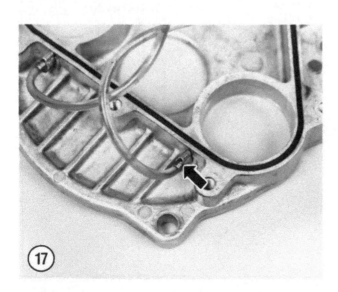

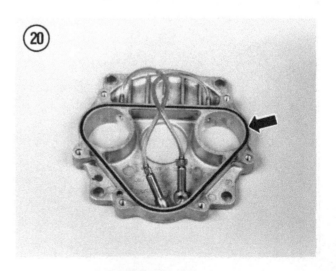

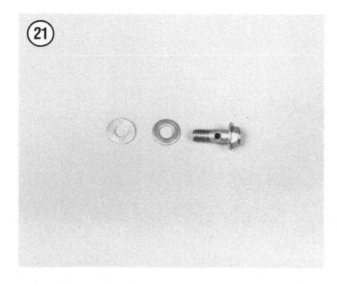

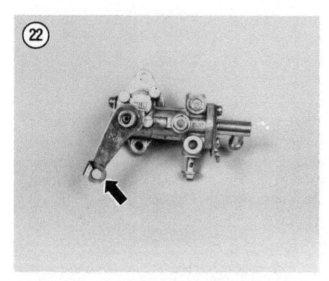

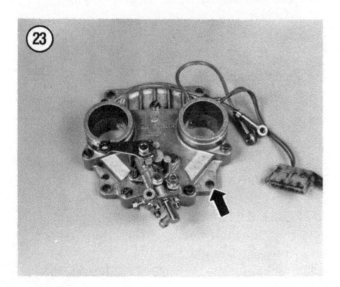

d. Install and tighten the nut (C, **Figure 8**). Use the same tool to hold the gear when tightening the nut as during disassembly.

3. Install the O-ring and assemble the rotary valve cover halves (**Figure 23**). Be sure to install the dowel pin (**Figure 15**). Install the cover screws and tighten securely.

4. If the rotary valve was removed, install it as described in Chapter Five.

5. Install the rotary valve cover (**Figure 7**). Install and tighten the cover screws securely.

6. Route the oil injection lines as shown in **Figure 24**. When installing the banjo bolts, make sure to place a washer on both sides of the hose valve as shown in **Figure 6**.

7. If the engine is installed in the frame, note the following:

    a. Connect the main oil hose at the oil pump.

    b. Connect the oil pump cable at the oil pump.

    c. Bleed the oil pump as described in this chapter.

    d. Install the carburetors as described in Chapter Six.

    e. Synchronize the carburetors, then adjust the oil pump as described in Chapter Three.

8

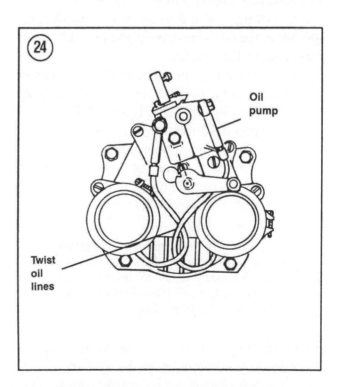

Oil pump

Twist oil lines

Chapter Nine

Liquid Cooling System

This chapter covers service procedures for the thermostat, water pump, drive belt and connecting hoses.

Cooling system flushing procedures are provided in Chapter Three.

The cooling system is a closed system. During operation, the coolant heats up and expands, thus pressurizing the system.

The liquid cooling system consists of a radiator cap, water pump, coolant reservoir tank, heat exchanger and hoses. See **Figures 1-4**.

This chapter describes repair and replacement of the liquid cooling system components.

Table 1, located at the end of this chapter, lists cooling system specifications.

> *WARNING*
> *Do not remove the radiator cap (**Figure 5**) when the engine is hot. The coolant is very hot and is under pressure. Severe scalding could result if the coolant comes in contact with your skin.*

The cooling system must be cooled before removing any component of the system.

THERMOSTAT

The thermostat blocks coolant flow to the heat exchanger when the engine is cold. As the engine warms up, the thermostat gradually opens, allowing coolant to circulate through the system.

> *CAUTION*
> *Do not operate the engine without a thermostat. This can lead to overheating or overcooling and serious engine damage.*

Removal

1. Open the shroud.

> *WARNING*
> *Make sure the engine is cold before proceeding with Step 2.*

2. Drain the cooling system as described in Chapter Three.
3. Loosen the hose clamp and disconnect the hose from the thermostat housing. Prop the hose up to prevent coolant loss.
4. Remove the thermostat. See **Figures 1-4**.

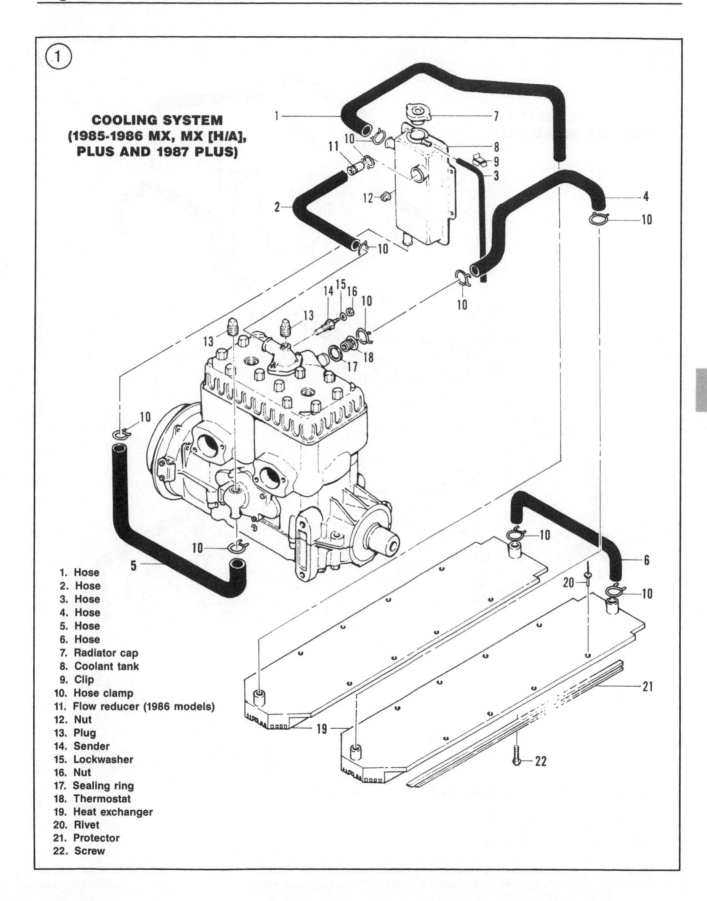

**COOLING SYSTEM
(1985-1986 MX, MX [H/A],
PLUS AND 1987 PLUS)**

1. Hose
2. Hose
3. Hose
4. Hose
5. Hose
6. Hose
7. Radiator cap
8. Coolant tank
9. Clip
10. Hose clamp
11. Flow reducer (1986 models)
12. Nut
13. Plug
14. Sender
15. Lockwasher
16. Nut
17. Sealing ring
18. Thermostat
19. Heat exchanger
20. Rivet
21. Protector
22. Screw

9

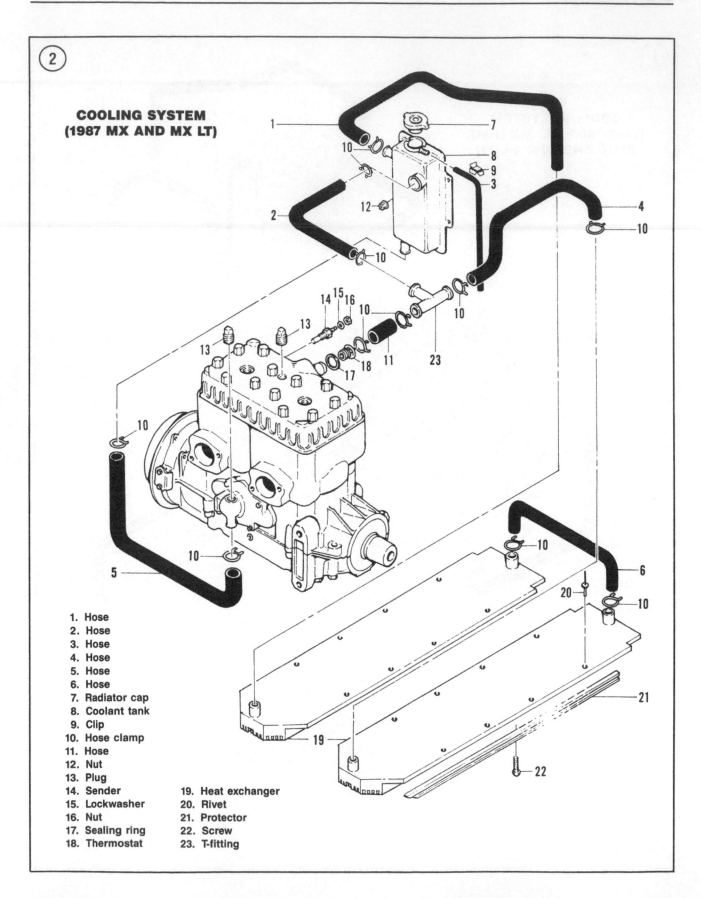

② COOLING SYSTEM
(1987 MX AND MX LT)

1. Hose
2. Hose
3. Hose
4. Hose
5. Hose
6. Hose
7. Radiator cap
8. Coolant tank
9. Clip
10. Hose clamp
11. Hose
12. Nut
13. Plug
14. Sender
15. Lockwasher
16. Nut
17. Sealing ring
18. Thermostat
19. Heat exchanger
20. Rivet
21. Protector
22. Screw
23. T-fitting

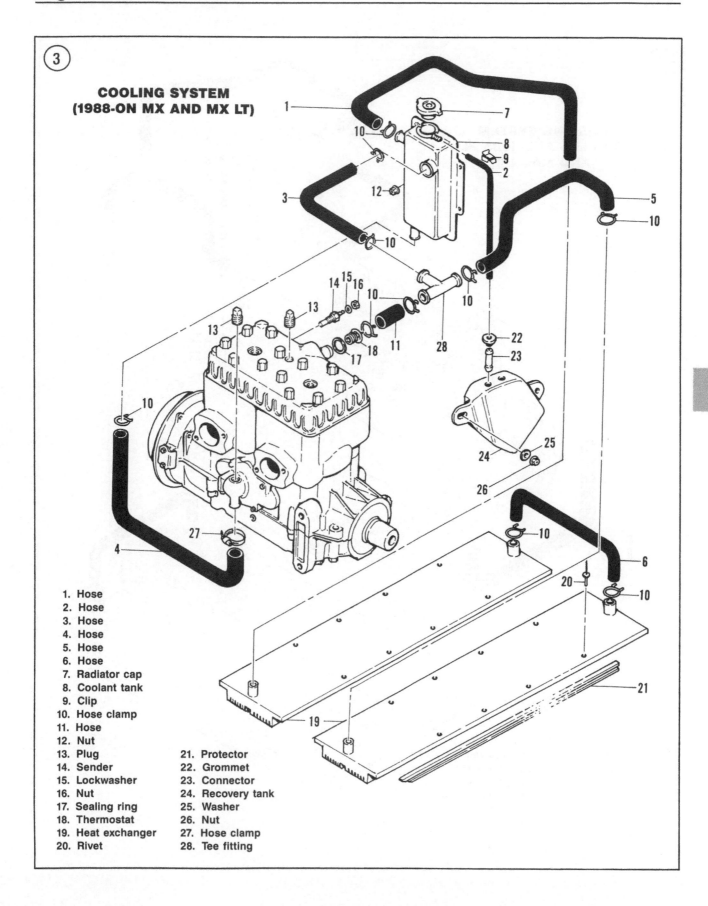

③ **COOLING SYSTEM (1988-ON MX AND MX LT)**

1. Hose
2. Hose
3. Hose
4. Hose
5. Hose
6. Hose
7. Radiator cap
8. Coolant tank
9. Clip
10. Hose clamp
11. Hose
12. Nut
13. Plug
14. Sender
15. Lockwasher
16. Nut
17. Sealing ring
18. Thermostat
19. Heat exchanger
20. Rivet
21. Protector
22. Grommet
23. Connector
24. Recovery tank
25. Washer
26. Nut
27. Hose clamp
28. Tee fitting

9

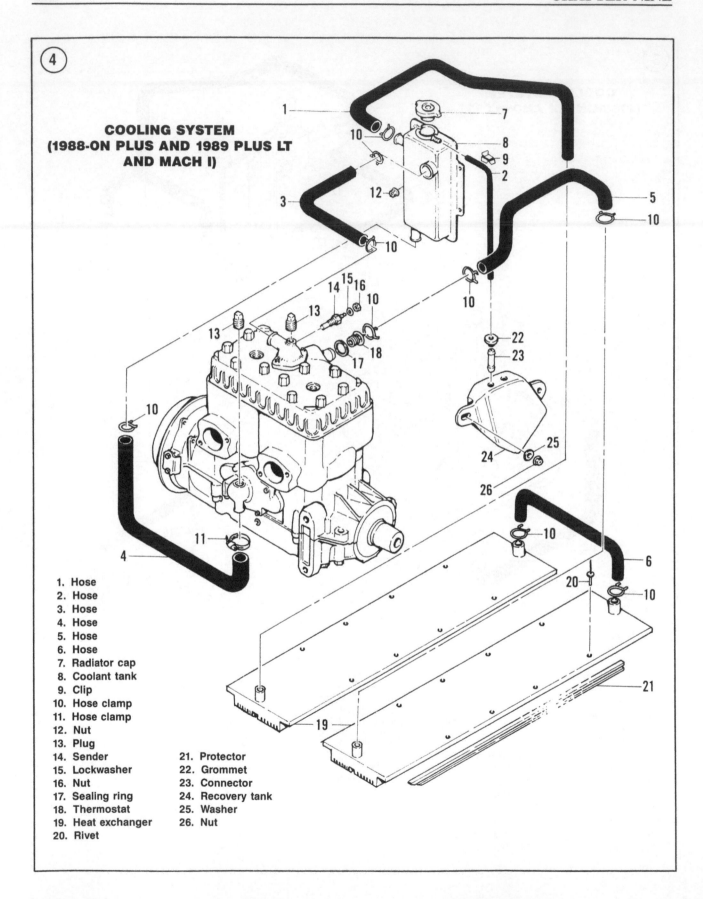

④

**COOLING SYSTEM
(1988-ON PLUS AND 1989 PLUS LT
AND MACH I)**

1. Hose
2. Hose
3. Hose
4. Hose
5. Hose
6. Hose
7. Radiator cap
8. Coolant tank
9. Clip
10. Hose clamp
11. Hose clamp
12. Nut
13. Plug
14. Sender
15. Lockwasher
16. Nut
17. Sealing ring
18. Thermostat
19. Heat exchanger
20. Rivet
21. Protector
22. Grommet
23. Connector
24. Recovery tank
25. Washer
26. Nut

Testing

Test the thermostat to ensure proper operation. The thermostat should be replaced if it remains open at normal room temperature or stays closed after the specified temperature has been reached during the test procedure.

1. Pour some water into a container that can be heated. Submerge the thermostat in the water and suspend a thermometer as shown in **Figure 6**. Use a thermometer that is rated higher than the test temperature (**Table 1**).

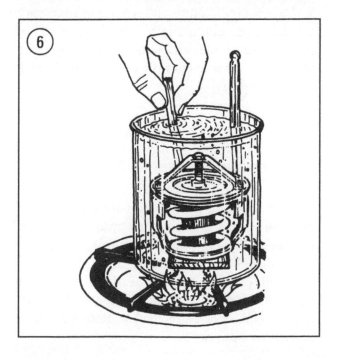

NOTE
Suspend the thermostat with wire so it does not touch the sides or bottom of the pan.

2. Heat the water until the thermostat starts to open. Check the water temperature on the thermometer. It should be approximately 107° F (42° C). If the thermostat valve did not start to open at the temperature, replace it.

3. Let the water cool to 10° *under* the thermostat's rated opening temperature. If the thermostat valve is not fully closed at this temperature, replace it.

4. Remove the thermostat from the water and let it cool to room temperature. Hold it close to a light bulb and check for leakage. If light can be seen at more than 1 or 2 tiny points around the edge of the valve, the thermostat is defective and should be replaced.

Installation

Refer to **Figures 1-4** for this procedure.

1. If a new thermostat is being installed, test it as described in this chapter.

2. Place the thermostat into the cylinder head as shown in **Figures 1-4**.

3. Install the hoses and secure with the hose clamps.

4. Refill the cooling system as described in Chapter Three.

WATER PUMP

The water pump impeller is mounted on the rotary valve shaft. Refer to Chapter Five for service procedures.

HEAT EXCHANGER

Removal/Installation

Refer to **Figures 1-4** for this procedure.

1. Drain the cooling system as described in Chapter Three.

2. Remove the engine as described in Chapter Five.

3. Remove the rear suspension, track and front axle as described in Chapter Fourteen.

4. Loosen the hose clamps and remove the 2 hoses from inside the engine compartment.

5. Remove the seat.

6. Loosen the hose clamps and remove the 2 hoses at the top of the tunnel.

7. Turn the machine on its side.

8. Remove the bolts holding the front heat exchanger to the frame and remove it.

9. Drill out the rivets holding the side and rear heat exchanger to the frame and remove them.

10. Inspect the hoses as described in this chapter.

11. Installation is the reverse of these steps. Note the following:

   a. Use new rivets when installing the side and rear heat exchanger.
   b. Replace worn or damaged water hoses.
   c. Replace fatigued or damaged hose clamps.
   d. Refill the cooling system as described in Chapter Three.

HOSES

Hoses deteriorate with age and should be replaced periodically or whenever they show signs of cracking or leakage. To be safe, replace the hoses every 2 years. Loss of coolant will cause the engine to overheat and result in severe damage.

Whenever any component of the cooling system is removed, inspect the hoses(s) and determine if replacement is necessary.

Inspection

1. With the engine cool, check the cooling hoses for brittleness or hardness. A hose in this condition will usually show cracks and must be replaced.

2. With the engine hot, examine the hoses for swelling along the entire hose length. Eventually a hose will rupture at this point.

3. Check area around hose clamps. Signs of rust around clamps indicate possible hose leakage.

Replacement

Hose replacement should be performed when the engine is cool.

1. Drain the cooling system as described under *Coolant Change* in Chapter Three.

NOTE
Make sure to note the routing and any clamps supported by the hoses.

2. Loosen the hose clamps from the hose to be replaced. Slide the clamps along the hose and out of the way.

3. Twist the hose end to break the seal and remove from the connecting joint. If the hose has been on for some time, it may have become fused to the joint. If so, cut the hose parallel to the joint connections with a knife or razor. The hose can then be carefully pried loose with a screwdriver.

CAUTION
Excessive force applied to the hose during removal could damage the connecting joint.

4. Examine the connecting joint for cracks or other damage. Repair or replace parts as required. If the joint is okay, remove rust with sandpaper.

5. Inspect hose clamps and replace as necessary.

6. Slide hose clamps over outside of hose and install hose to inlet and outlet connecting joint. Make sure hose clears all obstructions and is routed properly.

NOTE
If it is difficult to install a hose on a joint, soak the end of the hose in hot water for approximately 2 minutes. This will soften the hose and ease installation.

7. With the hose positioned correctly on joint, position clamps slightly back away from end of the hose. Tighten clamps securely, but not so much that the hose is damaged.

8. Refill cooling system as described under *Coolant Change* in Chapter Three. Start the engine and check for leaks. Retighten hose clamps as necessary.

Table 1 COOLING SYSTEM SPECIFICATIONS

Thermostat opening temperature	
1985	
MX	43° C (110° F)
Plus	42° C (108° F)
1986	
MX and MX (H/A)	37° C (98° F)
Plus	42° C (108° F)
1987-1988	
All models	43° C (109° F)
1989	
MX and MX LT	43° C (109° F)
Plus, Plus LT and Mach I	42° C (108° F)
Radiator cap pressure	
All models	0.9 kg/cm^2 (13 psi)

9

Chapter Ten

Recoil Starter

All models are equipped with a rope-operated recoil starter. The starter is mounted in a housing that is bolted onto the engine next to the flywheel. Pulling the rope handle causes the starter sheave shaft to rotate against spring tension, moving the drive pawl to engage the starter pulley on the flywheel and turn the engine over. When the rope handle is released, the spring inside the assembly reverses direction of the sheave shaft and winds the rope around the sheave.

Rewind starters are relatively trouble-free; a broken or frayed rope is the most common malfunction. This chapter covers removal and installation of the starter assembly, starter pulley, and rope and spring replacement.

Starter Housing
Removal/Installation

1. Open the shroud.
2. Pull the starter handle out (**Figure 1**). Untie the knot (**Figure 2**) and remove the handle. *Do not* release the rope yet. After removing the handle, tie a knot in the end of the rope to prevent the rope from retracting into the starter housing. Slowly release the rope.
3. Remove the bolts holding the starter housing to the engine. Remove the starter housing (**Figure 3**).
4. Installation is the reverse of these steps. Note the following.
5. Align the starter housing with the starter pulley and install the housing (**Figure 3**). Install the starter housing mounting bolts and tighten securely.
6. Feed the starter rope back through its original path and out the rear cowl hole.
7. Untie the knot at the end of the rope and feed the rope through the handle (**Figure 1**). Tie a knot in the end of the rope as shown in **Figure 4**. Operate the starter assembly to make sure it is working properly. Check the path of the rope through the engine compartment to make sure it is not kinked or interfering with any other component.
8. Close the shroud.

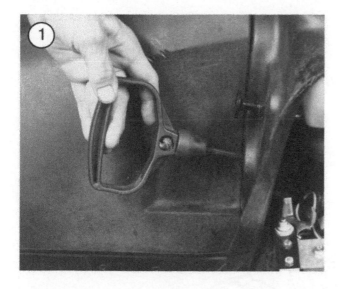

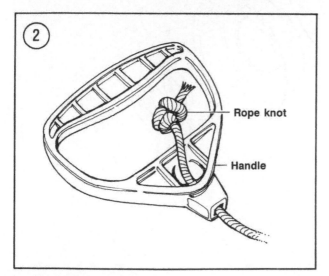

Rope knot

Handle

Starter Pulley
Removal/Installation

The starter pulley (**Figure 5**) is mounted onto the flywheel. The starter pulley can be removed with the engine installed in the frame. This procedure is shown with the engine removed for clarity.

1. Remove the shroud assembly.

2. Remove the recoil starter assembly as described in this chapter.

3. If the engine is installed in the frame, disconnect the fuel pump pulse hose at the crankcase.

CAUTION
Do not substitute the crankshaft locking tool described in Step 4. Substituting the tool with a softer piece of metal may cause crankshaft damage.

4. Insert the crankshaft locking tool (part No. 420 8766 40) through the pulse hose nozzle and engage the tool with the crankshaft (**Figure 6**). After installing the tool, rotate the crankshaft until the tool "locks" the crankshaft.

5. Remove the bolts and washers holding the starter pulley (**Figure 7**) to the flywheel. Remove the starter pulley.

6. Remove the flywheel counterweight (**Figure 5**), if so equipped.

10

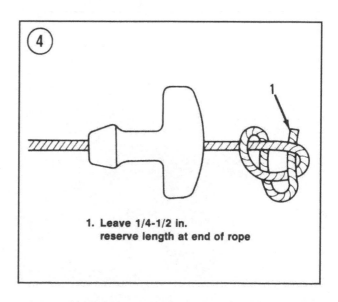

1. Leave 1/4-1/2 in. reserve length at end of rope

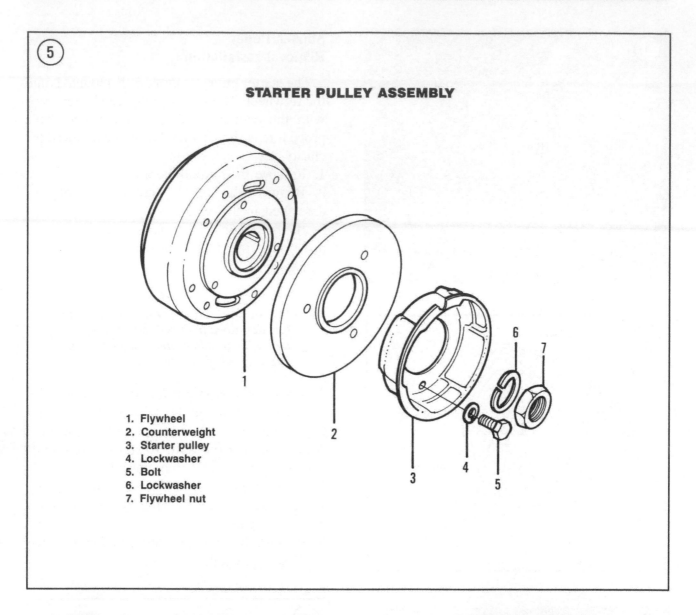

⑤

STARTER PULLEY ASSEMBLY

1. Flywheel
2. Counterweight
3. Starter pulley
4. Lockwasher
5. Bolt
6. Lockwasher
7. Flywheel nut

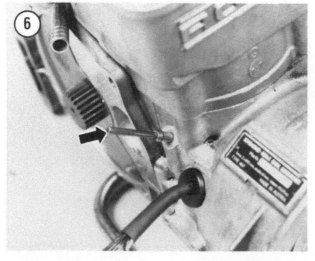

⑥

⑦

7. Installation is the reverse of these steps. Note the following.

8. Install the flywheel counterweight (if so equipped) by aligning the mark on the counterweight with the mark on the flywheel.

9. Install the starter pulley (**Figure 3**), washers and bolts. Tighten the bolts securely.

10. Remove the crankshaft locking tool from the pulse hose nozzle (**Figure 6**).

11. Reinstall the recoil starter housing as described in this chapter.

12. If the engine is installed in the frame, reconnect the pulse hose at the pulse hose nozzle on the crankcase. Secure the hose with the wire clip.

Starter Housing Disassembly

This procedure describes complete disassembly of the starter housing assembly. Refer to **Figure 8** for this procedure.

1. Remove the recoil starter housing as described in this chapter.

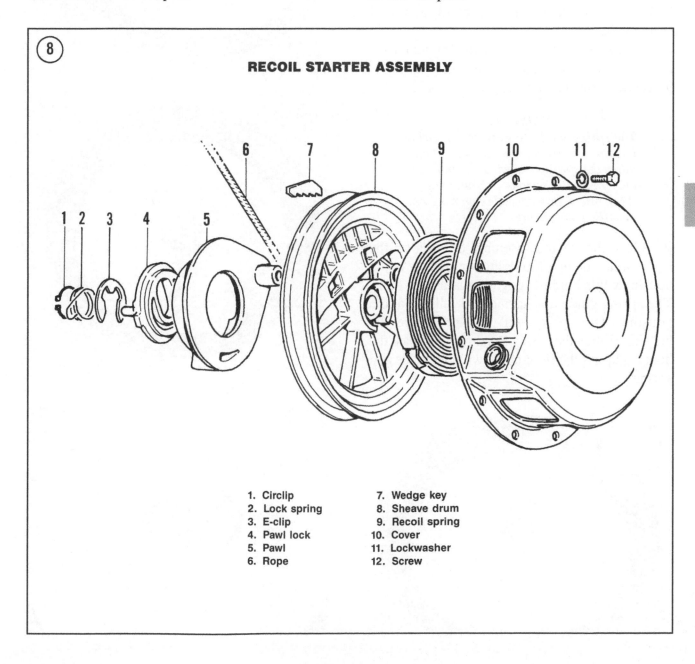

RECOIL STARTER ASSEMBLY

1. Circlip
2. Lock spring
3. E-clip
4. Pawl lock
5. Pawl
6. Rope
7. Wedge key
8. Sheave drum
9. Recoil spring
10. Cover
11. Lockwasher
12. Screw

10

2. Invert the recoil assembly on a clean work bench. Untie the knot in the starter rope. Slowly release the starter rope and allow the sheave to turn slowly until it stops.

3. Turn the recoil assembly over so that the sheave assembly faces up (**Figure 9**).

4. Remove the circlip from the housing post groove (**Figure 10**).

5. Remove the lock spring (**Figure 11**).

6. Remove the E-clip from the housing post groove (**Figure 12**).

7. Remove the pawl lock (**Figure 13**) and pawl (**Figure 14**).

8. Remove the sheave (**Figure 15**) and rope assembly from the starter housing.

9. Remove the key and rope.

> ### WARNING
> *The spring guide (**Figure 16**) is spring loaded. Failure to observe Step 10 may allow the spring to unwind violently and may result in serious personal injury. Wear safety glasses while removing and installing the spring.*

10. Place the rewind assembly housing on the floor right side up. Tap lightly on the top of the housing while holding it tightly against the floor. The spring guide and spring will fall out of the housing. The spring will unwind inside the housing. When the spring has unwound, pick the housing up off the floor along with the guide and spring.

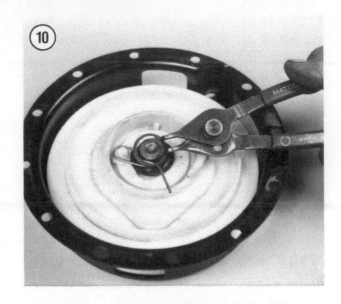

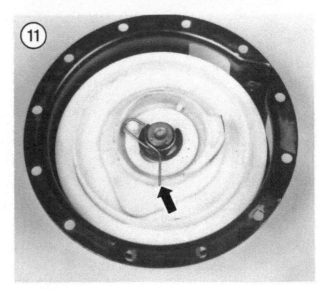

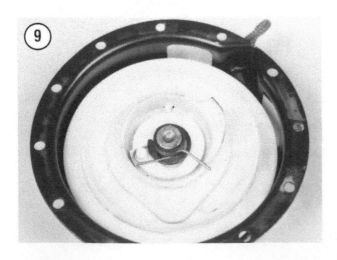

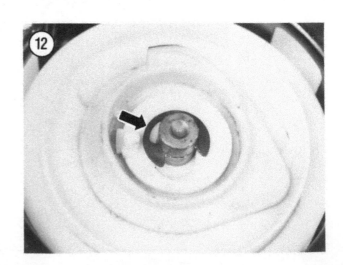

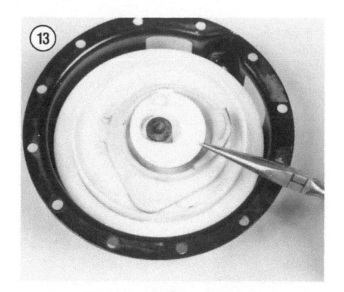

Starter Housing Inspection

NOTE
Before cleaning plastic components, make sure the cleaning agent is compatible with plastic. Some types of solvents can cause permanent damage.

1. Clean all parts thoroughly in solvent and allow to dry.

2. Visually check the starter post (**Figure 17**) in the housing for cracks, deep scoring or excessive wear. Check the post clip grooves for cracks or other damage. Replace the housing if the starter post is damaged.

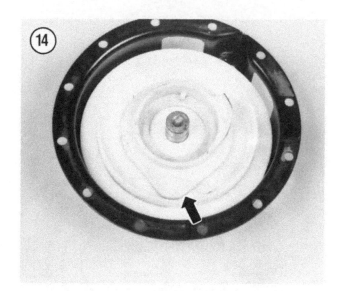

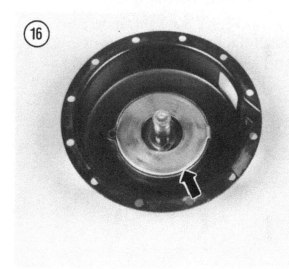

10

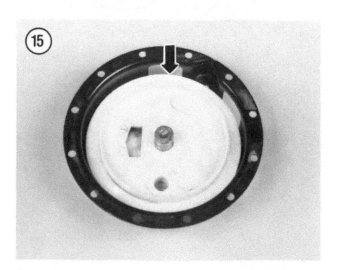

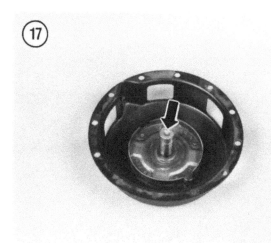

3. Check the pawl (A, **Figure 18**) and pawl lock (B, **Figure 18**) for cracks or other damage. Replace damaged parts as required.

4. Check the sheave drum (**Figure 19**) for cracks or damage.

5. Check the spring coil (**Figure 20**) for cracks or damage.

6. Check the spring guide (**Figure 21**) for cracks or damage.

7. Check the starter rope for fraying, splitting or breakage.

8. If there is any doubt as to the condition of any part, replace it with a new one.

Starter Housing Assembly

Refer to **Figure 8** for this procedure.

1. Install the spring coil (**Figure 20**) into the spring guide as follows:

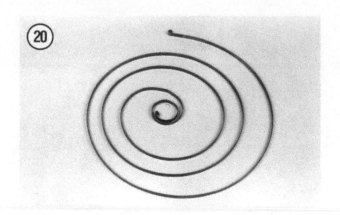

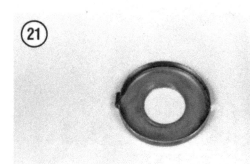

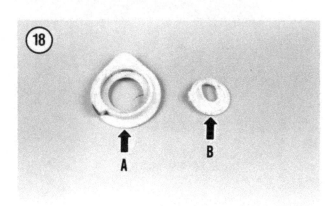

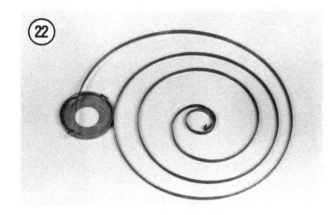

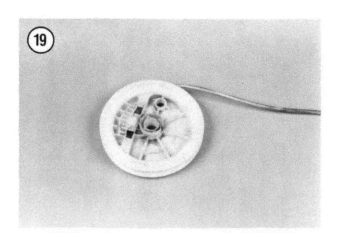

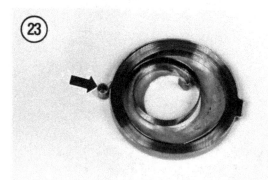

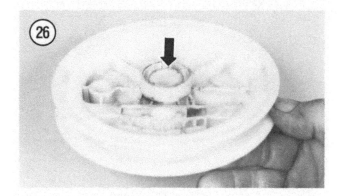

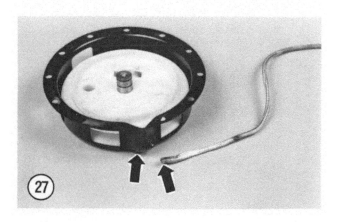

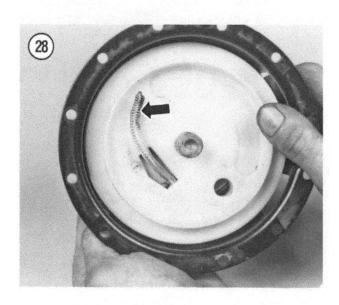

WARNING
Safety glasses should be worn while installing the spring coil.

a. Fit the outer spring loop over the spring guide notch as shown in **Figure 22**.

b. Wind the spring clockwise and fit spring into the pring guide. See **Figure 23**.

c. With the spring installed correctly (**Figure 23**), lubricate it with a low temperature grease.

d. Lubricate the starter post and spring guide in the starter housing with a low temperature grease. See **Figure 24**.

e. Carefully align the outer spring loop (**Figure 23**) with the guide notch in the starter housing and install the spring/guide assembly. See **Figure 25**. Make sure the outer spring loop (A, **Figure 25**) and the spring guide latch (B, **Figure 25**) engage as shown.

2. Align the inner hook on the spring coil (C, **Figure 25**) with notch on the sheave drum (**Figure 26**) and install the sheave drum. Twist the sheave drum slightly to make sure the drum and spring are engaged.

3. Install and tension rope as follows:

a. Align the rope with the rope hole in the starter housing (**Figure 27**) and install the rope. See **Figure 28**.

10

b. If the rope exited the sheave drum window, push it back into the sheave drum as shown in **Figure 29**.

c. Insert the key into the sheave drum. Lock the end of the rope with the key by wedging the key against the rope as shown in **Figure 30**. Pull the rope to seat the key.

d. With the sheave drum facing flywheel end up, wind the sheave drum clockwise until the spring coil binds. Then, release the sheave drum 1 turn. Pull the rope out part way and tie a knot at the end of the rope to hold it. See **Figure 31**.

4. Align the post on the pawl with the hole in the sheave drum and install the pawl (**Figure 32**).

5. Install the pawl lock (**Figure 33**).

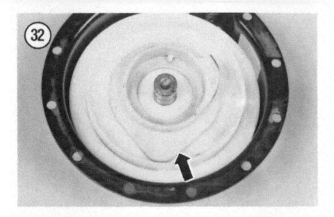

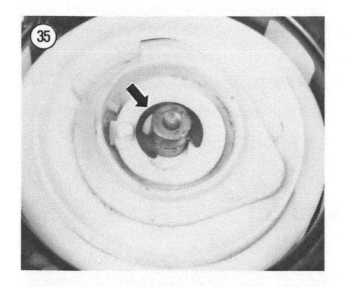

6. Install the E-clip (**Figure 34**) into the lower post groove. Make sure the E-clip engages the groove completely (**Figure 35**).

7. Install the lock spring (**Figure 36**) over the post so that the end of the spring fits over the pawl lock arm as shown in **Figure 37**.

8. Lubricate the lock spring with Molykote G-N paste.

9. Install the circlip (**Figure 38**) so that it seats in the upper post groove. Make sure the circlip seats in the post groove completely (**Figure 39**).

10. Operate the recoil starter to make sure the sheave assembly works smoothly and returns properly.

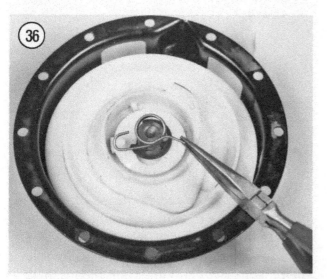

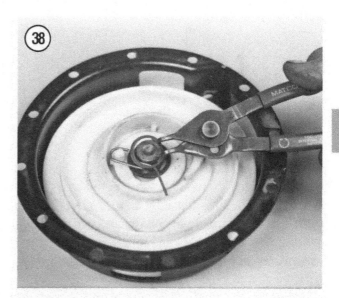

10

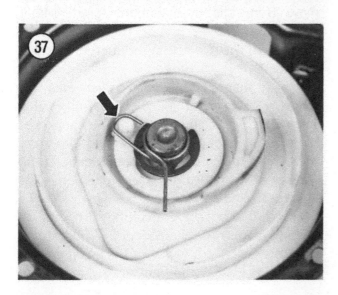

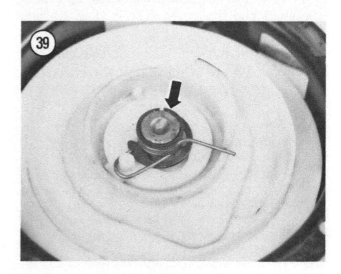

Chapter Eleven

Drive System

The drive train consists of a drive sheave on the end of the crankshaft, connected to a driven sheave on the chaincase by a belt, a drive chain and sprockets inside the chaincase, a drive shaft with 2 drive sprockets and a brake. This chapter describes complete service procedures for the primary and secondary sheave components. Service to the chaincase, chain and sprockets, jackshaft and brake are described in Chapter Twelve.

General drive specifications are listed in **Table 1**. **Tables 1-3** are at the end of the chapter.

> *WARNING*
> *Never lean into a snowmobile's engine compartment while wearing a scarf or other loose clothing when the engine is running or when the driver is attempting to start the engine. If the scarf or clothing should catch in the drive belt or drive system, severe injury or death could occur. Make sure the belt guard is in place.*

DRIVE UNIT

Torque is transferred from the engine crankshaft to the rear track unit by a centrifugally actuated sheave-type transmission. The transmission or drive unit automatically selects the proper drive ratio to permit the machine to move from idle to maximum speed. Major components are the primary sheave assembly, secondary sheave assembly and drive belt (**Figure 1**).

The primary and secondary sheaves are basically 2 sets of variable diameter pulleys that automatically vary the gear ratio. Gear changes are brought about by the primary and secondary sheave drums changing their diameter to correspond to prevailing track conditions. See **Figure 2**.

The shift sequence is determined by engine torque instead of engine rpm. When track resistance increases, such as when going up hill, the sheaves change the gear ratios; engine rpm will remain the same, but the vehicle's speed drops. When track resistance decreases, the sheaves automatically shift toward a higher ratio; engine rpm remains the same, but the vehicle speeds up.

PRIMARY SHEAVE ASSEMBLY

Major components of the primary sheave assembly are the sliding sheave, fixed sheave, weights, primary spring and V-belt. The V-belt connects the primary and secondary sheaves (**Figure 1**).

Fixed and Sliding Sheaves

The sheaves are precision made of mild steel and are balanced to prevent vibration induced problems with the crankshaft and its bearings. The sheave belt surfaces are machined to a

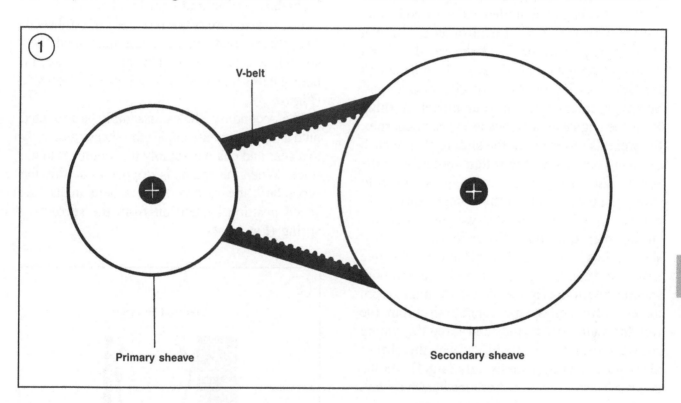

① V-belt

Primary sheave

Secondary sheave

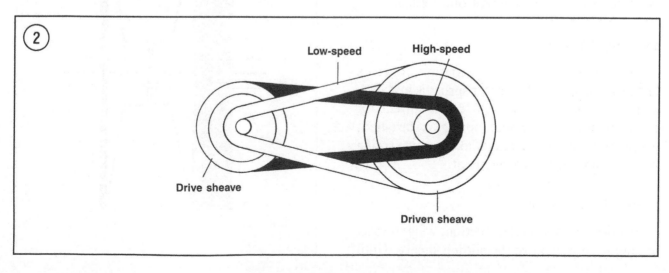

② Low-speed   High-speed

Drive sheave

Driven sheave

11

smooth tapered surface. The tapered surface on both sheaves matches the V-belt gripping surface (**Figure 3**).

The primary sheave assembly is mounted onto the engine crankshaft. When the engine is stopped or at idle, the fixed and sliding sheaves are held apart by the primary spring. This allows the V-belt to drop down between both sheaves. There is no engagement because the belt diameter is *less* than the space between the sheaves.

When the engine is started and allowed to idle, the primary spring is still pushing the sliding sheave away from the fixed sheave; the V-belt is not contacting either sheave. The engine will stay at idle until it reaches a specific engine rpm. See engagement rpm for your model in **Table 1**. As the engine approaches its engagement rpm, the weights mounted on the sliding sheave push or swing out against the rollers mounted on the spider. At this point, the weights begin to overcome the strength of the primary spring. As engine rpm increases, the weights force the sliding sheave to move closer toward the fixed sheave, thus increasing sheave diameter. The free space between the sheaves is reduced and both sheaves begin to grip the V-belt. As engine rpm increases from engagement rpm to shift rpm, the weights swing out further and force the sliding sheave closer to the fixed sheave (the sheave diameter is getting increasingly larger). As the primary sheaves come closer together, the V-belt is forced toward the sheaves outer edge.

Primary Sheave Spring

The primary sheave spring controls engagement speed. If a lighter spring is installed, a lower engine rpm will be required for engagement. If a heavier spring is installed, engine rpm must be increased to overcome spring pressure and allow engagement.

Centrifugal Weights

As noted in the previous section, weights react to engine rpm to move the sliding sheave. Until engine rpm is increased, the weights cannot move or pivot out and thus the sliding sheave cannot move. Weight movement is controlled by engine rpm. The faster the crankshaft rotates, the farther the weights pivot out.

SECONDARY SHEAVE

Major components of the secondary sheave assembly are the sliding sheave, fixed sheave, secondary spring and cam bracket.

The sheaves are precision made of mild steel. The sheave belt surfaces are machined to a smooth tapered surface. The tapered surface on both sheaves matches the V-belt gripping surface (**Figure 3**).

The secondary sheave assembly is mounted onto a jackshaft that couples the sheave assembly to a gear and chain assembly that connects to the track. When the engine is stopped or at idle, the secondary sheave assembly is held in its low speed position by tension from the secondary spring (**Figure 4**).

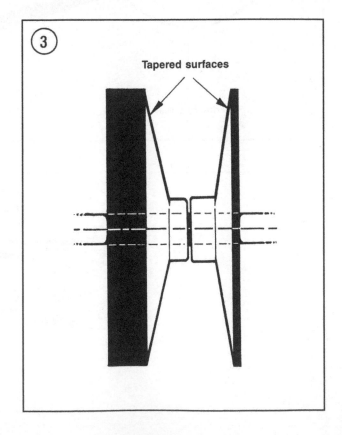

③ Tapered surfaces

The secondary sheave is a torque sensitive unit. If the snowmobile encounters an increased load condition while at wide open throttle, the cam bracket forces the secondary sheaves to downshift (secondary sheave halves move closer together). While the snowmobile will be running slower, the engine is still running at a high rpm. By sensing load conditions and shifting accordingly, engine rpm remains at its peak power range.

Secondary Sheave Spring

The secondary sheave spring determines engine rpm during the shift pattern. On the Ski-Doo secondary sheave, spring tension can be adjusted by repositioning the spring position in holes drilled in the cam. Note the following:

a. Increasing secondary spring tension increases engine rpm. If the secondary sheave is shifting into a higher ratio than the engine can pull, engine rpm will drop below the peak power range. Increasing spring tension will prevent the secondary sheave from upshifitng under the same riding conditions. By not shifting up, the engine will stay in a lower ratio at the peak power range.

b. Decreasing secondary spring tension decreases engine rpm. If the engine rpm is higher than the peak power range, decreasing spring tension will allow the clutch to shift into a higher ratio. Under the same load conditions, engine rpm will drop.

Torque Cam Angle

The secondary sheave spring and the torque cam angle work together to control how easily the driven clutch will shift up. For example, if the spring tension remains the same, but the cam angle is decreased, engine rpm will increase. When increasing the cam angle, engine rpm will decrease.

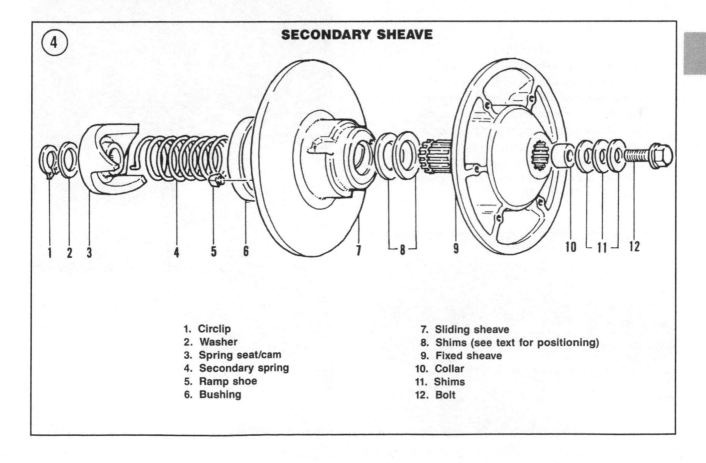

SECONDARY SHEAVE

1. Circlip
2. Washer
3. Spring seat/cam
4. Secondary spring
5. Ramp shoe
6. Bushing
7. Sliding sheave
8. Shims (see text for positioning)
9. Fixed sheave
10. Collar
11. Shims
12. Bolt

DRIVE BELT

The drive belt transmits power from the primary sheave to the secondary sheave and should be considered a vital link in the operation and performance of the snowmobile. To ensure top performance, the correct size drive belt must be matched to the primary and secondary sheaves. Belt width dimensions are critical. Belt wear affects clutch operation and its shifting characteristics. Because belt width will change with use, this dimension must be monitored and adjusted as described in this chapter. See **Table 1** for drive belt width specifications for your snowmobile.

With general use, there is no specific mileage or time limit on belt life. However, belt life is directly related to maintenance and snowmobile operation. The belt should be inspected at the intervals listed in Chapter Three. Premature belt failure (200 miles or less) is abnormal and the cause must be determined to prevent secondary damage.

WARNING
Never lean into the snowmobile's engine compartment while wearing a scarf or other loose clothing when the engine is running or when the driver is attempting to start the engine. If the scarf or clothing should catch in the drive belt or clutch, severe injury or death could occur. Do not run the engine with the belt guard removed (Figure 5).

Removal/Installation

1. Open the shroud.
2. Remove the pins (**Figure 6**) and remove the drive belt guard (**Figure 7**).

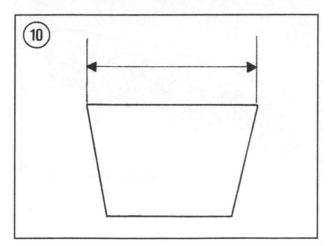

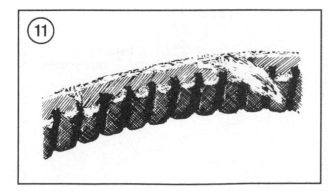

3. Check the drive belt for markings (**Figure 8**) so that during installation it will run in the same direction. If the belt is not marked, draw an arrow on the belt facing forward.
4. Push against the secondary sheave (A, **Figure 9**) and rotate it clockwise to separate the sheaves. Roll the belt over the secondary sheave and remove it (B, **Figure 9**).
5. Inspect the drive belt as described in this chapter.
6. Perform the *Drive Belt Alignment* as described in this chapter.
7. *1987-on:* If a new drive belt was installed, refer to *Drive Belt Tension Adjustment (1987-on)* in this chapter.
8. Reverse Steps 1-4 and install the drive belt. If installing the original belt, be sure to install it so that the arrow mark on the belt (made before removal) faces in the same direction (forward). When installing a new belt, install it so that you can read the belt identification marks while standing on the left-hand side of the machine and looking into the engine compartment (**Figure 8**).

Inspection

The drive belt should be inspected weekly or every 150 miles (240 km) of operation.

1. Remove the drive belt as described in this chapter.
2. Measure the width of the drive belt at its widest point (**Figure 10**). Replace the belt if the width meets or exceeds the wear limit in **Table 1**.
3. Visually inspect the belt for the following conditions:
   a. *Frayed edge:* Check the sides of the belt for a frayed edge cord (**Figure 11**). This indicates drive belt misalignment. Drive belt misalignment can be caused by incorrect sheave alignment and loose engine mounting bolts.
   b. *Worn narrow in one section:* Examine the belt for a section that is worn narrower in one section (**Figure 12**). This condition is

11

caused by excessive belt slippage due to a stuck track or a too high engine idle speed.

c. *Belt disintegration:* Drive belt disintegration (**Figure 13**) is caused by severe belt wear or misalignment. Disintegration can also be caused by the use of an incorrect belt.

d. *Sheared cogs:* Sheared cogs as shown in **Figure 14** are usually caused by violent drive sheave engagement. This is an indication of a defective or improperly installed drive sheave.

4. Replace a worn or damaged belt immediately. Always carry a spare belt (**Figure 7**) on your snowmobile for emergency purposes.

Drive Belt Deflection (1985-1986)

This procedure should be performed whenever a new drive belt is installed.

1. Check drive belt alignment as described in this chapter.

2A. A wooden stick and a spring scale will be required for this procedure. Check belt deflection as follows:

a. Position a straightedge on the drive belt for reference.

b. Using a stick and a spring scale, apply 15 lbs. (6.8 kg) of pressure at the center of the belt as shown in **Figure 15**. Belt deflection should be within specifications in **Table 1**.

c. If deflection is incorrect, perform Step 3.

2B. The Ski-Doo belt tension tester (**Figure 16**) is required for this procedure. Check belt deflection as follows:

a. Position a straightedge on the drive belt for a reference.

b. Position the lower tester O-ring to the belt deflection measurement for your model. See **Table 1**. Then, position the upper tester O-ring to zero on the force scale.

c. Place the end of the tester on the drive belt and push the tester down until the lower O-ring is flush with the edge of the ruler.

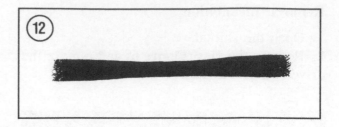

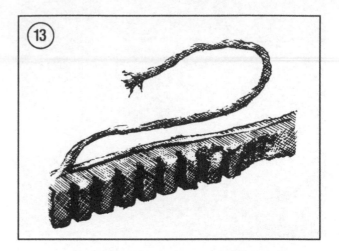

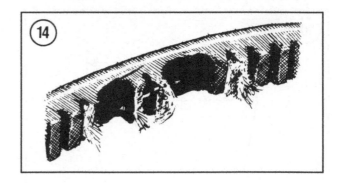

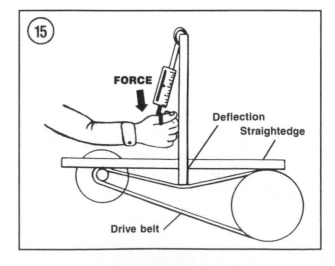

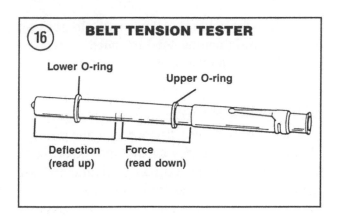

BELT TENSION TESTER

Lower O-ring

Upper O-ring

Deflection
(read up)

Force
(read down)

See **Figure 17**. Read the force scale reading at the top edge of the upper O-ring. It should read 15 lbs. (6.8 kg).

d. If force scale reading is incorrect, perform Step 3.

3. To change belt tension, turn Allen screws on the secondary sheave (**Figure 18**) 1/4 turn and recheck belt deflection. Note the following:

a. To tighten belt tension, turn the Allen screws counterclockwise.

b. To loosen belt tension, turn the Allen screws clockwise.

c. Turn Allen screws equally in 1/4 turn increments.

d. After making adjustment, rotate secondary sheave to help set belt in sheaves. Recheck adjustment.

Drive Belt Tension Adjustment (1987-on)

This procedure should be performed whenever a new drive belt is installed.

1. Check drive belt alignment as described in this chapter.

2. Drive belt must be flush with the top of the secondary sheave halves as shown in **Figure 19**. If belt is too high or too low, perform the following:

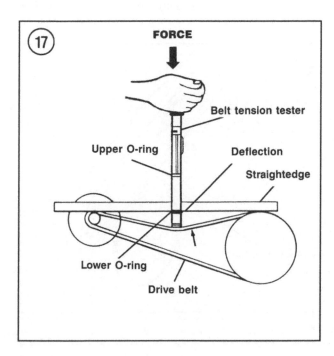

FORCE

Belt tension tester

Upper O-ring

Deflection

Straightedge

Lower O-ring

Drive belt

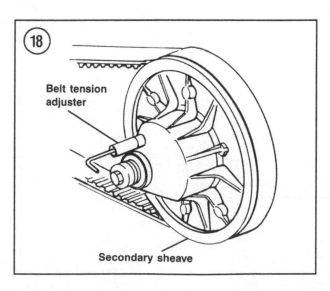

Belt tension adjuster

Secondary sheave

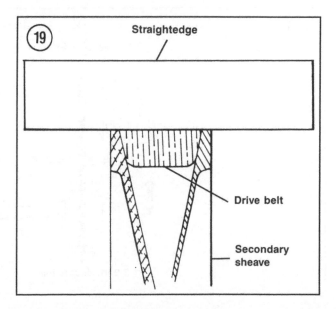

Straightedge

Drive belt

Secondary sheave

11

NOTE
Figure 20 *shows the secondary sheave removed for clarity.*

a. Loosen Allen screw jam nuts on the secondary sheave (**Figure 20**) and turn the Allen screws in 1/4 turn increments to align top of drive belt with top of secondary sheave halves. Loosening the Allen screws raises the belt; tightening the Allen screws lowers the belt.

b. Using the belt tension adjuster (**Figure 18**) (part No. 529 0087 00), and an Allen wrench, hold the Allen screws in position while tightening the jam nuts.

c. Recheck adjustment.

Drive Belt Alignment

The center-to-center distance from the primary sheave to the secondary sheave and the offset of the sheaves must be correctly maintained for good performance and long belt life.

Correct center-to-center distance assures correct belt tension and sheave ratio. A too short

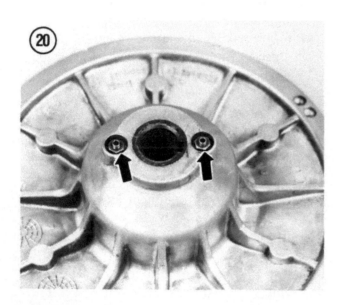

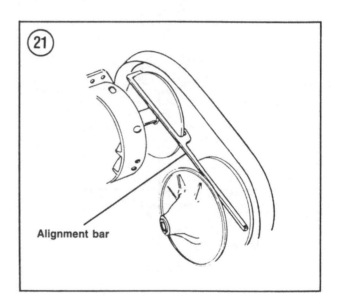

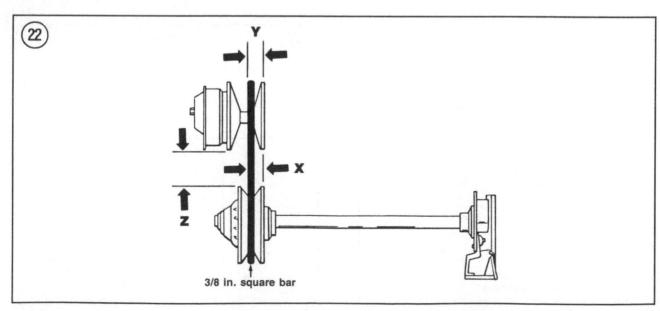

center-to-center distance will cause a narrow sheave shift ratio. If the center-to-center distance is too long, the drive belt will be pulled down between the secondary sheaves too soon; the machine cannot pull strongly because the sheaves are shifting too quickly towards the 1:1 ratio.

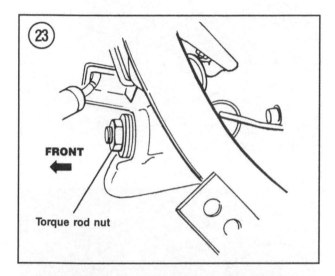

Torque rod nut

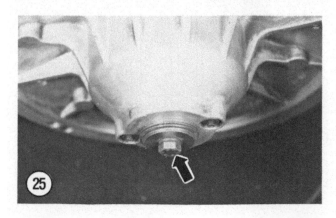

1. Adjust track tension as described in Chapter Three.

2. Remove the drive belt as described in this chapter.

3A. To check pulley alignment with Ski-Doo alignment bar (part No. 529 0071 00), perform the following:

    a. Place the alignment bar into the clutch assembly as shown in **Figure 21**. Clutch alignment is correct when the bar is easily slid over the primary sheave with slight contact against the secondary sheave.

    b. If alignment is incorrect, proceed to Step 4.

3B. To check pulley alignment with a 3/8 in. square bar 19 in. (48 cm) long, perform the following:

    a. Lay a 19 in. (48 cm) length of 3/8 in. square bar between the pulley halves as shown in **Figure 22**. It will be necessary to turn and push the secondary sheave pulley half to open it for placement of the square bar. When bar is in place, release the secondary sheave.

    b. Check pulley offset and distance as specified in **Figure 22**. Refer to **Table 2** for specifications.

    c. If alignment is incorrect, proceed to Step 4.

4A. *Pulley distance adjustment:* If distance "Z" in **Figure 22** is incorrect, perform the following:

    a. Loosen the engine torque rod nut (**Figure 23**) and the engine mount nuts (**Figure 24**).

    b. Reposition engine until distance "Z" (**Figure 22**) is correct.

    c. Tighten engine mount nuts to torque specifications in Chapter Five.

    d. Tighten torque rod nut so that nut seats against washer. Do not overtighten as overtightening will cause pulley misalignment.

4B. *Pulley alignment adjustment:* If pulley offset "X" in **Figure 22** is incorrect, perform the following:

    a. Remove the secondary sheave bolt (**Figure 25**) as described in this chapter. Don't lose the washers on the bolt (**Figure 26**).

11

b. Remove the secondary sheave (**Figure 27**).

c. Adjust offset by adding or removing shims placed between secondary sheave and the jackshaft bearing support (**Figure 28**).

d. Reinstall the secondary sheave and recheck the offset. If the offset is incorrect, repeat until the offset is correct.

e. After adjusting the sheave offset, perform the *Secondary Sheave Free Play Adjustment* as described in this chapter.

5. Reinstall the drive belt as described in this chapter.

PRIMARY SHEAVE SERVICE (1985 AND 1986 MX AND MX [H/A])

The primary sheave (**Figure 29**) is mounted onto the left end of the crankshaft.

Removal

The Ski-Doo crankshaft locking tool (part No. 420 8766 40) and the drive pulley puller (part No. 529 002 100) will be required to remove the primary sheave.

1. Remove the drive belt as described in this chapter.

2. Disconnect the fuel pump pulse hose at the crankcase.

> *NOTE*
> *In **Figure 30** the engine is removed for clarity.*

> *CAUTION*
> *Do not substitute the crankshaft locking tool described in Step 3. Substituting the tool with a softer piece of metal may cause crankshaft damage.*

3. Insert the crankshaft locking tool (part No. 420 8766 40) through the pulse hose nozzle and engage the tool with the crankshaft (**Figure 30**). After installing the tool, rotate the crankshaft slightly until the tool "locks" the crankshaft.

4. Loosen and remove the sheave bolt and washer.

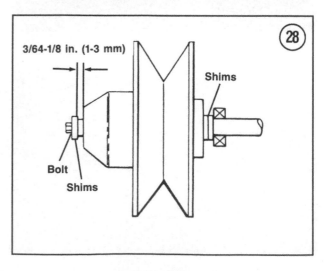

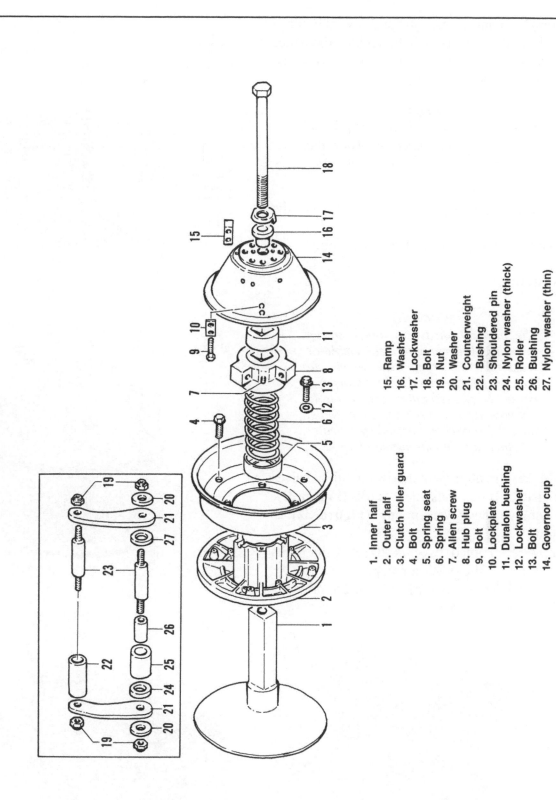

PRIMARY SHEAVE
(1985 MX AND PLUS; 1986 MX
AND MX [H/A])

1. Inner half
2. Outer half
3. Clutch roller guard
4. Bolt
5. Spring seat
6. Spring
7. Allen screw
8. Hub plug
9. Bolt
10. Lockplate
11. Duralon bushing
12. Lockwasher
13. Bolt
14. Governor cup
15. Ramp
16. Washer
17. Lockwasher
18. Bolt
19. Nut
20. Washer
21. Counterweight
22. Bushing
23. Shouldered pin
24. Nylon washer (thick)
25. Roller
26. Bushing
27. Nylon washer (thin)

11

5. Install the puller through the sheave and thread into the end of the crankshaft. Tighten the puller to break the primary sheave loose from the crankshaft taper.

> *NOTE*
> *It may be necessary to rap sharply on the head of the puller to shock the primary sheave loose from the crankshaft.*

6. When the primary sheave is loose, remove the puller.

7. Remove the primary sheave assembly.

Disassembly

> *WARNING*
> *The primary sheave is under spring pressure. Attempting to disassemble or reassemble the primary sheave without the use of the specified special tools may cause severe personal injury. If you do not have access to the necessary tools, have the service performed by a dealer or other snowmobile mechanic.*

1. Make sure alignment marks on pulley halves are visible. If not, make new marks (**Figure 31**).

2. Remove the governor cup (**Figure 32**) as follows:

   a. *1985:* Thread two 1/4-20 × 1 in. bolts through the governor cup.

   b. *1986:* Thread two M6 × 1 × 25 mm bolts through the governor cup.

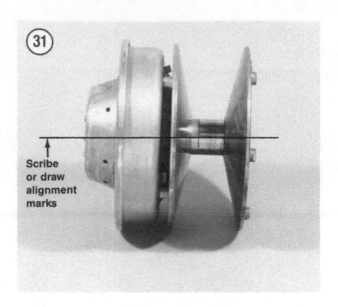

Scribe or draw alignment marks

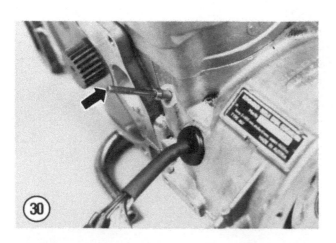

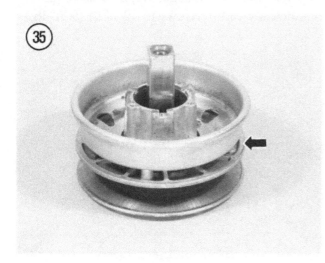

c. Then turn the bolts in equal amounts until the governor cup pulls out.

d. Remove the governor cup.

3. Remove the hub plug (**Figure 33**) as follows:

   a. The hub plug is under spring pressure.

   b. Loosen the 3 hub plug bolts in a crisscross pattern until the hub plug is free of spring tension.

   c. Remove the bolts and hub plug assembly.

4. Remove the spring (**Figure 34**).

5. Lift the outer half off of the inner half (**Figure 35**).

6. Remove the spring seat (**Figure 36**).

7. To remove the clutch roller guard, remove the bolts and washers and lift the roller guard off of the outer half.

8. Do not disassemble the counterweight assembly (A, **Figure 37**) unless necessary. Refer to *Inspection* in this section.

Inspection

CAUTION
Step 1 describes cleaning of the primary sheave assembly. Do not clean the rollers and weights with solvent as solvent will damage the Duralon bushings.

1. All parts, except the rollers and weights (A, **Figure 37**), should be cleaned thoroughly in solvent. Wipe the rollers and weights with a rag.

11

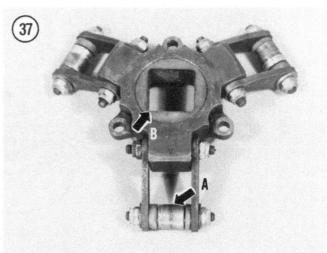

2. Remove Loctite residue from all threads.

3. Check the sliding and fixed sheaves for cracks or other damage.

4. Check the sheave drive belt surfaces for rubber or rust buildup. For proper operation, the sheave surfaces must be *clean*. Remove rubber or rust debris with a course grade sandpaper and finish with #400 wet-or-dry sandpaper.

5. Check the primary sheave spring for cracks or distortion. Replace the spring with the same color code.

6. Check the Duralon bushing (B, **Figure 37**) in the hub plug for wear or damage. Replace the bushing as follows:

   a. Remove the Allen screws (**Figure 38**) from the side of the hub plug.

   b. Remove the Duralon bushing with a hammer and punch or use a press.

   c. Clean the bushing area in the hub plug thoroughly. Remove all burrs with sandpaper.

   d. Apply Loctite 277 to the side of new Duralon bushing.

   e. Align the bushing with the hub plug so that the 3 Allen screw holes are properly aligned. In addition, install the hub plug so that the holes in the bushing are closest to the top of the bushing as shown in **Figure 39**.

   f. Install the bushing with a block of wood and hammer or use a press. Install the bushing so that it is flush with the hub plug as shown in **Figure 38**.

   g. Apply Loctite 242 (blue) to the 3 Allen screws. Then, install and tighten the 3 Allen screws (**Figure 38**).

7. Check each counterweight bushing for wear or damage. Refer to **Figure 29** and **Figure 40** for replacement. Apply Loctite 242 (blue) to the pin threads. Then, install the nuts and tighten to 16 N·m (12 ft.-lb.). Make sure each bushing rotates freely after reassembly.

8. Check the ramps (**Figure 41**) in the governor cup for wear or damage. Replace as required. Tighten the ramp screws to 12 N·m (9 ft.-lb.).

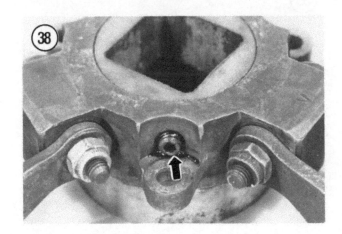

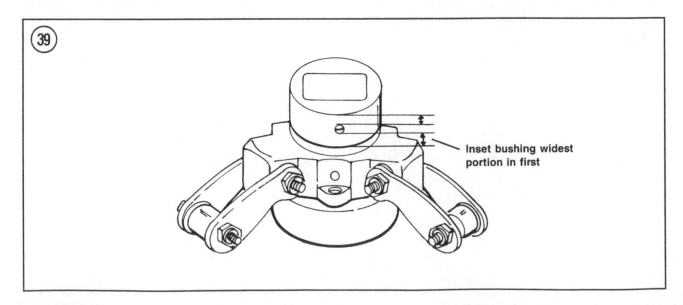
Inset bushing widest portion in first

9. If there is any doubt as to the condition of any part, replace it with a new one.

Reassembly

Refer to **Figure 29** for this procedure.

1. Make sure alignment marks are aligned during assembly.

2. If the clutch roller guard was removed, reposition it onto the outer half and secure with the mounting bolts.

3. Install the spring seat (**Figure 36**).

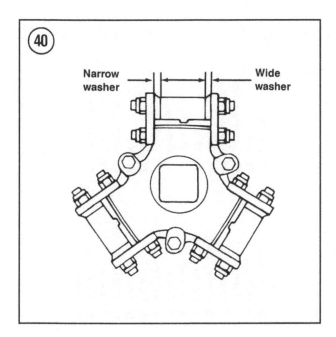

4. Place the outer half over the inner half (**Figure 35**).

5. Install the spring.

6. Position the hub plug into the outer half (**Figure 33**). Install the bolts and tighten evenly in a crisscross pattern.

7. Reinstall the governor cup (**Figure 32**).

Installation

1. Clean the crankshaft taper with lacquer thinner or electrical contact cleaner.

2. Slide the primary sheave onto the crankshaft.

3. Apply a light weight oil onto the primary bolt threads and thread the bolt into the end of the crankshaft with a new lockwasher.

4. Secure the crankshaft with the same tool used during disassembly.

5. Tighten the primary sheave bolt as follows:
   a. Tighten the primary sheave bolt to 85 N·m (53 ft.-lb.).
   b. Remove the crankshaft locking tool and reconnect the fuel pump pulse hose.
   c. Install the drive belt as described in this chapter.
   d. Close the shroud and jack the rear of the snowmobile so that the track clears the ground.
   e. Start the engine and run the engine. Apply the brake 2 or 3 times and turn the engine off.
   f. Reinstall the crankshaft locking tool.
   g. Retorque the primary sheave bolt to 85 N·m (53 ft.-lb.). Bend the lockwasher tab over the bolt to lock it.
   h. Remove the crankshaft locking tool and reconnect the fuel pump hose.
   i. Lower the snowmobile to the ground.

PRIMARY SHEAVE SERVICE (1986 PLUS AND ALL 1987 AND LATER MODELS)

The primary sheave is mounted onto the left end of the crankshaft. Refer to **Figure 42** when performing procedures in this section.

11

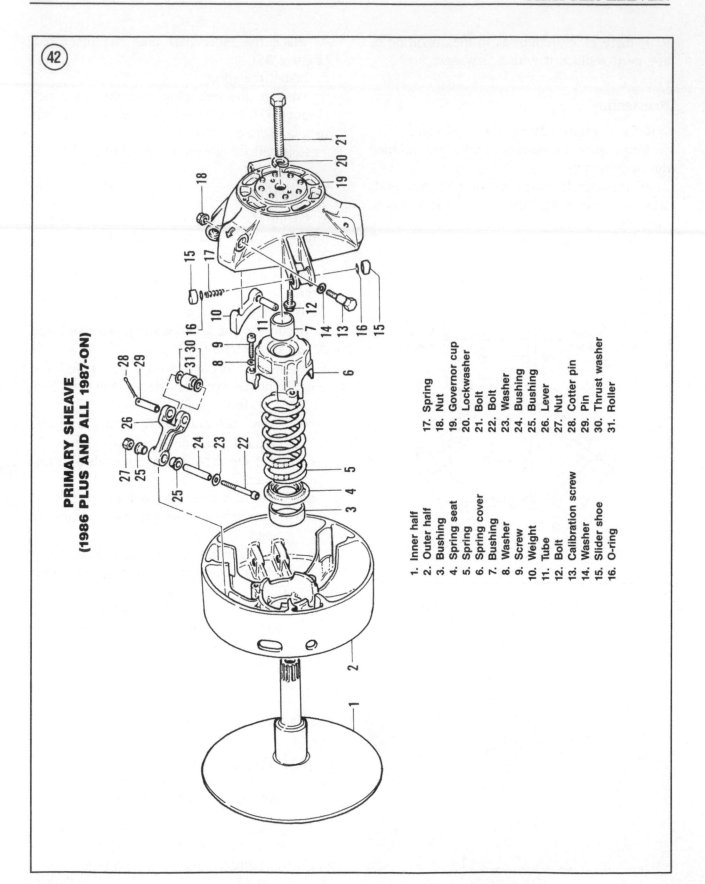

**PRIMARY SHEAVE
(1986 PLUS AND ALL 1987-ON)**

1. Inner half
2. Outer half
3. Bushing
4. Spring seat
5. Spring
6. Spring cover
7. Bushing
8. Washer
9. Screw
10. Weight
11. Tube
12. Bolt
13. Calibration screw
14. Washer
15. Slider shoe
16. O-ring
17. Spring
18. Nut
19. Governor cup
20. Lockwasher
21. Bolt
22. Bolt
23. Washer
24. Bushing
25. Bushing
26. Lever
27. Nut
28. Cotter pin
29. Pin
30. Thrust washer
31. Roller

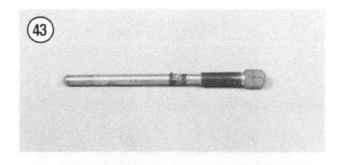

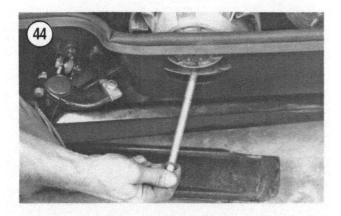

Removal

The Ski-Doo crankshaft locking tool (part No. 420 8766 40) and the drive pulley puller (part No. 420 4760 30) (**Figure 43**) will be required to remove the primary sheave.

1. Remove the drive belt as described in this chapter.
2. Disconnect the fuel pump pulse hose at the crankcase.

NOTE
Figure 30 shows the engine removed for clarity.

CAUTION
Do not substitute the crankshaft locking tool described in Step 3. Substituting the tool with a softer piece of metal may cause crankshaft damage.

3. Insert the crankshaft locking tool (part No. 420 8766 40) through the pulse hose nozzle and engage the tool with the crankshaft (**Figure 30**). After installing the tool, rotate the crankshaft slightly until the tool "locks" the crankshaft.
4. Loosen and remove the sheave bolt and washer (**Figure 44**).
5. Install the puller (**Figure 45**) through the sheave and thread into the end of the crankshaft. Tighten the puller to break the primary sheave loose from the crankshaft taper.

NOTE
It may be necessary to rap sharply on the head of the puller to shock the primary sheave loose from the crankshaft.

6. When the primary sheave is loose, remove the puller.
7. Remove the primary sheave assembly (**Figure 46**).

Disassembly

WARNING
The primary sheave is under spring pressure. Attempting to disassemble or

11

reassemble the primary sheave without the use of the specified special tools may cause severe personal injury. If you do not have access to the necessary tools, have the service performed by a dealer or other snowmobile mechanic.

1. Thread the puller (part No. 420 4760 30) approximately 13 mm (1/2 in.) into the inner half (**Figure 47**). Then, hold the sheave by the outer half and knock the puller to disengage the inner half. See **Figure 48**.

2. Remove governor cup (**Figure 49**) as follows:
   a. Lift the governor cup up until the slider shoes rise to their highest position in their guides.
   b. Place the fork (part No. 529 0055 00) over the slider shoes (**Figure 50**) and remove the governor cup assembly.

NOTE
The Ski-Doo puller (part No. 420 4760 30), spacer (part No. 529 0054 00) and

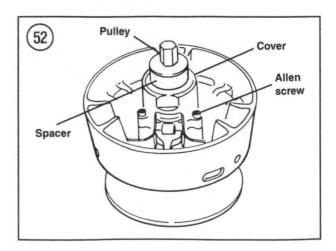

cover (part No. 529 0056 00) will be required for the following procedure.

3. Remove the spring cover assembly (**Figure 51**) as follows:

   a. Assemble the special tools as shown in **Figure 52**. Make sure the puller is screwed in all the way.

   b. Remove the 3 Allen screws evenly to slowly release spring pressure from the spring cover.

   c. When spring pressure is fully released, remove the special tools.

   d. Remove the spring cover (**Figure 53**) and spring (**Figure 54**).

4. Separate the inner and outer halve assemblies (**Figure 55**).

5. Remove the spring seat (**Figure 56**).

6. Disassemble the governor cup assembly (**Figure 57**) as follows:

   a. Remove the forks installed during removal (**Figure 58**).

11

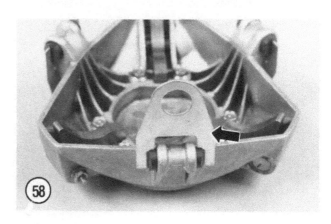

b. Remove a slider shoe (**Figure 59**).

c. Remove an O-ring (**Figure 60**).

d. Remove the spring (**Figure 61**).

e. Remove the opposite slider shoe (**Figure 62**) and O-ring (**Figure 63**).

f. Repeat for remaining slider shoe assemblies.

7. Remove the weights (**Figure 64**) from the outer half assembly as follows:

a. Remove the 2 bolts (**Figure 65**) holding the weight in place.

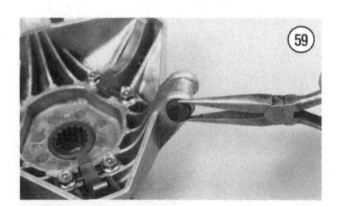

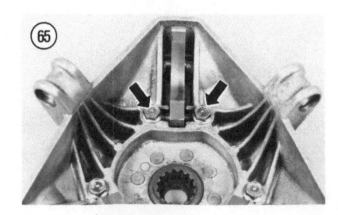

Figure 66

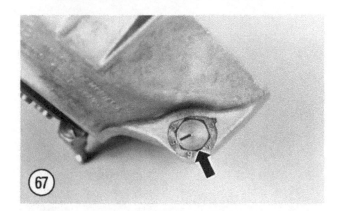

Figure 67

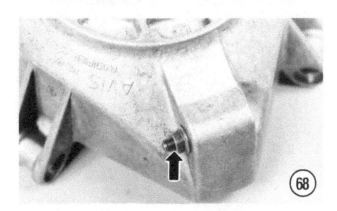

Figure 68

Figure 69

b. Remove the weight (**Figure 66**).

c. Repeat for the remaining weights.

8. Remove the calibration screw (**Figure 67**) as follows:

    a. Note the calibration screw position (**Figure 67**) and record it for reassembly.

    b. Loosen and remove the calibration screw nut (**Figure 68**).

    c. Remove the calibration screw and washer (**Figure 69**).

    d. Repeat for the remaining calibration screws.

9. Remove the lever and bushing assembly (**Figure 70**) as follows:

    a. Loosen the Allen bolt (A, **Figure 71**) and remove the nut (B, **Figure 71**).

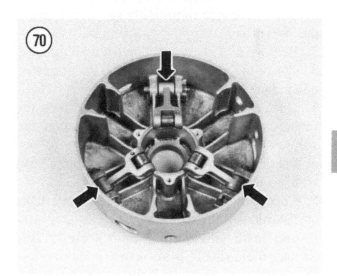

Figure 70

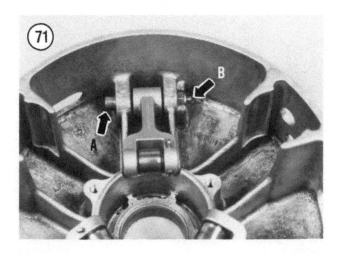

Figure 71

11

b. Slide the bolt (**Figure 72**) through the slot in the outer half assembly and remove it. See **Figure 73**.

c. Remove the lever and bushing assembly (**Figure 74**).

d. Repeat for the remaining lever and bushing assemblies.

Inspection

> *CAUTION*
> *Step 1 describes cleaning of the primary sheave assembly. Do not clean the rollers and weights with solvent as solvent will damage the Duralon bushings.*

1. All parts, except the rollers and weights, should be cleaned thoroughly in solvent. Wipe the rollers and weights with a rag.

2. Remove all Loctite residue from all threads.

3. Check the sheaves (**Figure 75**) for cracks or other damage.

4. Check the sheave drive belt surfaces for rubber or rust buildup. For proper operation, the sheave surfaces must be *clean*. Remove rubber or rust debris with a fine grade steel wool. Clean with a piece of lint-free cloth.

5. Check the primary sheave spring for cracks or distortion. If the spring appears okay, measure its free length with a vernier caliper (**Figure 76**). Replace the spring if its free length is shorter than the length specified in **Table 3**. Replace the spring with the same color code.

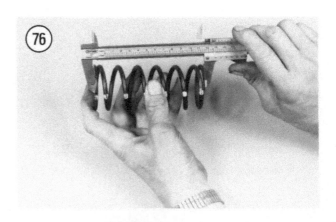

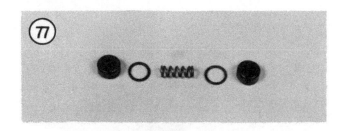

6. Check the slider shoe assembly (**Figure 77**) as follows:

    a. Check the slider shoes for wear. If there is no groove visible in the top of a slider shoe, replace the slider shoes as a set.

    b. Check the O-rings for cracks or other damage. Replace if necessary.

    c. Check the spring for cracks, breakage or fatigue.

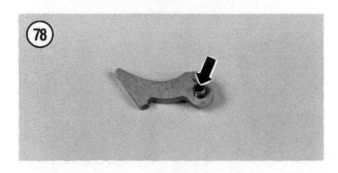

7. Check the weights (**Figure 78**) for cracks or other damage.

8. Check the calibration screws (**Figure 79**) for cracks or other damage.

9. Check the lever and roller assembly (**Figure 80**) for damage. If the roller and/or bushing are damaged, replace them as follows:

    a. Remove the cotter pin (**Figure 81**).

    b. Withdraw the bushing (**Figure 82**) and remove the bushing and thrust washers. See **Figure 83**.

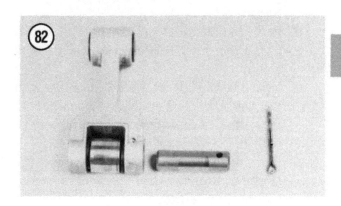

11

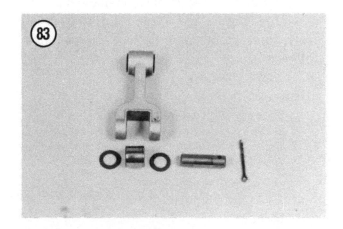

c. Replace worn or damaged parts.

d. Assemble by reversing these steps. Make sure to install a thrust washer on both sides of the roller (**Figure 84**).

10. Check 2 flange bushings in the top of the lever assembly (**Figure 85**). If the bushings appear worn or cracked, replace them. When installing new bushings, align bushing carefully with lever and press into place.

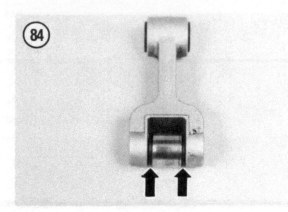

NOTE
If the replacement bushings are made of black plastic, ream bushings with a No. O letter drill after the new bushings have been installed in the lever assembly.

11. Inspect the Kahrlon bushing in the spring cover assembly (**Figure 86**). If the Kahrlon coating is worn from the bushing, replace the bushing as follows:

a. Place the spring cover in a press bed so that the arms on the cover face up as shown in **Figure 87**.

b. Press the bushing out of the spring cover.

c. Clean the outer half assembly with ethyl alcohol.

d. Coat the outside of a new bushing with Loctite 609. Then, press the bushing into the spring cover so that the bushing is flush with the cover as shown in **Figure 87**.

e. Flare bushing on outer spring cover side (**Figure 86**) to help prevent it from loosening in cover.

12. Inspect the Kahrlon bushing in the outer half assembly (**Figure 88**). If the Kahrlon coating is worn from the bushing, replace the bushing as follows:

NOTE
Special tools are required to correctly install the bushing. If you do not have access to the special tools, have the bushing replaced by a Ski-Doo dealer.

a. Press the old bushing out of the outer half assembly.

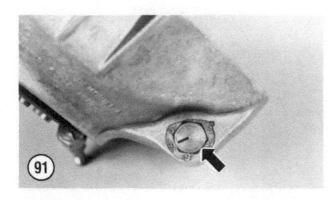

b. Clean the outer half assembly with ethyl alcohol.

c. Coat the outside of the new bushing with Loctite 609. Then, press the bushing into the outer half so that there is the same amount of clearance between the outer bushing edge and the outer half on both sides of the bushing.

d. Place the Ski-Doo outer flare tool (part No. 529 0060 00) on the bushing at the point shown in **Figure 89**. Then, press the flare tool with a press to flare the bushing. Knock the flare tool out of the bushing.

e. Place the Ski-Doo inner flare tool (part No. 529 0061 00) on the bushing at the point shown in **Figure 88**. Then, press the flare tool with a press to flare the bushing. Knock the flare tool out of the bushing.

13. If there is any doubt as to the condition of any part, replace it with a new one.

Reassembly

Refer to **Figure 42** for this procedure.

11

CAUTION
The primary sheave is assembled dry. Do not lubricate any component.

1. Install the calibration screw (**Figure 90**) as follows:

a. Install the washer on the inside of the governor cup and install the calibration screw (**Figure 91**).

b. Set the calibration screw to the same number setting recorded during removal.

c. Install the nut (**Figure 92**) and tighten securely.

d. Repeat for each calibration screw.

2. Install the weights (**Figure 78**) as follows:

a. Place the weight in the governor cup (**Figure 93**).

NOTE
*The bolts used to secure the weights have serrated flanges (**Figure 94**). Make sure to install the correct type bolts.*

NOTE
*If replacement dowel pins were installed in the weights (**Figure 95**), position the pins so that the open slot will face against the bolt flange when the bolts are tightened against the pins in sub-step b.*

    b. Install the bolts and tighten to 10 N·m (89 in.-lb.). See **Figure 96**.

    c. Repeat for each weight.

3. Install the lever and bushing assembly (**Figure 80**) as follows:

    a. Place the 3 lever assemblies in the outer half assembly so that the cotter pins face as shown in **Figure 97**. In addition, the head of the cotter pins must face up as shown in **Figure 98**.

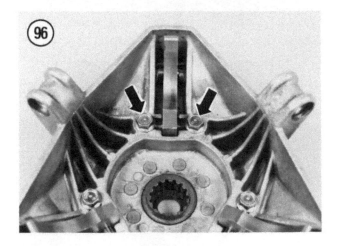

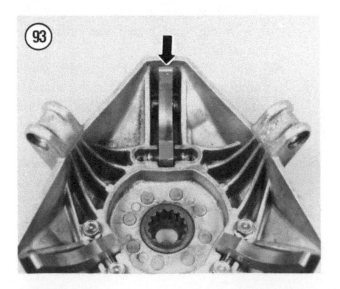

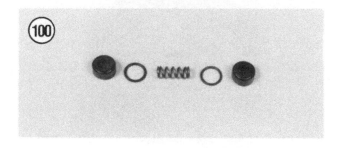

b. Install the Allen bolt (**Figure 99**) and nut. Tighten the Allen bolt to 12 N·m (106 in.-lb.).

c. Make sure the bushings move freely.

d. Repeat for each lever assembly.

4. Install the slider shoes (**Figure 100**) as follows:

NOTE
When installing the slider shoes, install them so that the groove in the shoe face is positioned vertically.

a. Install the first O-ring (**Figure 101**) and slider shoe (**Figure 102**).

b. Install the spring (**Figure 103**).

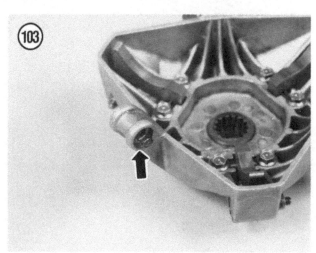

11

c. Install the second O-ring (**Figure 104**) and slider shoe (**Figure 105**).

d. Repeat for each slider shoe assembly.

e. Install the fork (part No. 529 0055 00) over each slider shoe assembly as shown in **Figure 106**.

5. Install the spring seat in the outer half assembly (**Figure 107**).

6. Install the spring (**Figure 108**).

7. Align the arrow on the spring cover (**Figure 109**) with arrow on outer half and install the spring cover. See **Figure 110**.

8. Slide the outer half assembly over the inner half assembly (**Figure 111**).

9. Install the 3 Allen screws (**Figure 112**) and tighten evenly until the spring cover is installed. Tighten the screws securely. If necessary, use the puller assembly to compress the spring cover when installing the Allen bolts.

10. With the forks installed over the slider shoes as shown in **Figure 106**, place the governor cup over the outer half assembly so that the arrow

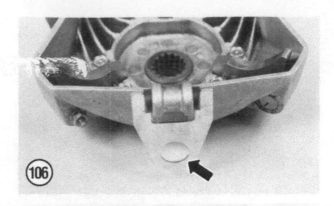

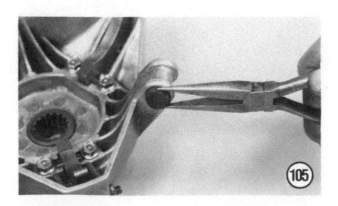

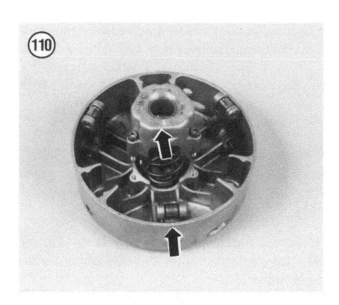

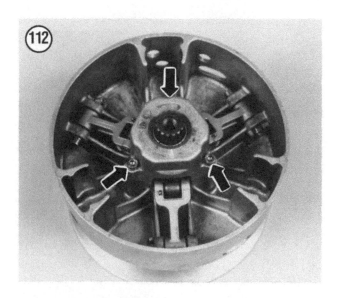

on the governor cup aligns with the arrow on the outer half assembly. See **Figure 113**.

11. Remove the forks and push the governor cup down so that its splines engage with the inner half shaft splines.

> *CAUTION*
> *Make sure the governor cup splines engage with the inner half shaft splines.*

Installation

> *CAUTION*
> *Do not install any antiseize lubricant onto the crankshaft taper when installing the primary sheave assembly.*

1. Clean the crankshaft taper with lacquer thinner or electrical contact cleaner.
2. Slide the primary sheave (**Figure 114**) onto the crankshaft.

11

3. Install the primary sheave bolt and lockwasher (**Figure 115**).

CAUTION
When replacing the primary sheave bolt lockwasher, always use a Ski-Doo factory replacement lockwasher.

4. Secure the crankshaft with the same tool used during disassembly or use a suitable holding tool as shown in **Figure 116**.

5. Tighten the primary sheave bolt as follows:
   a. Tighten the primary sheave bolt to 105 N·m (77 ft.-lb.).
   b. Remove the crankshaft locking tool and reconnect the fuel pump pulse hose.
   c. Install the drive belt as described in this chapter.
   d. Close the shroud and jack the rear of the snowmobile so that the track clears the ground.
   e. Start the engine and run the engine. Apply the brake 2 or 3 times and turn the engine off.
   f. Reinstall the crankshaft locking tool.
   g. Loosen the primary sheave bolt to 85 N·m (63 ft.-lb.). Then, retighten to 95 N·m (70 ft.-lb.).
   h. Remove the crankshaft locking tool and reconnect the fuel pump hose.
   i. Lower the snowmobile to the ground.

NOTE
After 10 hours of operation, recheck the primary sheave bolt tightness.

Drive Pulley Adjustment

The drive pulley calibration screw (**Figure 91**) can be changed to best suit high rpm depending on altitude, temperature and track conditions. Note the following:
   a. The stock calibration screw position is No. 3.
   b. Each number changes rpm by approximately 200 rpm.

   c. Lower numbers will decrease engine rpm whereas high numbers will increase engine rpm. For example, if the calibration screw is changed from No. 3 to No. 4, engine rpm is increased approximately 200 rpm.

NOTE
Do not remove the calibration screw locknut when performing the following procedure as the washer on the inside of the clutch assembly may fall out.

1. Loosen the calibration screw locknut (**Figure 92**) just enough to allow the calibration screw to be turned.

(115)

(116)

2. Turn the calibration screw to the desired position number.

3. Turn all 3 calibration screws to the same position number.

4. Tighten each locknut securely.

SECONDARY SHEAVE SERVICE

The secondary sheave is mounted onto the left-hand side of the jackshaft. Refer to **Figure 117**

or **Figure 118** when performing procedures in this section.

Removal

1. Remove the drive belt as described in this chapter.

2. Apply the parking brake to lock the jackshaft.

3. Loosen and remove the secondary sheave bolt (**Figure 119**) and extension. Don't lose the shims on the end of the bolt (**Figure 120**).

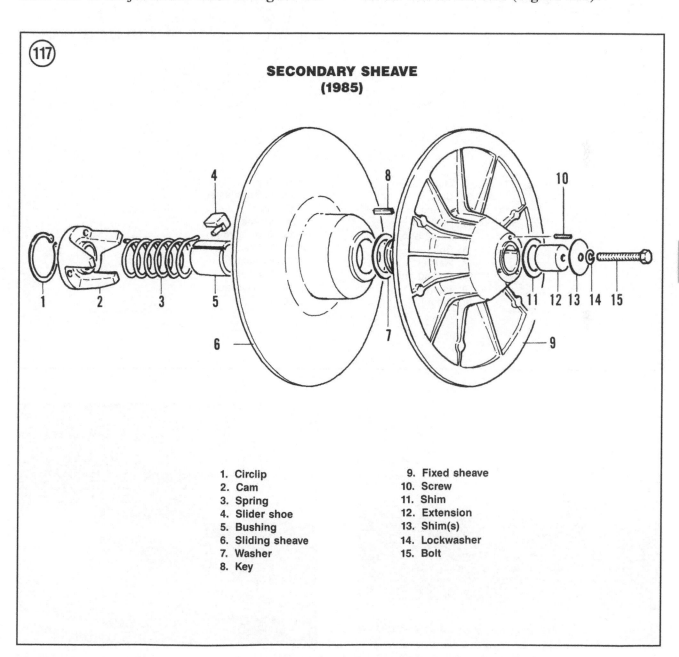

117

SECONDARY SHEAVE (1985)

1. Circlip
2. Cam
3. Spring
4. Slider shoe
5. Bushing
6. Sliding sheave
7. Washer
8. Key
9. Fixed sheave
10. Screw
11. Shim
12. Extension
13. Shim(s)
14. Lockwasher
15. Bolt

11

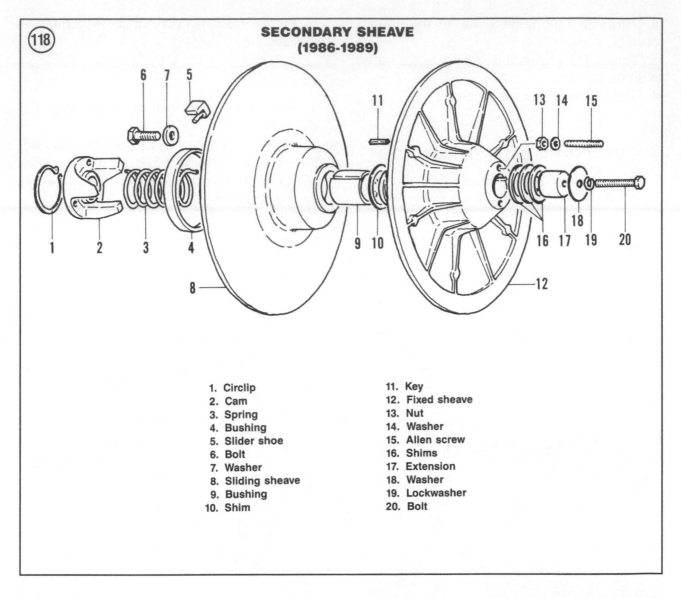

(118) **SECONDARY SHEAVE**
(1986-1989)

1. Circlip
2. Cam
3. Spring
4. Bushing
5. Slider shoe
6. Bolt
7. Washer
8. Sliding sheave
9. Bushing
10. Shim
11. Key
12. Fixed sheave
13. Nut
14. Washer
15. Allen screw
16. Shims
17. Extension
18. Washer
19. Lockwasher
20. Bolt

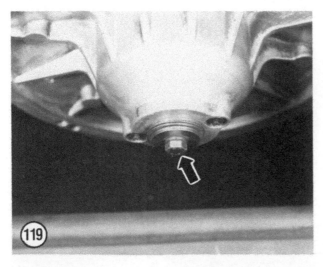

(119)

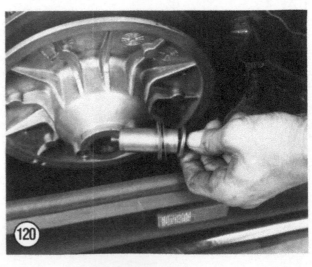

(120)

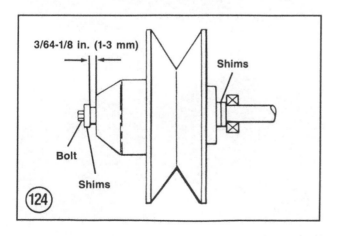

3/64-1/8 in. (1-3 mm)

Shims

Bolt

Shims

4. Remove the secondary sheave (**Figure 121**).

5. Remove the key (**Figure 122**).

NOTE
*The shim(s) (**Figure 123**) installed on the jackshaft behind the secondary sheave are used for sheave offset. Do not remove the shims unless jackshaft service is required. Make sure to install the same shims before secondary sheave installation.*

Installation

1. Make sure the offset shims are installed on the jackshaft (**Figure 123**).

2. Apply a low temperature grease to the splines in the secondary sheave.

3. Slide the secondary sheave onto the jackshaft (**Figure 121**).

4. Install the secondary sheave bolt and the original number of shims (**Figure 120**). Tighten the bolt to 25 N·m (18 ft.-lb.). Apply the parking brake to lock the jackshaft when tightening the secondary sheave bolt (**Figure 119**).

5. Check the secondary sheave free play adjustment as described in this chapter.

6. Install the drive belt as described in this chapter.

Secondary Sheave Free Play Adjustment

1. Remove the drive belt as described in this chapter.

2. Insert a 3/8 in. square bar between the secondary sheave pulley halves.

3. The secondary sheave must have side clearance when installed on the jackshaft and the sheave bolt tightened securely. Check free play by moving the sheave back and forth by hand or by measuring the clearance between the secondary sheave and washer with a feeler gauge. The sheave must have 1-3 mm (3/64-1/8 in.) axial clearance (**Figure 124**).

4. If the axial clearance is incorrect, proceed to Step 5.

5. Apply the parking brake to lock the jackshaft.

11

6. Loosen and remove the secondary sheave bolt (**Figure 119**).

7. Adjust the secondary sheave free play by adding or subtracting the number of shims on the extension (**Figure 120**). Adding shims decreases free play while removing shims increases it. Shims can be purchased through Ski-Doo dealers.

8. Install the secondary sheave bolt and the new number of shims (**Figure 120**). Tighten the bolt to 25 N·m (18 ft.-lb.).

9. Recheck the secondary sheave free play. If the free play is not within specifications, repeat this procedure until the free play is correct.

Disassembly

> *WARNING*
> *The secondary sheave is under spring pressure. Hold the secondary sheave securely when removing the circlip in Step 1.*

1. Compress the outer cam (A, **Figure 125**) and remove the circlip (B, **Figure 125**).

> *NOTE*
> *The outer cam is equipped with 6 holes for spring adjustment (A, **Figure 125**). Record the hole number before removing the outer cam in Step 2.*

2. Remove the cam (**Figure 126**).
3. Remove the spring (**Figure 127**).
4. Separate the sliding and fixed sheaves (**Figure 128**).
5. Remove the key (**Figure 129**).
6. Remove the washer (**Figure 130**).

Inspection

1. Clean the secondary sheave assembly in solvent.
2. Check the sheave surfaces (**Figure 128**) for cracks, deep scoring or excessive wear.
3. Check the sheave drive belt surfaces for rubber or rust buildup. For proper operation, the sheave surfaces must be *clean*. Remove rubber

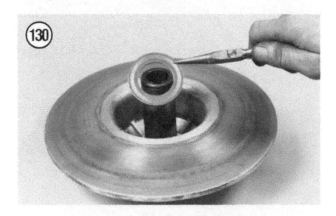

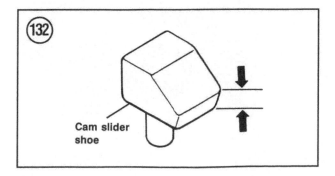

Cam slider shoe

or rust debris with a fine grade steel wool. Wipe off with a lint-free cloth.

4. The nylon ramp shoes (**Figure 131**) provide a sliding surface between the fixed and sliding secondary sheaves. Because the nylon pads are rubbing against aluminum, wear is normally minimal. However, if a ramp shoe is gouged or damaged, smooth the surface with emery cloth; do not use a file. If a ramp shoe cannot be repaired, replace all the ramp shoes as a set. If ramp shoes appear okay, measure slope thickness (**Figure 132**) with a vernier caliper. Replace the ramp shoes as a set if any one shoe measures 1 mm (0.039 in.) or less. Install the ramp shoes so that they face in the direction shown in **Figure 131**.

5. Check the secondary sheave spring (**Figure 133**) for cracks or distortion. Replace the spring if necessary. When replacing the secondary sheave spring, purchase a new spring with the same color code.

NOTE
Secondary sheave spring fatigue is due mainly to metal fatigue from the constant twisting action. As the spring weakens, the secondary sheave will open quicker. This condition can be noticed when riding in mountain areas or deep snow; the machine will drive slower and have quite a bit less pulling power. Because it is difficult to gauge spring wear, you should count on replacing the spring every 2 years or 2,000 miles, whichever comes first. In addition, if you have noticed that the sheave shift point is 500 rpm below the maximum torque rpm, replace the spring.

11

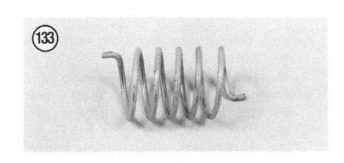

6. Check the cam ramps (**Figure 134**) for scoring, gouging or other signs of damage. Smooth the ramp area with a #400 wet-or-dry sandpaper. If the ramp area is severely damaged, replace the spring seat.

7. Inspect the sliding sheave bushing (A, **Figure 135**) for wear or damage. If necessary, replace the bushing as follows:

   a. Remove the 3 bolts holding bushing in place (B, **Figure 135**).

   b. Remove bushing with a press.

   c. Apply Loctite 609 onto the outside of bushing.

   d. Align new bushing bolt notches with sliding sheave and press bushing into sheave.

   e. Install and tighten bushing bolts.

8. Inspect the fixed sheave bushing (**Figure 136**) for wear or damage. If necessary replace bushing as follows:

   a. Press old bushing out of fixed sheave.

   b. Press new bushing into fixed sheave.

Assembly

1. Place the fixed sheave on the workbench so that the shaft faces up (**Figure 137**).

2. Install shim (**Figure 130**) over shaft.

3. Install key (**Figure 129**) in shaft.

4. Install the sliding sheave (**Figure 138**) over the fixed sheave shaft so that the belt surfaces face together.

5. Install the spring as follows:

   a. Install the spring (**Figure 127**) so that one end of spring engages the hole in the fixed sheave.

   b. The spring seat is equipped with 6 holes for spring adjustment (**Figure 126**).

   c. Hook the end of the secondary spring into the correct hole, as recorded before removal. See **Figure 125**.

   d. Attach a spring scale to the sliding sheave with the Ski-Doo spring scale hook (part No. 529 0065 00) or equivalent. See **Figure 139**. Then, hold the fixed sheave to prevent

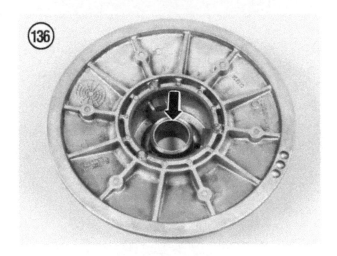

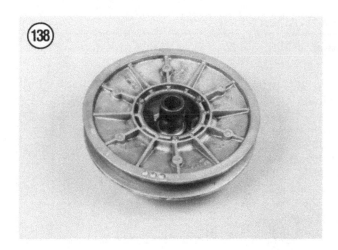

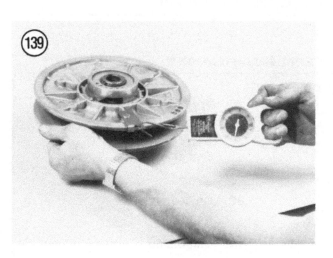

it from moving and pull the spring scale to measure spring preload. Preload should be 5.4-7.3 kg (12-16 lbs.). To adjust, reposition the spring in the cam holes (**Figure 140**) until preload is correct.

6. Compress the cam (A, **Figure 125**) and install the circlip (B, **Figure 125**). Make sure the circlip seats completely in the fixed sheave shaft groove.

7. Install the secondary sheave as described in this chapter.

11

Table 1 DRIVE SYSTEM SPECIFICATIONS

Engagement rpm	
1985-1986	
MX and MX (H/A)	3,100-3,400 rpm
Plus	3,600-3,900 rpm
1987-1988	
MX And MX LT	3,500-3,700 rpm
Plus	3,700-3,900 rpm
1989	
MX and MX LT	3,500-3,700 rpm
Plus and Plus LT	3,400-3,600 rpm
Mach I	3,000-3,200 rpm
Drive belt width	
1985-1987	34.92 mm (1 3/8 in.)
1988-on	34.5 mm (1 23/64 in.)
Drive belt deflection	
1985	30.2-38.1 mm (1 3/16-1 1/2 in.)
1986-1987	25.4-32.0 mm (1-1 1/4 in.)
1988-on	32.0 mm (1 1/4 in.)

Table 2 SHEAVE ALIGNMENT

Model	Pulley distance Z	Offset X	Y
1985	35 mm +3/-0	33.0 mm ±0.75	*
	(1.378 in. +0.12/-0)	(1.30 in. ±0.03)	*
1986			
MX and MX (H/A)	35 mm +3/-0	33.0 mm ±0.75	*
	(1.378 in. +0.12/-0)	(1.30 in. ±0.03)	*
Plus	26.5 mm +1/-0	33.0 mm ±0.75	*
	(1 3/64 in. +3/64/-0)	(1.30 in. ±0.03)	*
1987	26.5 mm +1/-0	36.0 mm ±0.50	*
	(1 3/64 in. +3/64/-0)	(1 27/64 in. ±1/64)	*
1988-on	27.0 mm	36.0 mm	*
	(1 1/16 in.)	(1 27/64 in.)	*

* Measurement Y must exceed measurement X by 0.75 mm (1/32 in.) to 1.5 mm (1/16 in.).

Table 3 PRIMARY SHEAVE SPRING SPECIFICATIONS

Model	Spring free length* mm (in.)	Spring color code
1985		
MX	77.7 (3.06)	Black
Plus	96.5 (3.80)	Orange
1986		
MX	77.7 (3.06)	Black
MX (H/A)	96.5 (3.80)	Orange
Plus	96.5 (3.80)	Blue/yellow
1987-1988		
All models	115.1 (4.53)	Yellow/blue
1989		
MX and MX LT	115.1 (4.53)	Blue/yellow
Plus and Plus LT	132.6 (5.22)	Blue/orange
Mach I	84.1 (3.31)	Red/blue

* ±1.5 mm (±0.060 in.)

Chapter Twelve

Brake, Chaincase, Jackshaft And Drive Axle

This chapter describes complete service procedures for the brake, chaincase, jackshaft and drive axle. **Tables 1-3** are at the end of the chapter.

DRIVE CHAIN AND SPROCKETS

Removal

Refer to **Figure 1** when performing this procedure.
1. Open the shroud.
2. Remove the exhaust pipe and muffler as described in Chapter Six.
3. Place a number of shops rags underneath the chaincase cover.

> *NOTE*
> *The chaincase is filled with oil and the cover is not equipped with a drain plug. When the chaincase cover is removed, try to absorb as much of the oil on the rags as possible.*

4. Remove the bolts and washers holding the chaincase cover (**Figure 2**) to the chaincase. Remove the cover and gasket. Discard the gasket.
5. Wipe up as much oil as possible with the rags.
6. Loosen chain tension as follows:
   a. Remove the hair pin (A, **Figure 3**).
   b. Turn the adjust bolt (B, **Figure 3**) counterclockwise and loosen chain tension.
   c. Check that chain tension is completely loose.
7. Remove cotter pin at end of jackshaft (**Figure 4**). Discard cotter pin.
8. Have an assistant apply the front brake. Then, loosen the drive sprocket nut (**Figure 6**).
9. Remove the driven sprocket circlip (**Figure 5**).
10. Remove the drive sprocket nut (**Figure 6**) and washer (**Figure 7**).
11. Remove the chain and sprockets (**Figure 8**) as an assembly. See **Figure 9**.
12. Remove the spacer from the jackshaft (**Figure 10**).
13. Remove the spacer from the drive axle (**Figure 11**).

①

CHAINCASE, CHAIN AND SPROCKETS

1. Bolt	9. Drive gear (pinion)	17. Flat washer	25. Roller
2. Chaincase cover	10. Shim	18. Circlip	26. Bearing
3. O-ring	11. Chaincase	19. Driven gear (sprocket)	27. Shim
4. Dipstick	12. Bearing	20. Shim	28. Chain tensioner
5. Chain	13. Circlip	21. Bolt	29. O-ring
6. Cotter pin	14. Oil seal	22. Washer	30. Brass washer
7. Nut	15. Bolt	23. Shim	31. Adjust bolt
8. Washer	16. Lockwasher	24. Shaft	32. Hair pin

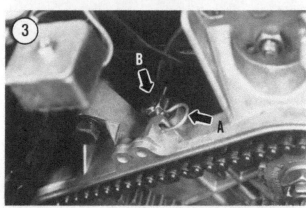

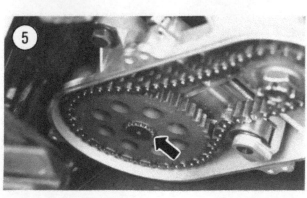

12

Inspection

Refer to **Figure 1** for this procedure. Drive chain specifications are listed in **Table 1**.

1. Clean all components thoroughly in solvent. Remove any gasket residue from the cover and housing machined surfaces.

2. Visually check the drive and driven sprockets (**Figure 12**) for cracks, deep scoring, excessive wear or tooth damage. Check the splines for the same abnormal conditions.

3. Check the chain (**Figure 13**) for cracks, excessive wear or pin breakage.

4. If the sprockets and/or chain are severely worn or damaged, replace all 3 components as a set.

5. Check the chain tensioner for wear and damage. If you have to replace the chain and sprockets because of severe wear or damage, the chain tensioner gear should also be replaced.

6. Disassemble the chain tensioner as shown in **Figure 1**. Check the bearing for damage. Replace if necessary. Reverse to assemble and install the tensioner.

7. Check the cover (**Figure 14**) for cracks, warpage or other damage.

Installation

Refer to **Figure 1** for this procedure.

1. Install the spacer on the drive axle (**Figure 11**).

2. Install the spacer on the jackshaft (**Figure 10**).

3. Lubricate the ends of the jackshaft and drive axle with the gear oil recommended in Chapter Three.

> *NOTE*
> *When installing the sprockets in Step 4, fit the sprockets onto the drive chain so that the stamped sprocket number faces out (Figure 15).*

4. Install the chain and sprockets (**Figure 9**) as an assembly. See **Figure 8**.

5. Install the drive sprocket washer (**Figure 7**) and nut (**Figure 6**).

6. Install the driven sprocket circlip (**Figure 5**). Make sure the circlip seats in the drive axle groove completely.

7. Lock the front brake. Then, tighten the drive sprocket nut to 53 N·m (39 ft.-lb.). Install a *new* cotter pin through the jackshaft and bend the ends over to lock it. See **Figure 4**.

8. Install the chaincase cover together with the O-ring. Make sure the O-ring fits in the cover groove completely. Install the cover and its mounting bolts. Tighten the bolts securely.

9. Adjust the chain tension as described in Chapter Three.

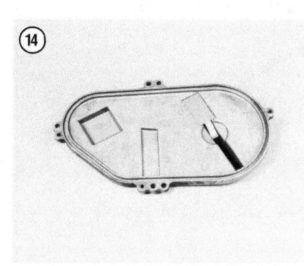

10. Fill the chaincase with the amount and type of gear oil specified in Chapter Three.

11. Install the exhaust pipe and muffler as described in Chapter Six.

12. Close and secure the shroud.

CHAINCASE, BRAKE CALIPER AND JACKSHAFT

A self-adjusting mechanical disc brake is installed on all models. The brake caliper is mounted onto the top of the chaincase. The brake disc rides on the jackshaft. The jackshaft is installed horizontally at the rear of the engine compartment directly over the drive axle. The secondary sheave is mounted onto the left-hand side of the jackshaft.

Refer to the following exploded views when performing procedures in this section:

a. **Figure 16**: 1985 models.
b. **Figure 17**: 1986-on models.
c. **Figure 18**: All models.

Removal

1. Open and secure the shroud.

2. Remove the air silencer as described in Chapter Six.

3. Remove the coolant reservoir.

4. Remove the rotary valve tank (if so equipped).

5. Remove the heat guard from the right-hand side.

6. Remove the chain and sprockets as described in this chapter.

7. Remove the secondary sheave as described in Chapter Eleven.

8. Pry the lock tab arms away from the caliper housing mounting nuts. Then, loosen and remove the nuts and lock tab. See **Figure 16** or **Figure 17**.

9. Remove the 2 caliper housing mounting bolts (**Figure 19**) and remove the outer caliper half (**Figure 20**).

10. Disconnect the brake cable (**Figure 21**) from the outer caliper half and remove the caliper half assembly.

12

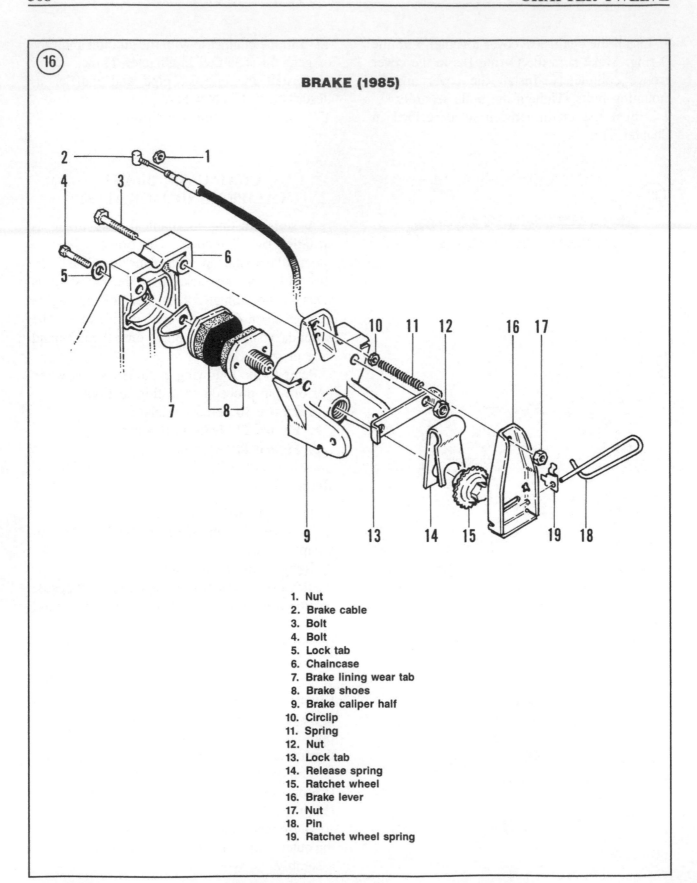

BRAKE (1985)

1. Nut
2. Brake cable
3. Bolt
4. Bolt
5. Lock tab
6. Chaincase
7. Brake lining wear tab
8. Brake shoes
9. Brake caliper half
10. Circlip
11. Spring
12. Nut
13. Lock tab
14. Release spring
15. Ratchet wheel
16. Brake lever
17. Nut
18. Pin
19. Ratchet wheel spring

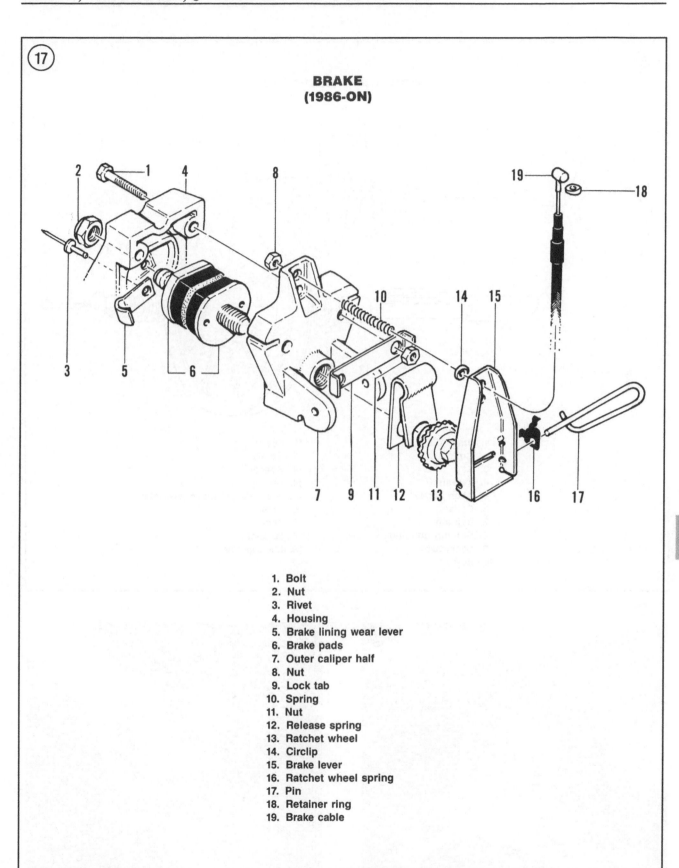

**BRAKE
(1986-ON)**

1. Bolt
2. Nut
3. Rivet
4. Housing
5. Brake lining wear lever
6. Brake pads
7. Outer caliper half
8. Nut
9. Lock tab
10. Spring
11. Nut
12. Release spring
13. Ratchet wheel
14. Circlip
15. Brake lever
16. Ratchet wheel spring
17. Pin
18. Retainer ring
19. Brake cable

12

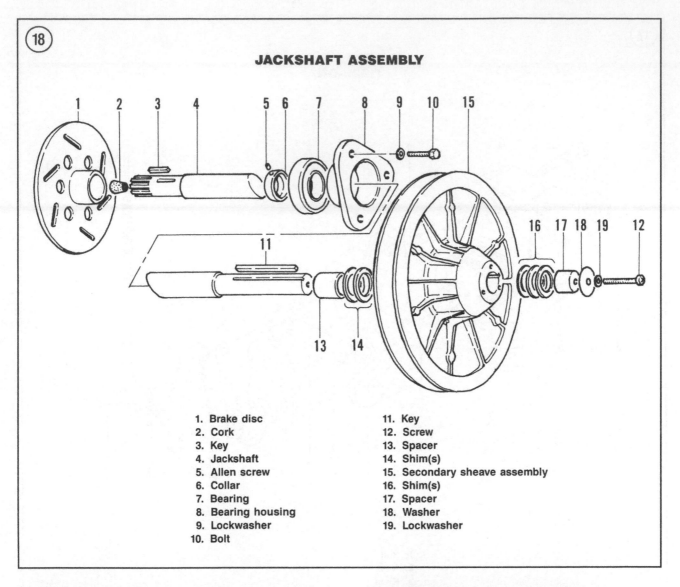

(18)

JACKSHAFT ASSEMBLY

1. Brake disc
2. Cork
3. Key
4. Jackshaft
5. Allen screw
6. Collar
7. Bearing
8. Bearing housing
9. Lockwasher
10. Bolt
11. Key
12. Screw
13. Spacer
14. Shim(s)
15. Secondary sheave assembly
16. Shim(s)
17. Spacer
18. Washer
19. Lockwasher

(19)

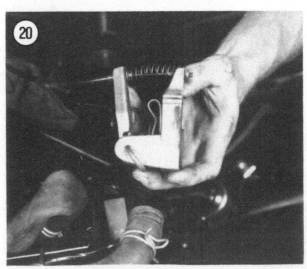

(20)

NOTE
When removing the chaincase mounting bolts in Step 11, note any shims placed between the chaincase and frame. These shims are used for chaincase alignment and must be installed during reassembly.

11. Remove the bolts securing the chaincase to the frame (**Figure 22** and **Figure 23**) and remove the chaincase. See **Figure 24**.

12. Loosen the lock collar Allen screw on the left-hand side of the vehicle (**Figure 25**). Use a punch and hammer and loosen the lock collar (**Figure 26**).

12

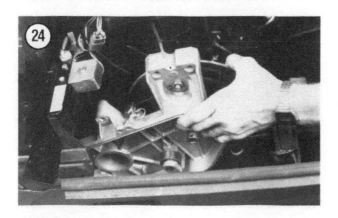

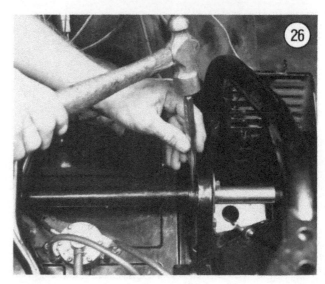

13. Remove the 3 bearing housing bolts and washers (**Figure 27**).

14. Loosen the jackshaft (**Figure 28**) and remove the bearing (**Figure 29**).

15. Remove the jackshaft/brake disc assembly from the right-hand side. See **Figure 30**.

16. Remove the bearing housing (**Figure 31**).

17. Inspect and service the different sub-assemblies as described in the following procedures.

Chaincase Inspection

1. Clean all components in solvent and dry thoroughly.

2. Remove all gasket residue from the chaincase and cover machined surfaces.

3. Check the chaincase oil seal (**Figure 32**) lip for cuts or damage. If necessary, replace the oil seal as follows:

   a. Before removing the oil seals, note and record the direction in which the lip of each seal faces for proper reinstallation.

   b. Carefully pry the oil seal out of the chaincase with a large flat-tipped screwdriver (**Figure 33**). Place a rag underneath the screwdriver to prevent case damage.

   c. Check the chaincase bearing as described in Step 4. If the bearing is worn or

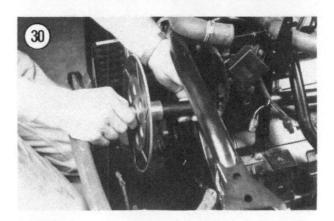

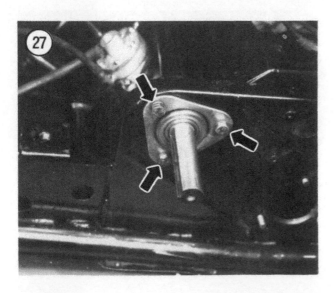

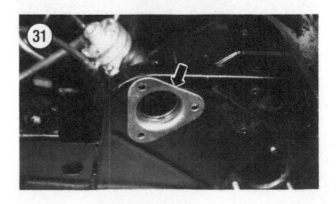

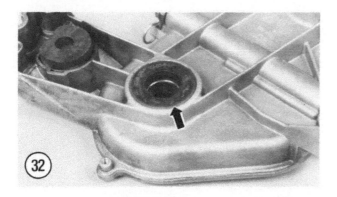

damaged, replace the bearing before installing the new oil seal.

d. The seal lips should face in the direction recorded during disassembly. After installation, pack the seal lip cavity with a low temperature grease.

e. Install the new seal by driving it squarely into the chaincase with a suitable size socket placed on the outer portion of the seal (**Figure 34**).

4. Check the chaincase bearings (**Figure 35**). Turn the bearing inner race by hand and check for roughness or excessive noise. The bearing should turn smoothly. Visually check the bearing for cracks or other abnormal conditions. If the bearing is worn or damaged, replace it as follows:

a. Remove the adjoining oil seal.

b. Drive the bearing out of the chaincase with a suitable size socket placed on the inner bearing race (**Figure 36**).

c. Clean the chaincase bearing area with solvent and dry thoroughly.

d. Install the new bearing by driving it squarely into the chaincase with a socket placed on the outer bearing race (**Figure 34**).

e. Install the oil seal as described in Step 3.

12

Jackshaft Inspection

Refer to **Figure 18** for this procedure.

1. Clean the jackshaft in solvent and dry thoroughly.

2. Check the jackshaft (**Figure 37**) for bending.

3. Turn the bearing (**Figure 38**) and check for excessive noise or roughness. Replace the bearing if worn or damaged.

4. Check both ends of the jackshaft for cracks, deep scoring or excessive wear. See **Figure 39** and **Figure 40**.

5. If there is any doubt as to the condition of the jackshaft, repair or replace it.

Brake Component Inspection

> *WARNING*
> *The brake system is an important part of machine safety. If you are unsure about the condition of any brake component or assembly, have it checked by a qualified snowmobile mechanic.*

Refer to **Figure 16** (1985) or **Figure 17** (1986-on) for this procedure.

1. Clean all brake components (except brake pads) in solvent.

2. Check the brake pads for scoring, uneven wear, cracks or pad breakage. Replace the brake pads if necessary.

   a. Replace the fixed pad (**Figure 41**) by removing the nut (**Figure 42**).

   b. Replace the movable pad (**Figure 43**) by removing the nut.

   c. Reverse to install.

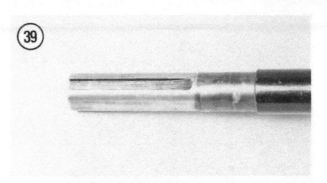

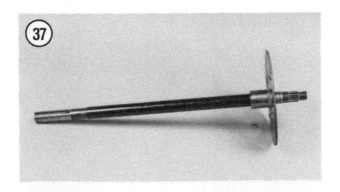

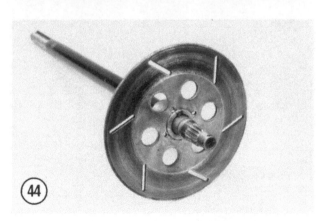

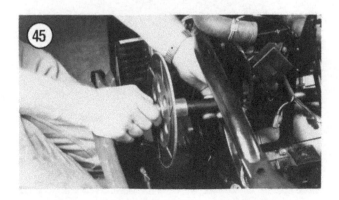

3. Visually check the brake disc (**Figure 44**) for cracks, deep scoring, heat discoloration or checking. If necessary, replace the brake disc. A press will be required to remove and install the brake disc. Refer service to a Ski-Doo dealer or machine shop.

CAUTION
Do not resurface the brake disc. If the brake disc surface is damaged or worn, replace the brake disc.

4. Check the ratchet wheel for wear or damage. Refer to **Figure 16** or **Figure 17** to replace the ratchet wheel. Note the following:
  a. Apply a low temperature grease to the ratchet threads and spring seat.
  b. Install ratchet and turn until it is tight. Then, back off 1/2 (1985-1986) or 1 (1987-on) turn.
5. If there is any doubt as to the condition of any part or assembly, replace it.

Installation

1. Pack the seal lip cavity (**Figure 32**) with a low temperature grease.
2. Install the bearing housing (**Figure 31**). Do not install the bolts at this time.
3. Insert the jackshaft (**Figure 45**) part way into the frame.
4. Install the lock collar over the jackshaft (**Figure 46**).

12

5. Install the bearing so that the shoulder faces in. See **Figure 47**.

6. Push the jackshaft (**Figure 48**) all the way to the right-hand side.

7. Position the chaincase over the jackshaft (**Figure 49**).

> *NOTE*
> *When installing the chaincase in Step 8, make sure to install any shims recorded during removal that were placed between the chaincase and frame. These shims are used for chaincase alignment and must be installed in their exact position.*

8. Install the chaincase mounting bolts and washers (**Figure 50** and **Figure 51**). Tighten the bolts securely.

9. Insert the 2 caliper mounting bolts (**Figure 52**) through the chain housing.

10. Reconnect the brake cable at the holder (**Figure 53**).

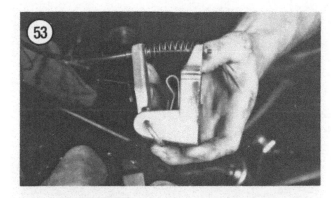

11. Install the holder (**Figure 54**) onto the chain housing. Install the lock tab and the 2 nuts. Tighten the nuts securely and bend the lock tabs over the nuts to lock them.

12. Adjust the jackshaft bearing as follows:

  a. Check that the jackshaft is centered in the bearing housing. Then, install the 3 bearing housing bolts and washers (**Figure 55**). Tighten the bolts securely.

> *NOTE*
> *If the jackshaft is not centered in the bearing housing, perform the **Chaincase Alignment Check/Adjustment** in this chapter.*

  b. Install the chain and sprockets as described in this chapter.

> *NOTE*
> *The drive sprocket nut must be tightened before adjusting the jackshaft bearing.*

  c. Slide the lock collar (**Figure 46**) toward the bearing. Then, turn the lock collar to engage the collar and bearing. Engagement will probably require 1/4 turn. See **Figure 56**.

  d. Turn the lock collar in the direction of rotation until the collar and inner race lock. Then, insert a punch in the lock collar (**Figure 57**) and tighten it by tapping the punch with a hammer.

12

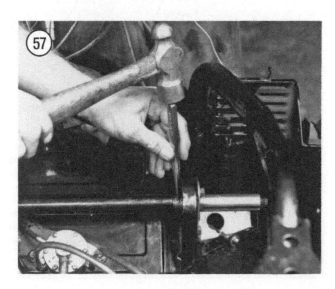

e. Apply Loctite 242 (blue) to the lock collar Allen screw. Install and tighten the screw securely (**Figure 58**).

13. Install the chaincase cover as described in this chapter.

14. Reinstall the secondary sheave as described in Chapter Eleven.

15. Reinstall the heat guard on the right-hand side.

16. Install the oil injection tank.

17. Reinstall the coolant reservoir.

18. Reinstall the air silencer as described in Chapter Six.

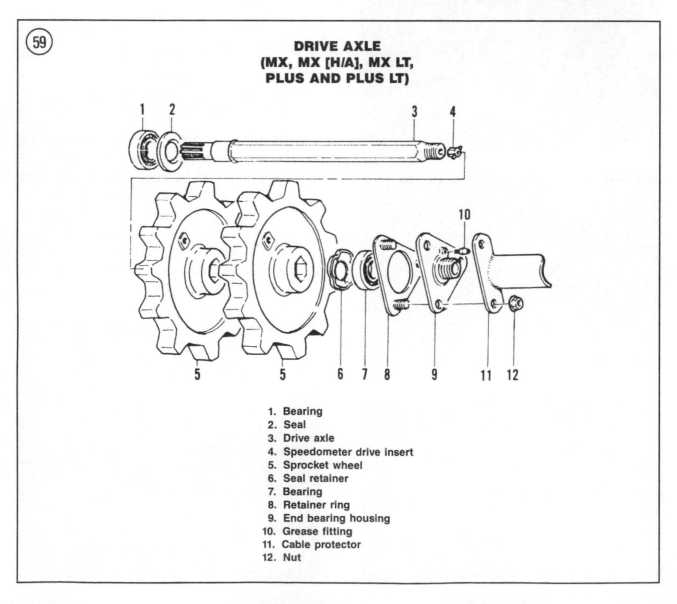

DRIVE AXLE
(MX, MX [H/A], MX LT,
PLUS AND PLUS LT)

1. Bearing
2. Seal
3. Drive axle
4. Speedometer drive insert
5. Sprocket wheel
6. Seal retainer
7. Bearing
8. Retainer ring
9. End bearing housing
10. Grease fitting
11. Cable protector
12. Nut

DRIVE AXLE

The drive axle receives power from the jackshaft through the drive chain assembly on the right-hand side. Refer to **Figure 59** or **Figure 60** when performing procedures in this section.

Removal

1. Remove the chain and sprockets as described in this chapter.

2. Remove the secondary sheave as described in Chapter Eleven.

3. Disconnect the speedometer cable at the drive axle on the right-hand side. Then, remove the end bearing housing and retainer ring (**Figure 61**).

4. Remove the rear suspension as described in Chapter Fourteen.

5. Lift up the drive axle and remove it (**Figure 62**).

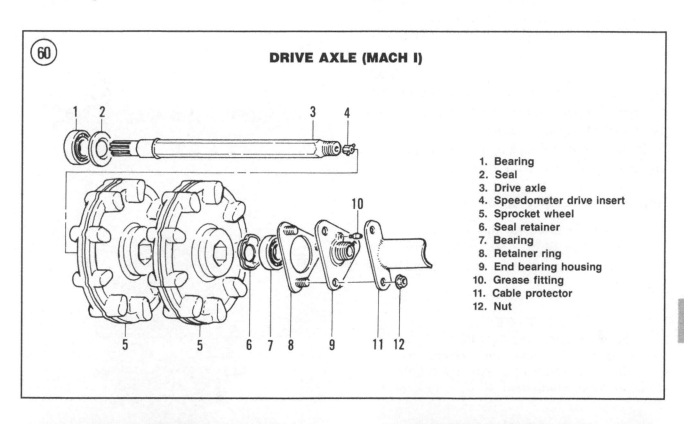

DRIVE AXLE (MACH I)

1. Bearing
2. Seal
3. Drive axle
4. Speedometer drive insert
5. Sprocket wheel
6. Seal retainer
7. Bearing
8. Retainer ring
9. End bearing housing
10. Grease fitting
11. Cable protector
12. Nut

12

Inspection

This section describes drive axle inspection. A press is required to disassemble and reassemble the drive axle. Refer to **Figure 59** or **Figure 60**.

1. Hold the bearing by its outer race and spin the inner race by hand. Check for roughness or excessive noise. Also check the bearing seal for damage. Replace the bearing if necessary. Repeat for both bearings. See **Figure 63** and **Figure 64**.

2. Check the drive axle for bending (**Figure 65**).

3. Check the drive axle splines (**Figure 64**) for cracks or other damage.

4. Visually inspect the sprocket wheels (**Figure 66**) for excessive wear, cracks, distortion or other damage. If necessary, replace the sprocket wheels as described in this chapter.

5. If there is any doubt as to the condition of any part, replace it with a new one.

Sprocket Wheel Replacement

A press is required for this procedure. Refer to **Figure 59** or **Figure 60** for this procedure.

1. Purchase all new parts to have on hand before disassembly.

> *NOTE*
> *Before removing the right-hand oil seal (**Figure 67**), note and record the direction in which the seal lip faces for proper reinstallation.*

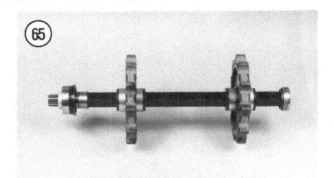

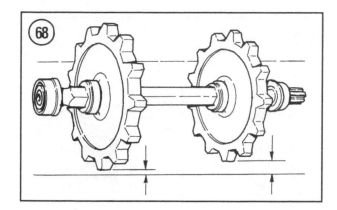

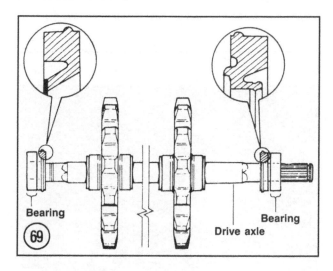

Bearing

Bearing

Drive axle

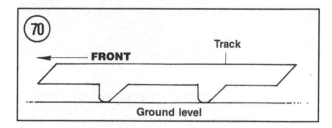

Track

FRONT

Ground level

2. Remove the outer bearings with a suitable bearing puller. See **Figure 63** and **Figure 64**.

3. Place the drive axle in a press and press off each sprocket wheel.

4. Clean the drive axle thoroughly. Remove all nicks with a file.

5. Note the following when installing the sprocket wheels:

a. Both sprocket wheels have an index mark that must align when both wheels are installed on the drive axle. After installing the first sprocket wheel, install the second so that the index marks align.

b. Press the sprocket wheels onto the jackshaft to the dimensions shown in **Table 2**.

6. After installing the sprocket wheels, check synchronization as follows:

a. Place the drive axle on a perfectly flat surface (**Figure 68**).

b. Measure the gap between the sprocket teeth and the flat surface as shown in **Figure 68**.

c. If the gap exceeds 1.5 mm (1/16 in.), the sprocket wheels are out of synchronization. Remove one of the sprocket wheels and realign during assembly.

d. When sprocket synchronization is within specifications, proceed to Step 7.

7. When installing bearings and oil seals, note the following:

a. Install new oil seals during installation.

b. Install oil seals so that they face in the direction shown in **Figure 69**.

c. Install both bearings with their shields facing toward the sprocket wheels.

d. Install bearing on splined side (**Figure 67**) until it seats on drive axle shoulder.

e. Install the end bearing so that it is flush with the drive axle as shown in **Figure 63**.

Installation

1. When installing the track, orient the track lugs to run in the direction shown in **Figure 70**.

2. Align the drive axle (**Figure 71**) so that the splined end faces toward the right-hand side, then install the drive axle. See **Figure 62**.

12

3. Position the drive axle oil seal (**Figure 72**) so that a 2 mm (5/64 in.) gap separates seal and the bearing housing. See **Figure 73**.

4. Install the chain, sprockets and cover as described in this chapter.

5. Install the end bearing housing and retainer ring (**Figure 61**). Reconnect the speedometer cable.

6. Perform the procedures under *Drive Axle Axial Play*.

7. Adjust the chain and refill the chain housing oil as described in Chapter Three.

8. Install the secondary sheave. See Chapter Eleven.

9. Install the rear suspension as described in Chapter Fourteen.

DRIVE AXLE AXIAL PLAY

Chaincase Alignment Check/Adjustment

Before checking drive axial play, perform the following.

1. Remove the secondary sheave as described in Chapter Eleven.

2. Loosen the lock collar Allen screw on the left-hand side of the vehicle (**Figure 74**). Then, use a punch and hammer and loosen the lock collar (**Figure 75**).

3. Turn the lock collar clockwise (**Figure 76**) and release the bearing (**Figure 77**).

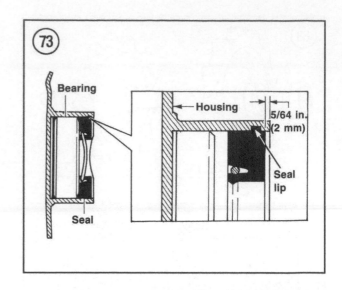

4. Now reinstall the bearing into its support (**Figure 78**). Interpret the ease of bearing installation as follows:

   a. If the bearing was easy to install, chaincase alignment is okay. Proceed to Step 6.

   b. If the bearing was difficult to install, chaincase alignment is incorrect. Proceed to Step 5.

5. If the bearing was difficult to install, refer to **Figure 79** and note at what position the difficulty was more noticeable. Then, refer to the chart in **Figure 79** and place shims between the chaincase

12

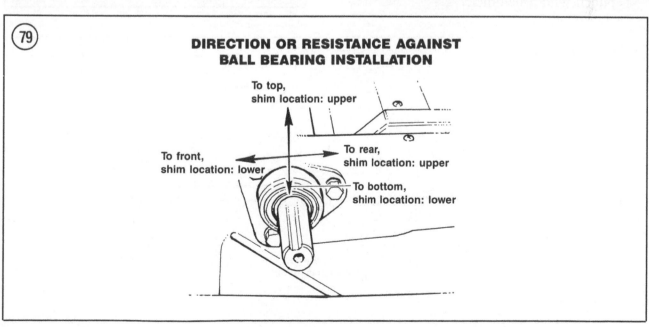

DIRECTION OR RESISTANCE AGAINST BALL BEARING INSTALLATION

To top, shim location: upper

To front, shim location: lower

To rear, shim location: upper

To bottom, shim location: lower

and frame to correct alignment. See **Figure 80** for shim positions. Recheck alignment after installing shims. Refer to *Chaincase, Brake Caliper and Jackshaft* to remove and install the chaincase.

> *NOTE*
> *When installing 1 or more shims to align chaincase, install longer chaincase bolts.*

6. Perform the following:
   a. Install the chain and sprockets as described in this chapter, if previously removed.

> *NOTE*
> *The drive sprocket nut must be tightened before adjusting the jackshaft bearing.*

   b. Install the bearing so that the bearing shoulder faces toward the chaincase (**Figure 77**).
   c. Slide the lock collar (**Figure 76**) toward the bearing. Turn the lock collar to engage the collar and bearing. Engagement will probably require 1/4 turn.
   d. Turn the lock collar in the direction of rotation until the lock collar and inner race lock. Insert a punch in the lock collar (**Figure 75**) and tighten it by tapping the punch with a hammer.

   e. Apply Loctite 242 (blue) to the lock collar Allen screw. Install and tighten the screw securely (**Figure 74**).
   f. Do not install the secondary sheave at this time. Proceed to *Axial Play Adjustment*.

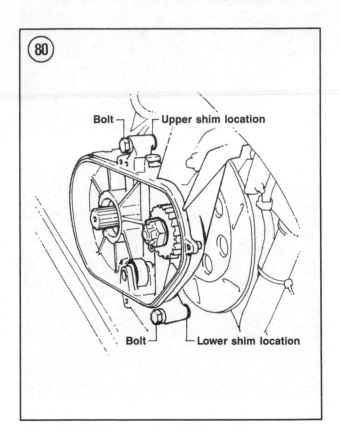

Figure 80 — Bolt, Upper shim location, Bolt, Lower shim location

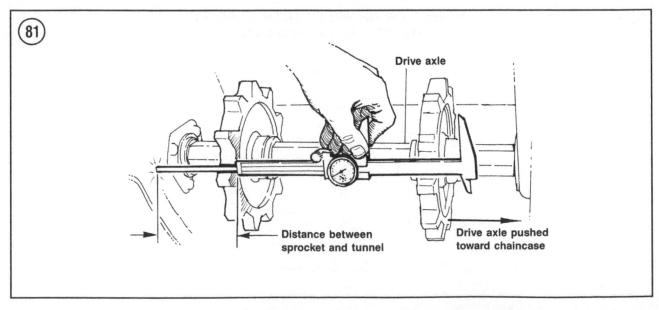

Figure 81 — Drive axle, Distance between sprocket and tunnel, Drive axle pushed toward chaincase

Axial Play Adjustment

1. Once the chaincase is properly aligned, perform the following.

2. Remove the rear suspension as described in this chapter. Then, reinstall the drive axle without the track.

3. Push the drive axle toward the chaincase by hand. Measure the distance between the tunnel and sprocket as shown in **Figure 81**.

> *NOTE*
> *Make sure the drive axle bearing contacts the chaincase bearing bore completely when making the measurement in Step 3.*

4. Now push the drive axle toward the left-hand side end bearing housing by hand. Measure the distance between the tunnel and sprocket (**Figure 81**).

5. The difference in the free play recorded in Step 3 and Step 4 is drive axle axial play. The axial play should be within 0-1.5 mm (0-0.060 in.). If the axial play is excessive, proceed to Step 6. If axial play is correct, proceed to Step 8.

6. Shim drive axle as follows:

　a. Remove the drive axle.

　b. Determine the number of shims required to obtain the correct drive axle axial play. Then, refer to **Table 3** for shim position and quantity. For example, if only 1 shim is required, place the shim on the end bearing housing side. If 2 shims are required, place 1 shim on the end bearing housing side and 1 shim on the chaincase side. If 3 shims are required, place 2 shims on the end bearing side and 1 shim on the chaincase side.

　c. Install the required number of shims on the drive axle. When installing shims between the chaincase and drive axle bearing, install the same number of shims between the drive chain sprocket and spacer. See **Figure 82**.

7. Reinstall drive axle (without track) and recheck axial play. Remove drive axle and readjust shims as required.

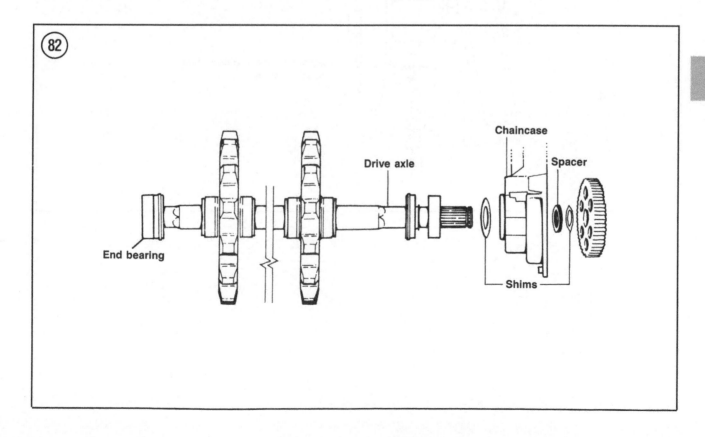

(82)

End bearing · Drive axle · Chaincase · Spacer · Shims

8. When drive axle axial free play is correct, install the track, drive axle and rear suspension. Adjust track tension and alignment as described in Chapter Fourteen.

9. Raise the track off of the ground. Support the vehicle securely. Then, perform the following:

   a. Center the track as follows: Push the slide suspension to the right, then to the left. Note the difference in both movements and center the track at this point by hand.

   b. Check track alignment as follows: Measure the gap between the guide cleat and the slider shoe (behind front axle) on both sides of snowmobile. See **Figure 83**. Record the distance for each side. If the distance between the left- and right-hand sides exceeds 3 mm (1/8 in.), remove the drive axle and reposition shims as follows. If the difference recorded was 3-4.5 mm (1/8-3/16 in.), remove one shim from the larger gap side and add one shim to the smaller gap side. If the difference recorded was 4.5-6 mm (3/16-1/4 in.), remove 2 shims from the larger gap side and add 2 shims to the smaller gap side.

   c. Reinstall drive axle and recheck alignment.

10. Reinstall parts as required after adjustment is complete.

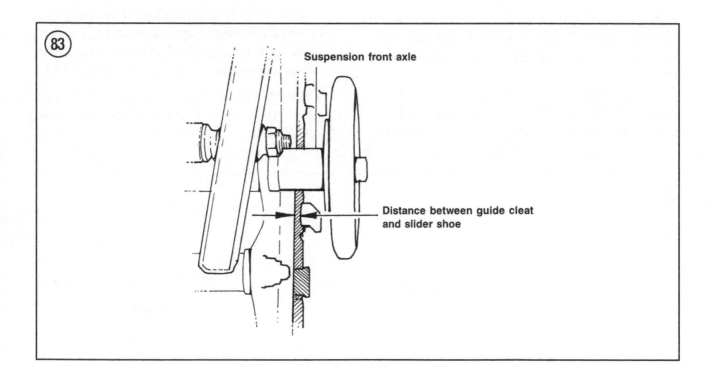

Suspension front axle

Distance between guide cleat and slider shoe

Table 1 DRIVE CHAIN SPECIFICATIONS

Chain pitch	
All models	3/8 in. silent
Chain drive ratio	
1985	
MX	26/40
Plus	26/40
1986	
MX	26/40
MX (H/A)	26/44
Plus	20/38
1987-on	
MX and MX LT	22/44
Plus and Plus LT	20/38
Mach I	22/40

Table 2 SPROCKET WHEEL INSTALLATION SPECIFICATIONS

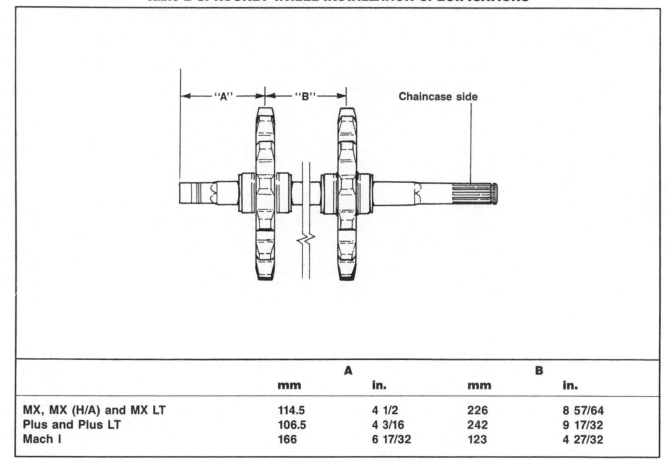

	A		B	
	mm	in.	mm	in.
MX, MX (H/A) and MX LT	114.5	4 1/2	226	8 57/64
Plus and Plus LT	106.5	4 3/16	242	9 17/32
Mach I	166	6 17/32	123	4 27/32

12

Table 3 DRIVE AXLE SHIM POSITION AND QUANTITY

Shim(s)	Chaincase side	End bearing housing side
1		1
2	1	1
3	1	2

Chapter Thirteen

Front Suspension and Steering

All models are equipped with a telescopic strut suspension system. This chapter describes service procedures for the skis, handlebar, steering assembly, steering column and tie rods.

Table 1 and **Table 2** are at the end of the chapter.

SKIS

The skis are equipped with wear bars, or skags, that aid in turning the machine and protect the bottoms of the skis from wear and damage caused by road crossings and bare terrain. The bars are expendable and should be checked for wear and damage at periodic intervals and replaced when they are worn to the point they no longer protect the skis or aid in turning.

Removal/Installation

Refer to **Figure 1** for this procedure.

1. Support the front of the machine so both skis are off the ground.

NOTE
Mark the skis with a "L" (left-hand) or "R" (right-hand). The skis should be installed on their original mounting side.

2. Remove the cotter pin from the end of the ski pivot bolt. Loosen and remove the nut, bolt (A, **Figure 2**) and ski (B, **Figure 2**). Don't lose the collars from both sides of the ski pivot bolt hole.

3. If necessary, remove the ski boot (A, **Figure 3**) and stop binding from the ski.

4. Inspect the skis as described in this chapter.

5. Installation is the reverse of these steps. Note the following.

6. Wipe all old grease from the pivot bolt, collar and washers.

7. Apply a low temperature grease to the pivot bolt and collar before installation.

8. Reinstall the stop binding and ski boot, if previously removed.

9. When installing the skis, refer to the alignment marks made before removal.

10. Hold the ski in place. Install the collar, washers and pivot bolt.

11. Install the ski pivot bolt nut and tighten to the torque specification in **Table 1**.

13

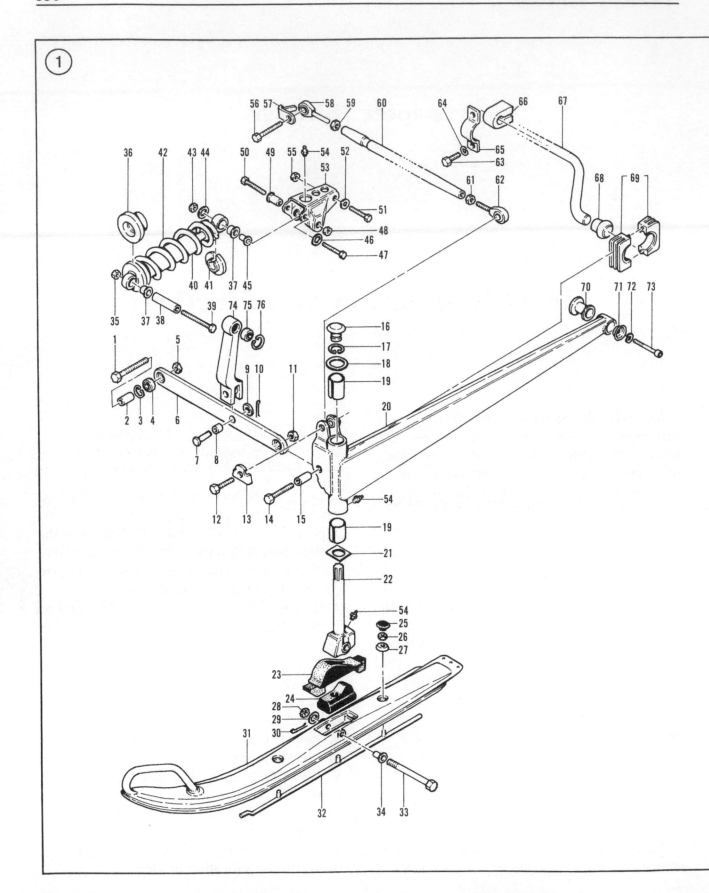

FRONT SUSPENSION AND SKI

1. Bolt		39. Bolt	
2. Bushing		40. Shock absorber	
3. Circlip		41. Spring stopper	
4. Ball joint		42. Spring	
5. Nut		43. Nut	
6. Lower bracket		44. Washer	
7. Clevis pin		45. Bushing	
8. Bushing		46. Washer	
9. Washer		47. Bolt	
10. Cotter pin		48. Nut	
11. Nut		49. Bushing	
12. Bolt		50. Bolt	
13. Lock tab		51. Bolt	
14. Bolt		52. Washer	
15. Bushing		53. Rocker arm	
16. Cap		54. Grease nipple	
17. Circlip		55. Nut	
18. Thrust washer		56. Bolt	
19. Housing		57. Screw stopper	
20. Swing arm		58. Ball joint	
21. Wear plate		59. Nut	
22. Ski leg		60. Upper bracket	
23. Ski boot		61. Nut	
24. Stop bounding		62. Ball joint	
25. Cap		63. Bolt	
26. Nut		64. Lockwasher	
27. Cup		65. Clamp	
28. Nut		66. Rubber clamp	
29. Washer		67. Stabilizer bar	
30. Cotter pin		68. Slider joint	
31. Ski		69. Slider	
32. Wear bar (skag)		70. Rubber damper	
33. Bolt		71. Cup	
34. Bushing		72. Lockwasher	
35. Nut		73. Bolt	
36. Stopper ring		74. Crank rod	
37. Bushing		75. Ball joint	
38. Spacer		76. Circlip	

13

12. Check that the ski pivots up and down with slight resistance. If the ski is tight, remove the ski and check parts.

13. Install a new cotter pin through the pivot bolt. Bend the ends of the cotter pin over to lock it.

Inspection

Refer to **Figure 1** for this procedure.

1. Check the rubber stoppers for wear, cracking or deterioration. Replace if necessary.

2. Check the skis for metal fatigue or damage. Repair or replace the ski(s) as required.

3. Check the wear bars (skags) on the bottom of the skis for severe wear or damage. If necessary, replace the wear bars as follows:

   a. Remove the rubber plugs from the top of the ski (B, **Figure 3**).

   b. Remove the nuts (**Figure 4**) holding the wear bars to the ski.

   c. Remove the cups from inside the ski.

   d. Remove the wear bar.

   e. Install new wear bar by reversing these steps. Tighten the wear bar nuts to the tightening torque in **Table 1**.

FRONT SUSPENSION

The Ski-Doo progressive reaction system (PRS) is used on all models covered in this manual. Refer to **Figure 1** when performing procedures in this section. Because of the number of parts used in the suspension, identify parts during removal as well as storing them in separate boxes.

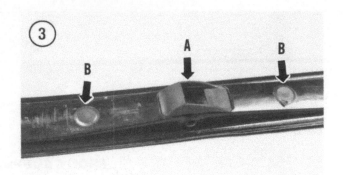

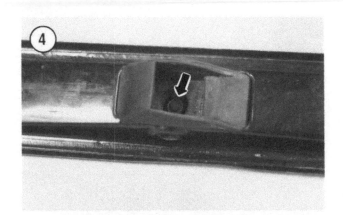

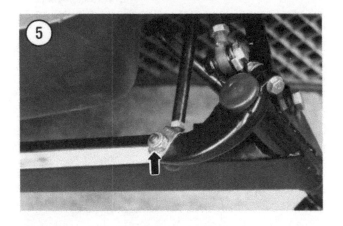

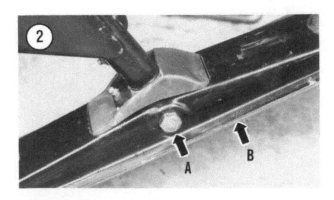

Steering Arm, Ski Leg and Swing Arm Removal/Installation

1. Support the front of the machine so both skis are off the ground.

2. Remove the skis as described in this chapter.

3. Disconnect the tie rod at the steering arm (**Figure 5**).

4. Remove the steering arm and ski leg as follows:

   a. Remove the cap (**Figure 6**).

   b. Remove the circlip (**Figure 7**) from the groove in the ski leg.

   c. Remove steering arm pinch bolt.

   d. Remove the steering arm (**Figure 8**).

   e. Remove the circlip and washer (**Figure 9**).

   f. Remove the ski leg (**Figure 10**).

5. Disconnect the upper control arm at the swing arm (**Figure 11**).

6. Remove the swing arm as follows:

   a. Remove the front swing arm-to-lower control arm bolt (**Figure 12**).

13

b. Remove the rear swing arm to stabilizer bar bolt.

c. Remove the swing arm (**Figure 13**).

7. Perform the *Steering Arm, Ski Leg and Swing Arm Inspection* procedure in this section.

8. Installation is the reverse of these steps. Note the following:

a. Tighten the front and rear swing arm bolts to the torque specification in **Table 1**.

b. Tighten the steering arm pinch bolt to the torque specification in **Table 1**.

c. Tighten the upper control arm bolt at the swing arm to the torque specification in **Table 1**.

Steering Arm, Ski Leg and Swing Arm Inspection

1. Clean all components thoroughly in solvent. Remove all dirt and other residue from all surfaces.

> *NOTE*
> *If paint has been removed from parts during cleaning, touch up areas as required before assembly or installation.*

2. Visually check the steering arm (A, **Figure 14**) for cracks, excessive wear or other damage. Check splines (B, **Figure 14**) for damage. Check mating splines in ski leg (A, **Figure 15**).

3. Check the ski leg (B, **Figure 15**) for cracks or other damage.

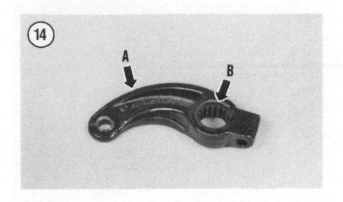

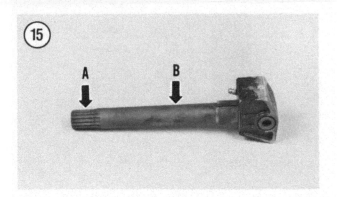

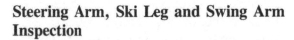

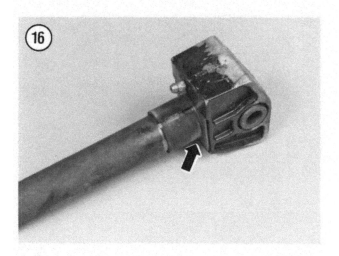

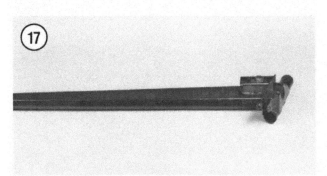

4. Check the ski leg housing and wear plate for damage (**Figure 16**).

5. Check the swing arm (**Figure 17**) for cracks, excessive wear or other damage. Check the housing (**Figure 18**) for cracks or other damage. Check the damper and bushing for wear or damage.

6. Replace worn or damaged parts as required.

Lower Control Arm
Removal/Installation

Refer to **Figure 1** for this procedure.

1. Support the front of the machine so both skis are off the ground.

2. Open the shroud.

3. Remove the lower control arm bolt at the swing arm (**Figure 19**).

4. Remove the clevis cotter pin and remove the clevis pin from the bell crank rod (**Figure 20**).

5. Remove the lower control arm bolt at the frame (**Figure 21**).

6. Remove the lower control arm.

7. Perform the *Lower Control Arm Inspection* as described in the following procedure.

8. Installation is the reverse of these steps. Note the following:

    a. Tighten the lower control arm bolts to the torque specification in **Table 1**.

    b. Install the clevis pin assembly as shown in **Figure 1**. Secure the clevis pin with a new cotter pin.

13

Lower Control Arm
Inspection

1. Clean the lower control arm in solvent and dry thoroughly.

2. If paint has been removed from the lower control arm during cleaning or through use, touch up areas before installation.

> *CAUTION*
> *Do not paint over radial ball-joints in lower control arm.*

3. Check the lower control arm (**Figure 22**) for cracks or other damage.

4. Check the radial ball-joints (**Figure 23**) at both ends of the lower control arm for excessive wear or damage. Replace the radial ball-joint by first removing the circlip and driving the ball-joint out of the lower arm. Install new ball-joint with a press or suitable tool.

5. Check the radial ball-joint bushings for wear or damage.

6. Replace worn or damaged parts as required.

Stabilizer Bar
Removal/Installation

Refer to **Figure 1** for this procedure.

1. Remove the stabilizer bar mounting bolts and brackets and remove the stabilizer bar.

2. Perform the following *Stabilizer Bar Inspection* procedure.

3. Reverse to install. Tighten the stabilizer bar bolts to the torque specification in **Table 1**. Make sure the stabilizer bar has movement in its bushings.

Stabilizer Bar
Inspection

1. Clean the stabilizer bar in solvent and dry thoroughly.

2. If paint has been removed from the stabilizer bar during cleaning or through use, touch up areas before installation.

3. Check the stabilizer bar for wear, cracks or other damage.

4. Check the slider and slider joint for damage.

5. Replace worn or damaged parts as required.

Upper Control Arm
Removal/Installation

Refer to **Figure 1** for this procedure.

1. Open the shroud.

2. Disconnect the upper control arm at the steering arm (**Figure 24**).

3. Disconnect the upper control arm at the frame (**Figure 25**).

4. Remove the upper control arm.

5. Perform the following *Upper Control Arm Inspection* procedure.

6. Installation is the reverse of these steps. Tighten the upper control arm bolts to the torque specification in **Table 1**.

Upper Control Arm
Inspection

1. Clean the upper control arm in solvent and dry thoroughly.

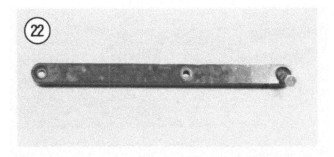

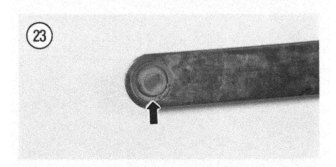

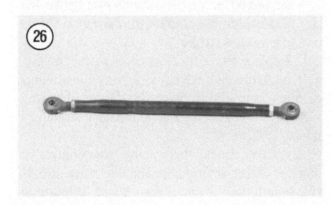

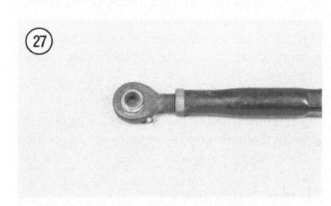

2. If paint has been removed from the upper control arm during cleaning or through use, touch up areas before installation.

3. Check the upper control arm (**Figure 26**) for wear, cracks or other damage.

4. Check the ball-joints (**Figure 27**) for excessive wear or damage. If damaged, loosen the nut and unscrew the ball-joint. Reverse to install. When installing new ball-joints, make sure the exposed thread length between the ball-joint and tie rod does not exceed 15 mm (19/32 in.).

5. Replace worn or damaged parts as required.

6. If one or more ball-joints were loosened or replaced, check steering adjustment as described in this chapter.

Rocker Arm
Removal/Installation

Disconnect all steering components at the rocker arm (**Figure 1**). Remove the rocker arm. Reverse to install. Apply Loctite 272 (red) to the rocker arm bolt during installation and tighten the bolt to the torque specification in **Table 1**.

Rocker Arm
Inspection

1. Clean the rocker arm in solvent and dry thoroughly.

2. If paint has been removed from the rocker arm during cleaning or through use, touch up areas before installation.

3. Check the rocker arm for wear, cracks or other damage.

4. Replace worn or damaged parts as required.

SHOCK ABSORBER

Refer to **Figure 1** for this procedure.

Removal/Installation

1. Open the shroud.

2. Remove the 2 shock absorber mounting bolts (**Figure 28**) and remove the shock absorber.

13

3. Perform the *Shock Inspection* as described in this section.

4. Installation is the reverse of these steps. Tighten the shock absorber bolts to the torque specifications in **Table 1**.

Spring
Removal/Installation

Refer to **Figure 1** for this procedure.

1. Remove the shock absorber as described in this chapter.

2. Secure the bottom of the shock absorber in a vise with soft jaws.

3. Slide the rubber bumper down the shock shaft (**Figure 29**).

> *WARNING*
> *Do not attempt to remove or install the shock spring without the use of a spring compressor. Use the Ski-Doo spring remover (part No. 414 5796 00) or a suitable equivalent. Attempting to remove the spring without the use of a spring compressor may cause severe personal injury. If you do not have access to a spring compressor, refer spring removal to a Ski-Doo dealer.*

4. Install a spring compressor onto the shock absorber following the manufacturer's instructions. Compress the spring with the tool and remove the spring stopper from the top of the shock. Remove the spring.

> *CAUTION*
> *Do not attempt to disassemble the shock damper housing.*

5. Inspect the shock and spring as described in the following procedure.

6. Installation is the reverse of these steps. Make sure the spring stopper secures the spring completely.

Shock Inspection

1. Remove the shock spring as described in this chapter.

2. Clean all components thoroughly in solvent and allow to dry.

3. Check the damper rod as follows:
   a. Check the damper housing for leakage. If the damper is leaking, replace it.
   b. Check the damper rod for bending or damage. Check the damper housing for dents or other damage.
   c. Operate the damper rod by hand and check its operation. If the damper is operating correctly, a small amount of resistance should be felt on the compression stroke and a considerable amount of resistance felt on the return stroke.
   d. Replace the damper assembly if necessary.

4. Check the damper housing for cracks, dents or other damage. Replace the damper housing if damaged. Do not attempt to repair or straighten it.

5. Visually check the spring for cracks or damage. If the spring appears okay, measure its free length with a tape measure and compare to the specification in **Table 2**. Replace the spring if it has sagged significantly.

6. Check the spring stopper for cracks, deep scoring or excessive wear. Replace if necessary.

STEERING ASSEMBLY

This section describes service to the handlebar, steering column and tie rods.

Handlebar
Removal/Installation

Refer to **Figure 30** for this procedure.

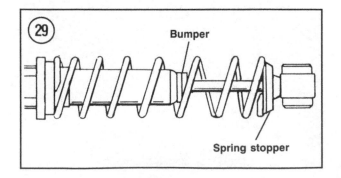

29 Bumper

Spring stopper

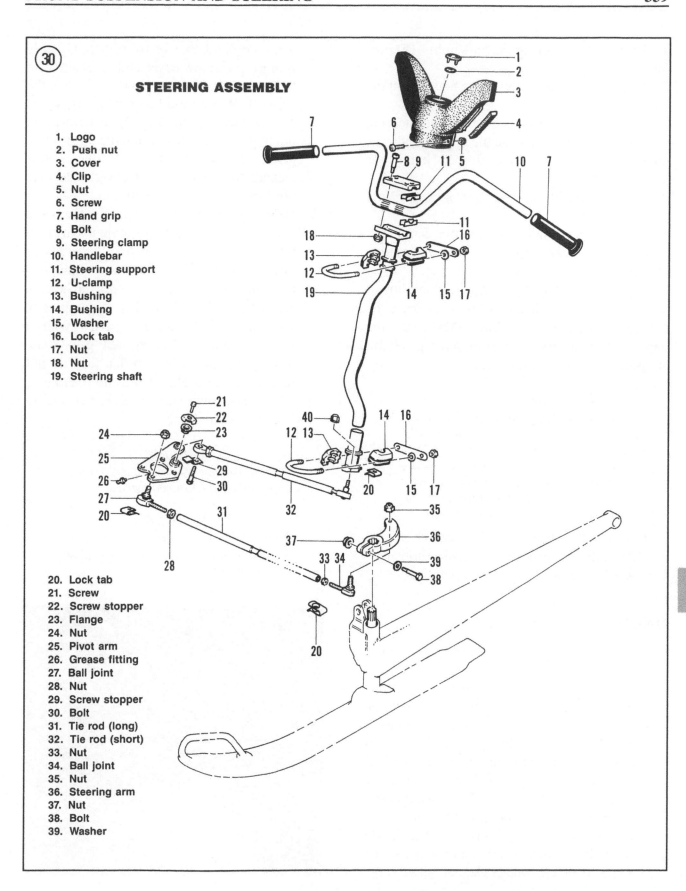

STEERING ASSEMBLY

1. Logo
2. Push nut
3. Cover
4. Clip
5. Nut
6. Screw
7. Hand grip
8. Bolt
9. Steering clamp
10. Handlebar
11. Steering support
12. U-clamp
13. Bushing
14. Bushing
15. Washer
16. Lock tab
17. Nut
18. Nut
19. Steering shaft

20. Lock tab
21. Screw
22. Screw stopper
23. Flange
24. Nut
25. Pivot arm
26. Grease fitting
27. Ball joint
28. Nut
29. Screw stopper
30. Bolt
31. Tie rod (long)
32. Tie rod (short)
33. Nut
34. Ball joint
35. Nut
36. Steering arm
37. Nut
38. Bolt
39. Washer

13

1. Remove the steering pad assembly (**Figure 30**).

2. Remove the throttle and brake housings at the handlebar.

3. Remove the steering clamp bolts and steering clamp (**Figure 31**). Remove the upper handlebar supports, handlebar and lower handlebar supports.

4. Installation is the reverse of these steps. Note the following.

5. Install lower handlebar supports, handlebar and upper handlebar supports. Install the steering clamp and the 4 mounting bolts.

6. Position the handlebar and tighten the 4 steering clamp bolts in a crisscross pattern to the torque specification in **Table 1**. Make sure the gap between the steering clamp steering column support is equal on all 4 sides.

> *WARNING*
> *If the handlebar is positioned too high, the brake lever may contact the windshield during turning. Check handlebar position before riding snowmobile and adjust so that you have total handlebar movement from side-to-side without brake lever contact.*

7. Reinstall the steering pad assembly.

8. Reinstall the throttle and brake housings. Check all controls for proper operation.

Handlebar Grips

Note the following when replacing handlebar grips.

a. *Grips without heating element:* These grips can be removed and installed easily by blowing compressed air into the opposite handlebar end while covering the hole in opposite grip. To install grips with air, align the grip with the handlebar and direct air from the opposite side through the handlebar. If you do not have access to compressed air, insert a thin-tipped screwdriver underneath the grip and squirt some electrical contact cleaner between the grip and handlebar or twist grip. Quickly remove the screwdriver and twist the grip to break its hold on the handlebar; slide the grip off. When installing new grips, squirt contact cleaner into the grip as before and quickly twist it onto the handlebar or twist grip. Allow plenty of time for the contact cleaner to evaporate and the grip to take hold before riding snowmobile.

b. *Grips with heating element:* To remove a grip, locate the heating element wires on the grip. Cut the grip on the opposite side away from the wires. Slowly peel the grip back and locate the gap between the heating element. Cut along this gap to completely remove grip. Install new grips with a hammer. When installing new grip, route heating element wires so that they do not interfere with brake or throttle operation.

c. *All models:* Make sure grips are *tight* before riding snowmobile.

> *WARNING*
> *Do not use any type of grease, soap or other lubricant to install grips.*

> *WARNING*
> *Loose grips can cause you to crash. Always replace damaged or loose grips before riding.*

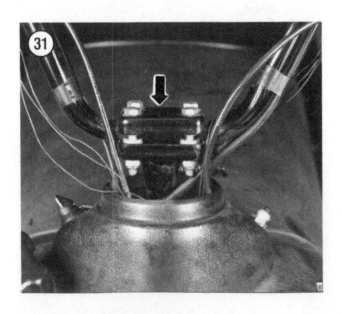

Steering Column
Removal/Installation

Refer to **Figure 30** for this procedure

NOTE
The following procedure is shown with the engine removed for clarity.

1. Remove the handlebar as described in this chapter.
2. Disconnect the tie rod at the steering column (**Figure 32**).

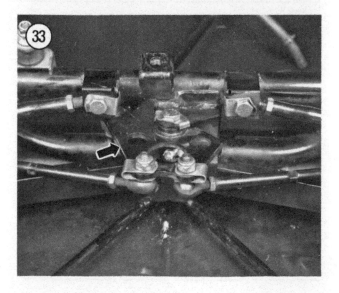

3. Remove the upper and lower U-clamp and remove the steering column. Store the upper and lower steering bushings separately so that they can be reinstalled in their original position.
4. Inspect the steering column as described under *Steering Column Inspection* in this chapter.
5. Installation is the reverse of these steps. Note the following.
6. Apply a low temperature grease to the inner bearing halves before assembly.
7. If reinstalling the original steering bushings, install them in their original positions.
8. Assemble the steering bushings and U-clamps as shown in **Figure 30**. Tighten the U-clamps to the torque specifications in **Table 1**.
9. After tightening the bearing half mounting bolts, bend the lockwasher tabs over the bolts.
10. Check ski alignment as described in this chapter.

Steering Column
Inspection

Refer to **Figure 30** for this procedure.
1. Clean all components thoroughly in solvent. Remove all grease residue from the bearing halves.
2. Visually check the steering column surfaces for cracks or deep scoring. Also check the welds at the top and bottom of the shaft for cracks or breakage.
3. Check the steering bushings for cracks, deep scoring or excessive wear.

Tie Rod
Removal/Installation

Refer to **Figure 30** for this procedure.
1. Disconnect the tie rod at the steering column (**Figure 32**).
2. Disconnect the tie rod at the pivot arm (**Figure 33**).
3. Remove the tie rod.

13

4. Inspect the tie rod as described in the following procedure.

5. Installation is the reverse of these steps. Note the following.

6. After installing the tie rods, check ski alignment as described in this chapter. When ski alignment is correct, tighten tie rod bolts to the torque specification in **Table 1**. Secure bolts with lockwasher tab.

Tie Rod
Inspection

1. Clean the tie rod in solvent and dry thoroughly.

2. If paint has been removed from the tie rod during cleaning or through use, touch up areas before installation.

3. Check the tie rod for wear, cracks or other damage.

4. Check the ball-joints (**Figure 34**) for excessive wear or damage. If damaged, loosen the nut and unscrew the ball-joint. Reverse to install. When installing new ball-joints, make sure the exposed thread length between the ball-joint and tie rod does not exceed 15 mm (19/32 in.).

5. Replace worn or damaged parts as required.

6. If one or more ball-joints were loosened or replaced, check steering adjustment as described in this chapter.

STEERING ADJUSTMENT

Steering adjustment is a four-part procedure. Perform the following adjustments in order:
   a. Pivot arm center adjustment.
   b. Camber angle adjustment.
   c. Handlebar alignment.
   d. Toe-out adjustment.

Incorrect ski alignment can cause difficult steering.

Pivot Arm Center Adjustment

1. Park the snowmobile on a level surface.

2. Open the shroud.

3. Center handlebars so that they face straight ahead.

4. See **Figure 35**. Measure the distance from the center of the tie rod ball joints to the centerline shown in **Figure 35**. Repeat for both tie rod ball-joints. The distance should be the same for both tie rods. If not, perform the following:
   a. Turn the handlebar to the left-hand side.
   b. Loosen the tie rod jam nut at the pivot arm (**Figure 33**). Remove the tie rod mounting bolt and pivot the tie rod away from the pivot arm.
   c. Turn the ball-joint (**Figure 34**) as required and reinstall the tie rod. Continue adjustment until the center distance of both tie rods is equal.

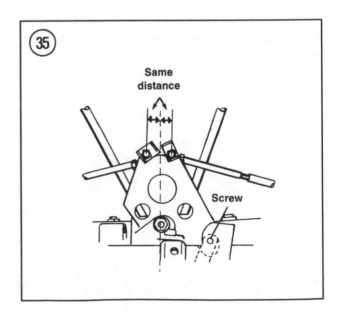

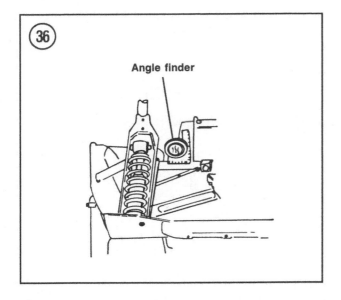

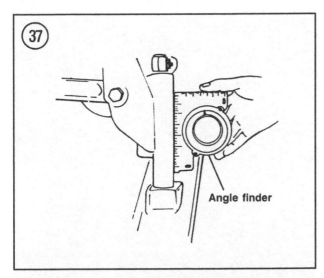

NOTE
When adjusting ball-joint, make sure the exposed thread length between the ball-joint and tie rod does not exceed 15 mm (19/32 in.).

d. When adjustment is correct, tighten tie rod mounting bolt to the torque specification in **Table 1**. Bend lockwasher tab over bolt to lock it.

Camber Angle Adjustment

1. Perform pivot arm center adjustment as described in this chapter.
2. Raise and support the snowmobile so that the skis are approximately 25 mm (1 in.) off the ground.

NOTE
Camber angle can only be measured when the front suspension is fully extended. Do not measure angle with skis on ground.

3. Place an angle finder (available at well equipped hardware or building supply stores) on the main horizontal frame member as shown in **Figure 36** to check vehicle level. If necessary, block machine so that both sides are level.

NOTE
When using the angle finder in Step 4, do not place the angle finder on a decal or weld. This will give a false reading and result in an incorrect camber adjustment.

4. Place the angle finder on one of the swing arms as shown in **Figure 37**. The camber reading should be 0° ±0.5°. If the adjustment is incorrect, loosen the upper control arm inner and outer (**Figure 38**) locknuts. Then, turn the arm to adjust the camber. Tighten the locknuts securely and recheck the adjustment.
5. Repeat for the opposite side.
6. Lower skis to ground.

13

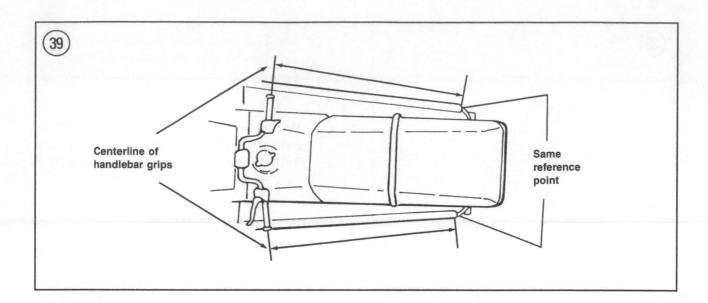

Centerline of
handlebar grips

Same
reference
point

Handlebar Alignment

1. Place the snowmobile on a level surface.
2. Turn handlebar until both skis face forward.
3. Check handlebar alignment as follows:
   a. Measure from the end of one handlebar grip to a point at the back of the snowmobile (**Figure 39**). Record the length.
   b. Repeat measurement for opposite side (**Figure 39**). Be sure to use the same reference points. Record the length.
   c. The distance recorded in sub-step b and sub-step c must be approximately the same. If adjustment is required, proceed to Step 4.
4. Loosen the 2 locknuts on each tie rod (**Figure 40**). Determine from Step 3 which direction the skis have to be turned to correctly align handlebar. Turn 1 tie rod to shorten its length and turn the opposite tie rod to lengthen it. Make sure to turn both tie rods the *same* amount so that you do not change the toe out adjustment. Tighten locknuts and recheck adjustment.

> *NOTE*
> *When adjusting ball-joint, make sure the exposed thread length between the ball-joint and tie rod does not exceed 15 mm (19/32 in.).*

5. Perform the following *Toe-Out Adjustment* procedure.

Toe-Out Adjustment

1. Check handlebar alignment as described in previous section.
2. Park snowmobile on a level surface.
3. Turn handlebar so that skis face forward. Grasp skis and turn inward to remove all slack from steering assembly. If skis want to spring outward, pull them together with a tie down strap or rope.
4. Working from the inner edge of each ski, measure the distance between the skis at the front and back as shown in **Figure 41**. When

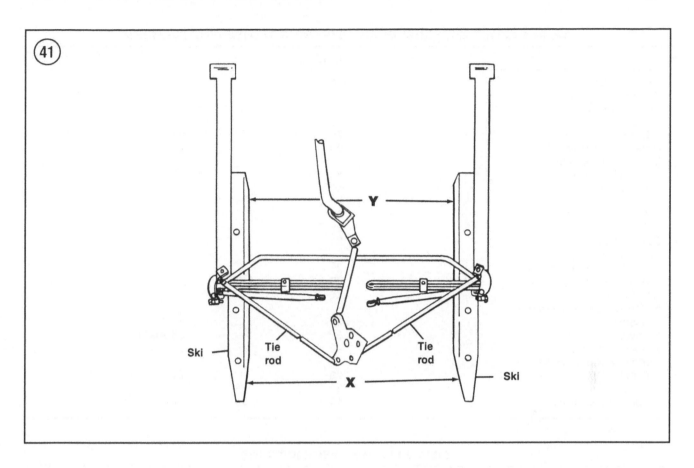

measuring skis, try to measure as far forward and backward on the skis as possible. Pick a reference point or make chalk marks so that the same measuring points are used for all measurements. Interpret results as follows:

a. Subtract the front measurement from the rear measurement.

b. For the toe-out to be correct, the front measurement must be 3 mm (1/8 in.) larger than the rear measurement.

c. If measurement is incorrect, loosen tie rod locknuts (**Figure 40**) and rotate tie rods to increase or reduce toe out as needed. Tighten locknuts and recheck toe out.

NOTE
When adjusting ball-joint, make sure the exposed thread length between the ball-joint and tie rod does not exceed 15 mm (19/32 in.).

Tables are on the following page.

Table 1 FRONT SUSPENSION AND STEERING TIGHTENING TORQUES

	N·m	ft.-lb.
Ski pivot bolt	40	30
Ski wear bar nuts	18	13
Swing arm bolts		
Front	29	21
Rear	25	18
Upper control arm bolts	29	21
Lower control arm		
At swing arm	85	63
At frame	52	38
Tie rod		
At steering arm	29	21
At pivot arm	29	21
Steering arm pinch bolt	25	18
Tie rod (center)		
At steering column	29	21
At pivot arm	48	35
Handlebar	26	19
Steering column U-clamps		
Upper	28	19
Lower	9.5	85 in.-lb.
Rocker arm bolt	48	35
Shock absorber bolts		
Front	29	21
Rear	35	26
Stabilizer bar bolts	15	11

Table 2 SPRING SPECIFICATIONS

Free length*	
1985-1986	215.9 mm (8.50 in.)
1987-1988	290.1 mm (11.42 in.)
1989	241.3 mm (9.50 in.)
Color code	
1985-1986	White/white
1987-1988	Blue/yellow
1989	Green/green
* ±3 mm (±0.12 in.)	

Chapter Fourteen

Track and Rear Suspension

All models are equipped with a slide-rail suspension. Suspension adjustment is found in Chapter Four.

Track specifications are found in **Table 1**. **Tables 1-4** are at the end of the chapter.

REAR SUSPENSION

Removal

1. Loosen the locknuts on the track adjusting bolts (**Figure 1**) and back the bolts out to relieve track tension.

> *NOTE*
> *If you do not have access to a suitable snowmobile jack, you will have to turn the snowmobile on its side when performing the following steps. If the snowmobile is going to be turned on its side during rear suspension removal, plug the oil injection reservoir cap and the chaincase filler cap vent holes to prevent leakage. In addition, place a*

large piece of cardboard next to the snowmobile before turning it on its side.

2A. *MX, MX (H/A), MX LT, Plus and Plus LT models:* Jack up the snowmobile so that the track is clear of the ground. Remove the bolts in the order shown in **Figure 2**. Repeat for the opposite side.

2B. *Mach I:* Jack up the snowmobile so that the track is clear of the ground. Remove the bolts in the order shown in **Figure 3**. Repeat for the opposite side.

3. Slide the track/suspension assembly away from the tunnel and lift the suspension assembly (**Figure 4**) out of the track and remove it.

Inspection

1. Inspect the suspension mounting bolts for thread damage or breakage. Replace damaged bolts as required.

2. Clean all bolts thoroughly in solvent and remove all Loctite residue.

14

3. Visually inspect the long bolts for any signs of wear, cracks, breakage or other abnormal conditions.

4. Check the long bolts for bending.

5. If there is any doubt as to the condition of a bolt, replace it.

6. Lightly grease the long bolt with a low temperature grease before installation.

Installation
(MX, MX [H/A], MX LT, Plus and Plus LT)

NOTE
If you do not have access to a suitable snowmobile jack, you will have to turn the snowmobile on its side when performing the following steps. If the snowmobile is going to be turned on its side during rear suspension installation, plug the oil injection reservoir cap and the chaincase filler cap vent holes to prevent leakage. In addition, place a large piece of cardboard next to the snowmobile before turning it on its side.

1. Pull the track away from the tunnel and open it.

2. Install the rear suspension into the track, starting with the front of the suspension and working toward the rear.

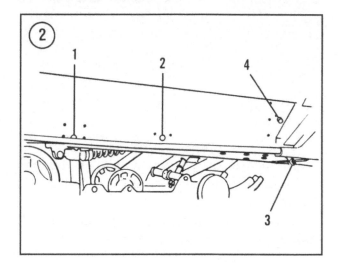

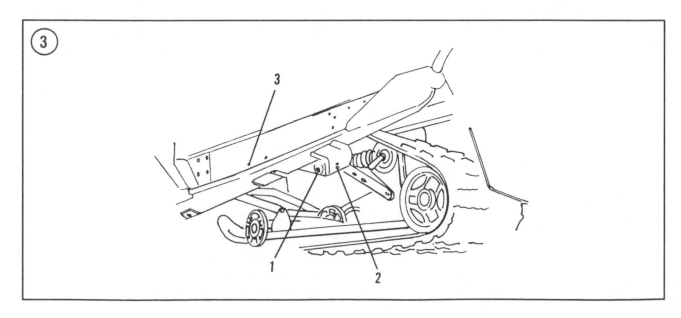

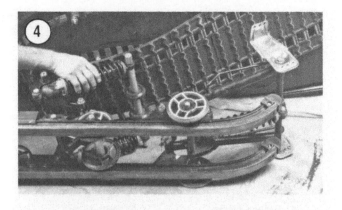

3. Install the rear suspension mounting bolts in the following order:

> *NOTE*
> *When installing the rear suspension mounting bolts in the following steps, do not tighten any one bolt until all of the bolts are installed.*

a. Raise the front arm and align the retainer plate bolt holes with the frame. Install the bolts and washers finger-tight.

b. Install a spacer on each shock pivot (**Figure 5**). Then, install the shock pivot bolt and washer.

c. Align the rear arm with the frame (**Figure 6**) and install the rear arm bolts and washers (**Figure 7**).

4. After all bolts have been installed, reverse the bolt sequence in **Figure 2** and tighten the bolts to the torque specifications in **Table 2**.

5. Lower the track to the ground.

6. If the snowmobile was placed on its side, bleed the oil pump as described in Chapter Eight.

Installation
(Mach I)

> *NOTE*
> *If you do not have access to a suitable snowmobile jack, you will have to turn the snowmobile on its side when performing the following steps. If the snowmobile is going to be turned on its side during rear suspension installation, plug the oil injection reservoir cap and the chaincase filler cap vent holes to prevent leakage. In addition, place a large piece of cardboard next to the snowmobile before turning it on its side.*

1. Pull the track away from the tunnel and open it.

2. Install the rear suspension into the track, starting with the front of the suspension and working toward the rear.

14

3. Install the rear suspension mounting bolts in the following order:

NOTE
When installing the rear suspension mounting bolts in the following steps, do not tighten any one bolt until all of the bolts are installed.

a. Raise the front arm and align the retainer plate bolt holes with the frame. Install the bolts and washers finger-tight.

b. Install a spacer on each shock pivot. Then, install the shock pivot bolt and washer.

c. Align the rear arm with the frame and install the rear arm bolts and washers.

d. Install the rear shock pivot bolt and washer.

4. After all bolts have been installed, reverse the bolt sequence in **Figure 3** and tighten the bolts to the torque specifications in **Table 3**.

5. Lower the track to the ground.

6. If the snowmobile was placed on its side, bleed the oil pump as described in Chapter Eight.

SLIDE SHOES

Inspection/Replacement

The slide shoes are mounted at the bottom of the runner and held in position with a single screw. The slide shoes should be replaced when worn or if they show signs of abnormal wear or damage.

Refer to **Figure 8** (1985), **Figure 9** (1986-on MX, MX [H/A], MX LT, Plus and Plus LT) or **Figure 10** (Mach I).

1. Remove the rear suspension as described in this chapter.

2. Visually inspect the slide shoes for cracks or other damage. See A, **Figure 11**. If a crack is detected, replace the slide shoes as described in this procedure.

NOTE
The slide shoes must be replaced as a set.

REAR SUSPENSION (1985)

1. Bolt	50. Bushing
2. Washer	51. Stopper
3. Nut	52. Nut
4. Runner	53. Retainer pin
5. Slider shoe	54. Cotter pin
6. Bolt	55. Center axle
7. Lockwasher	56. Washer
8. Grease fitting	57. Wheel cap
9. Wheel cap	58. Spacer
10. Washer	59. Rear shackle
11. Idler wheel	60. Grease fitting
12. Bolt	61. Rear shackle lower axle
13. Bearing	62. Rear arm
14. Circlip	63. Grease fitting
15. Housing	64. Rear shackle upper axle
16. Front axle	65. Cup
17. Front shackle	66. Screw
18. Grease fitting	67. Cross pivot
19. Front swing arm	68. Flat washer
20. Grease fitting	69. Lockwasher
21. Bolt	70. Screw
22. Spacer	71. Axle
23. Nut	72. Flat washer
24. Bolt	73. Bearing
25. Front swing arm axle	74. Spacer
26. Nut	75. Lockwasher
27. Front arm	76. Screw
28. Grease fitting	77. Shock absorber
29. Front arm upper axle	78. Spring stopper
30. Retainer plate	79. Spring
31. Screw	80. Spring seat
32. Screw	81. Cam
33. Lockwasher	82. Screw
34. Flat washer	83. Spacer
35. Front arm lower axle	84. Rear shock pivot
36. Flat washer	85. Inner spacer
37. Nut	86. Rear axle
38. Cam	87. Outer spacer
39. Spring stopper	88. Idler wheel
40. Spring	89. Rubber stopper
41. Spring seat	90. Rivet
42. Shock absorber	91. Push nut
43. Screw	92. Tensioner stopper
44. Spacer	93. Nut
45. Cushion	94. Screw
46. Nut	95. Nut
47. Stopper	96. Washer
48. Flat washer	97. Washer
49. Spring	

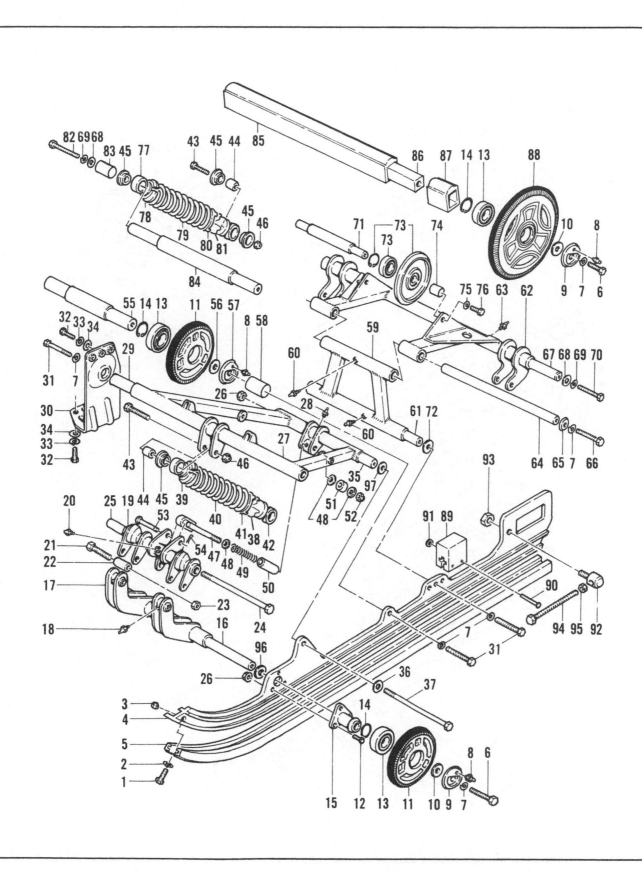

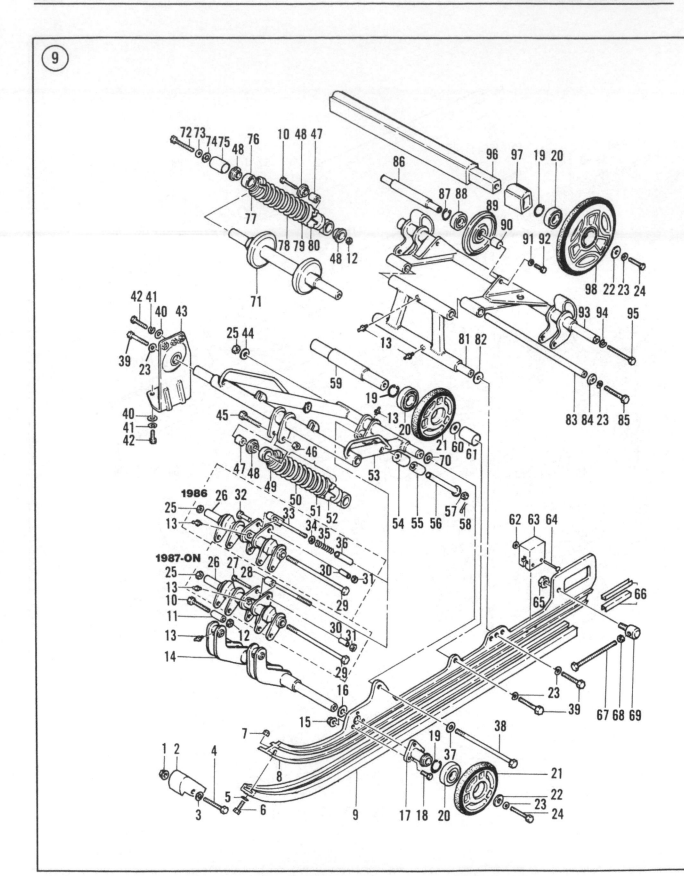

REAR SUSPENSION
(1986-ON MX, MX [H/A],
MX LT, PLUS AND PLUS LT)

1. Nut
2. Protector
3. Washer
4. Bolt
5. Washer
6. Bolt
7. Nut
8. Runner
9. Slider shoe
10. Bolt
11. Spacer
12. Nut
13. Grease nipple
14. Front shackle
15. Nut
16. Washer
17. Housing
18. Bolt
19. Circlip
20. Bearing
21. Idler wheel
22. Washer
23. Lockwasher
24. Bolt
25. Nut
26. Front swing arm
27. Bolt
28. Stopper
29. Axle
30. Spacer
31. Nut
32. Retainer pin
33. Stopper
34. Washer
35. Spring
36. Bushing
37. Washer
38. Bolt
39. Bolt
40. Washer
41. Lockwasher
42. Bolt
43. Shackle
44. Washer
45. Bolt
46. Nut
47. Spacer
48. Cushion
49. Spring seat
50. Spring
51. Spring seat
52. Adjuster cam
53. Front arm
54. Cup
55. Stopper
56. Pin
57. Nut
58. Cotter pin
59. Center axle
60. Washer
61. Spacer
62. Push nut
63. Rubber stopper
64. Pin
65. Nut
66. Cover
67. Bolt
68. Nut
69. Tensioner stopper
70. Washer
71. Rear shock pivot assembly
72. Bolt
73. Washer
74. Washer
75. Spacer
76. Damper assembly
77. Spring seat
78. Spring
79. Spring seat
80. Adjuster cam
81. Rear shackle
82. Washer
83. Axle
84. Cup
85. Bolt
86. Axle
87. Circlip
88. Bearing
89. Idler wheel
90. Spacer
91. Washer
92. Bolt
93. Cross pivot
94. Lockwasher
95. Bolt
96. Inner spacer and rear axle
97. Outer spacer
98. Idler wheel

14

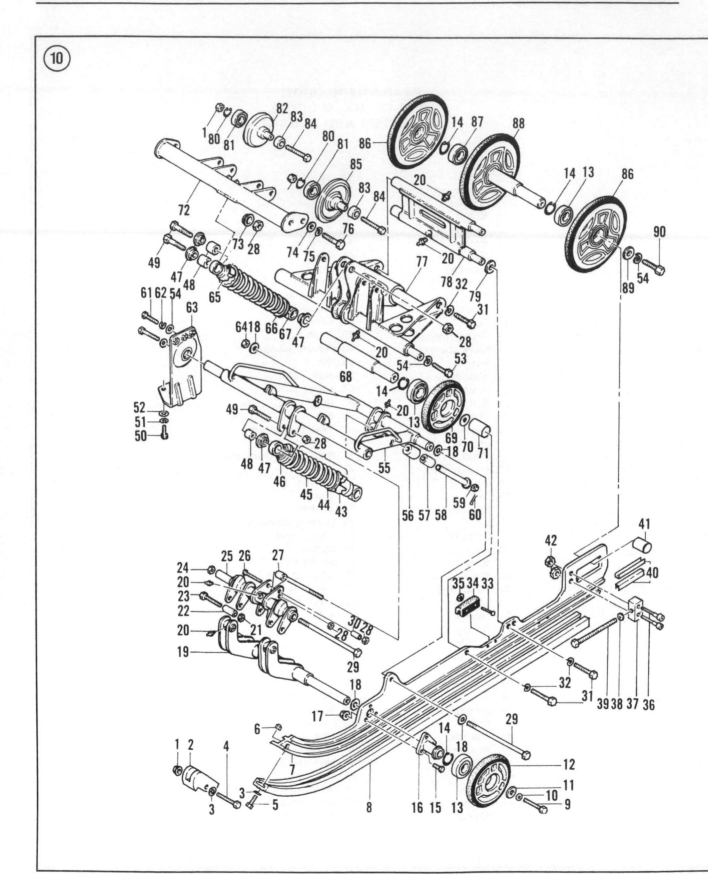

REAR SUSPENSION
(1989 MACH I)

1. Nut	46. Spring seat
2. Protector	47. Cushion
3. Washer	48. Spacer
4. Bolt	49. Bolt
5. Bolt	50. Bolt
6. Nut	51. Lockwasher
7. Runner	52. Washer
8. Slider shoe	53. Bolt
9. Bolt	54. Washer
10. Lockwasher	55. Front arm
11. Washer	56. Cup
12. Idler wheel	57. Stopper
13. Bearing	58. Pin
14. Circlip	59. Nut
15. Bolt	60. Cotter pin
16. Housing	61. Bolt
17. Nut	62. Lockwasher
18. Washer	63. Shackle
19. Front shackle	64. Nut
20. Grease fitting	65. Adjuster/spring seat
21. Nut	66. Spring/spring seat
22. Bushing	67. Damper
23. Bolt	68. Center axle
24. Nut	69. Idler wheel
25. Front swing arm and axle	70. Washer
26. Bolt	71. Spacer
27. Stopper	72. Shock pivot
28. Nut	73. Cushion
29. Bolt	74. Washer
30. Spacer	75. Lockwasher
31. Bolt	76. Bolt
32. Lockwasher	77. Rear arm and cross pivot
33. Pin	78. Rear shackle
34. Rubber stopper	79. Washer
35. Push nut	80. Circlip
36. Bolt	81. Bearing
37. Stopper	82. Wheel support
38. Nut	83. Spacer
39. Bolt	84. Bolt
40. Protector	85. Idler wheel and axle
41. Outer spacer	86. Idler wheel
42. Nut	87. Bearing
43. Cam	88. Idler wheel
44. Spring seat/damper body	89. Washer
45. Spring	90. Bolt

14

3. Remove the slide shoes as follows:
   a. Turn the rear suspension over and rest it upside down on the workbench.
   b. Remove the screws at the front of each shoe.
   c. Working at the front of the slide shoe, drive the shoe to the rear of the runner. If necessary, use a block of wood and hammer (**Figure 12**).

> *CAUTION*
> *Do not use a steel punch to remove the slide shoes or the runner may become damaged.*

4. Inspect the runner as follows:
   a. Clean the runner with solvent and dry thoroughly.
   b. Place a straightedge alongside the runner and check for bends. If a runner is bent, it must be straightened. If the bend is severe or if the runner is cracked or damaged, replace it.
   c. Check the runner for gouges or cracks along the runner path. Smooth rough surfaces with a file or sandpaper.
5. Install the slide shoes as follows:
   a. Lightly grease the runner mating surface to ease installation.
   b. Working from the back of the runner, align the slide shoe with the runner so that the hole in the slide shoe faces to the front. Drive the slide shoe onto the runner with the same wood block and hammer used during removal.
   c. Continue to drive the slide shoe onto the runner until the screw hole in the slide shoe aligns with the threads in the runner. Install the screw and lockwasher and tighten securely.

RUNNER PROTECTOR

Inspect the runner protector (B, **Figure 11**) for wear, cracks or other damage. Replace the protector by removing the nut, bolt and washer. Reverse to install.

REAR SHOCK ABSORBERS

Front Shock
Removal/Installation

Refer to **Figure 8**, **Figure 9** or **Figure 10** for this procedure.
1. Remove the rear suspension as described in this chapter.

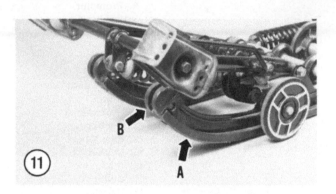

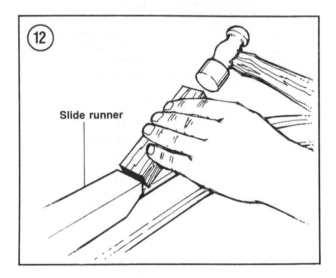

Slide runner

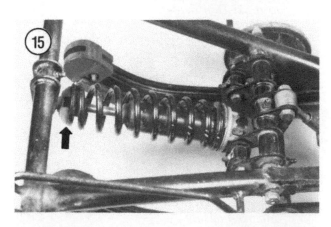

2. Loosen limiter screw (**Figure 13**) as follows:
   a. Remove the cotter pin at the bottom of the limiter screw.

> *NOTE*
> *Because the limiter screw controls vehicle weight during acceleration, note the number of threads visible from the base of the limiter screw to the end of the threads. Record this number so that you can reinstall the limiter screw to the same setting. Adjustment of the limiter screw is covered in Chapter Four.*

   b. Loosen the limiter screw nut (**Figure 14**) until free play is obtained in screw.

3. Remove the shock bolts and nuts and remove the front shock (**Figure 15**).

4. Remove the shock bushings and inspect them for wear or damage. Replace the bushings if necessary.

5. Inspect the shock absorbers as described in this chapter.

6. Installation is the reverse of these steps. Note the following.

7A. *MX and Plus:* Tighten the shock bolts to the torque specification in **Table 2**.

7B. *Mach I:* Tighten the shock bolts to the torque specification in **Table 3**.

8. Tighten the limiter screw nut (**Figure 14**) to secure the limiter screw at the original setting recorded during disassembly. Install a new cotter pin through the end of the limiter screw to prevent the nut from backing all the way off. Bend the cotter pin out to lock it.

**Rear Shock Absorbers
Removal/Installation (MX, MX [H/A],
MX LT, Plus and Plus LT)**

Refer to **Figure 8** or **Figure 9** for this procedure.

1. Remove the rear suspension as described in this chapter.

2. Remove the spacer and cushion from the shock absorber (**Figure 16**).

3. Remove the rear shock bolt and washers (A, **Figure 17**) and remove the shock (B, **Figure 17**).

14

4. Repeat for the opposite shock.

5. Inspect the shock absorbers as described in this chapter.

6. Installation is the reverse of these steps. Note the following:

7. Tighten the shock bolt to the torque specification in **Table 2**.

Rear Shock Absorbers
Removal/Installation
(Mach I)

Refer to **Figure 10** for this procedure.

1. Remove the rear suspension as described in this chapter.

2. Remove the bolts securing the shock absorber in place and remove the shock absorber.

3. Repeat for the opposite shock.

4. Inspect the shock absorbers as described in this chapter.

5. Tighten the shock bolts to the torque specification in **Table 3**.

Spring
Removal/Installation

1. Remove the shock absorber as described in this chapter.

2. Secure the bottom of the shock absorber in a vise with soft jaws.

3. Slide the rubber bumper down the shock shaft (**Figure 18**).

4. Turn the spring adjuster to its softest position.

WARNING
Do not attempt to remove or install the shock spring without the use of a spring compressor. Use the Ski-Doo spring remover (part No. 414 5796 00) or a suitable equivalent. Attempting to remove the spring without the use of a spring compressor may cause severe personal injury. If you do not have access to a spring compressor, refer spring removal to a Ski-Doo dealer.

5. Install a spring compressor onto the shock absorber following the manufacturer's instructions. Compress the spring with the tool and remove the spring stopper from the top of the shock. Remove the spring.

6. Remove the thrust washers installed underneath the spring.

CAUTION
Do not attempt to disassemble the shock damper housing.

7. Inspect the shock and spring as described in the following procedure.

8. Installation is the reverse of these steps. Note the following:

  a. Lightly grease the thrust washers with a low temperature grease. Then slide the thrust washers over the shock body until they seat against the spring adjuster.

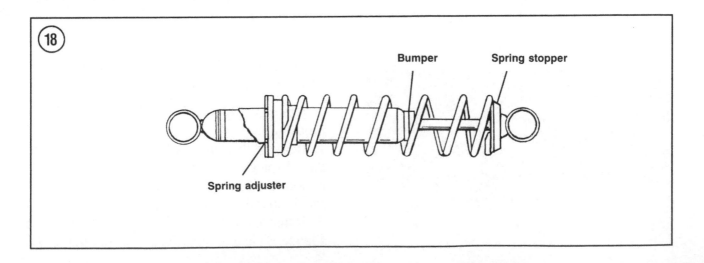

18

Bumper   Spring stopper

Spring adjuster

b. When installing the spring, make sure the spring stopper secures the spring completely.

c. Reposition the spring adjuster (**Figure 18**).

Shock Inspection

1. Remove the shock spring as described in this chapter.

2. Clean all components thoroughly in solvent and allow to dry.

3. Check the damper rod as follows:

   a. Check the damper housing for leakage. If the damper is leaking, replace it.

   b. Check the damper rod for bending or damage. Check the damper housing for dents or other damage.

   c. Operate the damper rod by hand and check its operation. If the damper is operating correctly, a small amount of resistance should be felt on the compression stroke and a considerable amount of resistance felt on the return stroke.

   d. Replace the damper assembly if necessary.

4. Check the damper housing for cracks, dents or other damage. Replace the damper housing if damaged. Do not attempt to repair or straighten it.

5. Visually check the spring for cracks or damage. If the spring appears okay, measure its free length with a tape measure and compare to the specification in **Table 4**. Replace the spring if it has sagged significantly.

6. Check the spring stopper for cracks, deep scoring or excessive wear. Replace if necessary.

FRONT SHACKLE, FRONT SWING ARM AND FRONT ARM

Removal/Installation

Refer to **Figure 8, Figure 9** or **Figure 10** for this procedure.

1. Remove the rear swing arm as described in this chapter.

2. Loosen limiter screw (**Figure 13**) as follows:

   a. Remove the cotter pin at the bottom of the limiter screw.

> *NOTE*
> *Because the limiter screw controls vehicle weight during acceleration, note the number of threads visible from the base of the limiter screw to the end of the threads. Record this number so that you can reinstall the limiter screw to the same setting. Adjustment of the limiter screw is covered in Chapter Four.*

   b. Loosen the limiter screw nut (**Figure 14**) until free play is obtained in screw.

3. Remove the nuts and bolts connecting the front shackle to the front swing arm (**Figure 19**).

4. Remove the shock absorber as described in this chapter.

5. Loosen the front swing arm pivot shaft nut. Remove the nut and slide the pivot shaft out of the front swing arm.

6. Remove the front swing arm pivot shaft nut. Remove the nut and slide the pivot shaft out of the front swing arm.

7. If necessary, remove the retainer plates at the front swing arm (**Figure 20**).

8. Lift the sub-assemblies out of the rear suspension unit.

14

9. Installation is the reverse of these steps. Note the following.

10. When installing the retainer plates, turn the rear suspension assembly over. Position the retainer plates so that they face in the same angle as shown in **Figure 21**.

11. Refer to **Table 2** or **Table 3** for critical tightening torques. If a fastener is not identified in **Table 2** or **Table 3**, tighten it securely.

12. Tighten the limiter screw nut (**Figure 14**) to secure the limiter screw at the original setting recorded during disassembly. Install a new cotter pin through the end of the limiter screw to prevent the nut from backing all the way off. Bend the cotter pin out to lock it.

Inspection

1. Clean all parts in solvent and dry thoroughly.

2. Visually inspect all welded joints for cracks or bends.

3. Check all pivot bolt surfaces for cracks, deep scoring or excessive wear.

4. If there is any doubt as to the condition of any part, repair or replace it as required.

5. Apply a low temperature grease to all pivot shafts and bushings.

REAR SHACKLE
AND REAR ARM

Removal/Installation
(MX, MX [H/A], MX LT,
Plus and Plus LT)

Refer to **Figure 8** or **Figure 9** when performing this procedure.

1. Remove the rear suspension as described in this chapter.

2. Remove the rear shock absorbers (A, **Figure 22**) as described in this chapter.

3. Remove the rear shackle upper axle bolts and washers (B, **Figure 22**). Withdraw the upper axle.

4. Remove the rear shackle bolts at the runner. Note the bolt hole position (**Figure 23**) so that

the rear shackle can be reinstalled to the same position.

5. Remove the rear arm bolts and washers.

6. Remove the rear shackle and rear arm sub-assemblies.

7. Installation is the reverse of these steps. Note the following:

   a. Install the rear shackle so that the grease fittings face forward. See **Figure 24**.

   b. Make sure to install the washers on the inside of the rear shackle as shown in **Figure 8** or **Figure 9**.

   c. Install the rear shackle in the same adjustment hole as recorded during removal (**Figure 23**).

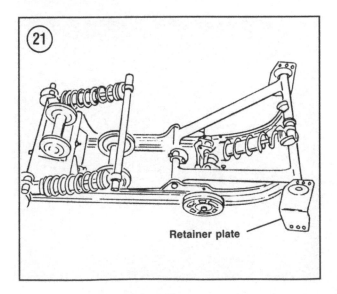

Retainer plate

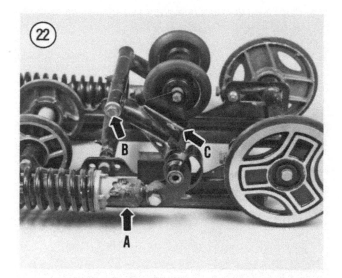

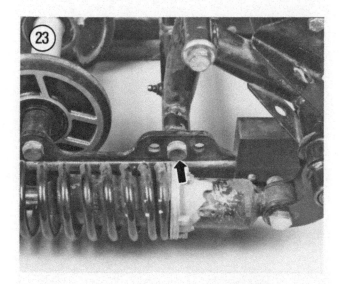

d. Install the rear arm so that the angle brackets face in the direction shown in C, **Figure 22**. Refer to **Figure 8** or **Figure 9** also.

e. Tighten the rear arm bolts to the torque specification in **Table 2**.

f. Apply Loctite 242 (blue) to the rear shackle upper axle bolts and tighten the bolts securely.

Removal/Installation (Mach I)

Refer to **Figure 10** when performing this procedure.

1. Remove the rear suspension as described in this chapter.

2. Remove the rear shock absorbers as described in this chapter.

3. Disconnect the shock pivot at the rear arm.

4. Remove the rear shackle bolts. Note the lower bolt hole position so that the rear shackle can be reinstalled in the same position.

5. Remove the rear arm bolts and washers.

6. Remove the rear shackle and rear arm sub-assemblies.

7. Installation is the reverse of these steps. Note the following:

a. Install the rear shackle so that the grease fittings face forward.

b. Make sure to install the washers on the inside of the rear shackle as shown in **Figure 10**.

c. Install the rear shackle in the same adjustment hole as recorded during removal.

d. Install the rear arm so that the angle brackets face in the direction shown in **Figure 10**.

e. Apply Loctite 242 (blue) to all mounting bolts. Install bolts and washers and tighten the bolts securely.

Inspection

1. Clean all parts in solvent and dry thoroughly.

2. Visually inspect all welded joints for cracks or bends.

3. Check all pivot bolt surfaces for cracks, deep scoring or excessive wear.

4. If there is any doubt as to the condition of any part, repair or replace it as required.

5. Apply a low temperature grease to all pivot shafts and bushings.

REAR AXLE

Removal/Installation
(MX, MX [H/A], MX LT,
Plus and Plus LT)

1. Remove the rear suspension as described in this chapter.

2. Remove the idler wheel bolt and washers and remove the idler wheel (**Figure 25**). Repeat for opposite wheel.

3. Remove the outer spacer (**Figure 26**).

4. Remove the runner from one side.

5. Remove the rear axle and its inner spacer.

6. Installation is the reverse of these steps. Note the following:

   a. Install the rear axle so that the notch in the axle faces as shown in **Figure 26**.

   b. Install the outer spacer so that the hole in the spacer faces toward the rear axle as shown in **Figure 26**.

   c. Tighten the idler wheel bolt to the torque specification in **Table 2**.

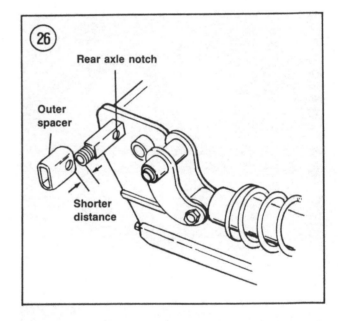

Removal/Installation
(Mach I)

1. Remove the rear suspension as described in this chapter.

2. Remove the idler wheel bolt and washers and remove the idler wheel. Repeat for opposite wheel.

3. Remove the outer spacer (**Figure 27**).

4. Remove the runner from one side.

5. Remove the rear axle and its inner spacer.

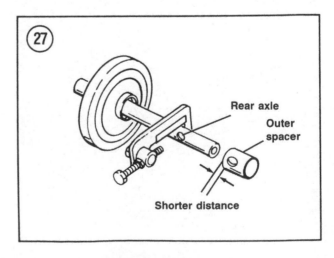

6. Installation is the reverse of these steps. Note the following:

   a. Install the rear axle so that the notch in the axle faces as shown in **Figure 27**.

   b. Install the outer spacer so that the hole in the spacer faces toward the rear axle as shown in **Figure 27**.

   c. Tighten the idler wheel bolt to the torque specification in **Table 3**.

Inspection

1. Clean all components thoroughly in solvent.

2. Check the rear axle for cracks or other damage.

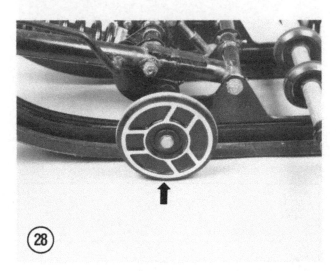

3. Check the inner spacer for cracks or other damage.

4. Check the outer spacer for cracks or other damage. Check the adjust bolt hole in the spacer for cracks.

5. If there is any doubt as to the condition of any part, repair or replace it as required.

6. Apply a low temperature grease to all pivot shafts and bushings.

SUSPENSION AND GUIDE WHEELS

Refer to the following exploded views for your model when servicing the suspension and guide wheels:

   a. **Figure 8**: 1985 MX and Plus.

   b. **Figure 9**: 1986-on MX, MX (H/A), MX LT, Plus and Plus LT.

   c. **Figure 10**: Mach I.

Inspection

1. Remove the rear suspension as described in this chapter.

2. Spin the suspension and guide wheels. See **Figure 28** and **Figure 29**, typical. The wheels should spin smoothly with no signs of roughness, binding or noise. These abnormal conditions indicate worn or damaged bearings. Replace worn or damaged bearings as described in this chapter.

3. Check the wheel hubs for cracks.

4. Check the outer wheel surface for cracks, deep scoring or excessive wear.

5. If there is any doubt as to the condition of any part, replace it with a new one. Note the following:

   a. Remove pressed on wheels with a universal type bearing puller.

   b. When installing pressed on wheels, install the wheel by driving the bearing onto the shaft with a socket placed on the inner bearing race. Do not install by driving on the outer bearing race or bearing and/or wheel damage may occur.

14

6. When installing wheels, note the following:

   a. Coat all sliding surfaces with a low temperature grease.

   b. When installing circlips, make sure the circlip seats in the shaft groove or wheel groove completely.

   c. Spin each wheel and check its operation. If a wheel is tight or binding, remove the wheel and check the bearing.

Wheel Bearing Replacement

1. Remove the wheel from its shaft.

2. Remove the circlip from the groove in the wheel.

3. Using a socket or bearing driver, remove the wheel bearing.

4. Clean the wheel in solvent and dry thoroughly.

5. Check the circlip groove in the wheel hub (**Figure 30**) for cracks. Check the hub area for breakage or other damage. Replace the wheel if the hub is damaged.

6. Install the new bearing as follows:

   a. Align the new bearing with the hub.

   b. Drive the bearing into the hub with a socket or bearing driver placed on the *outer* bearing race (**Figure 31**).

   c. Drive the bearing squarely into the hub until it bottoms out.

7. Secure the bearing with the circlip. Make sure the bearing seats in the hub groove completely.

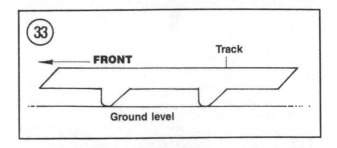

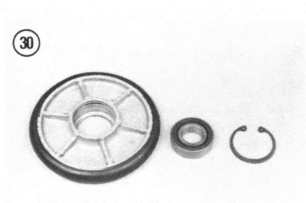

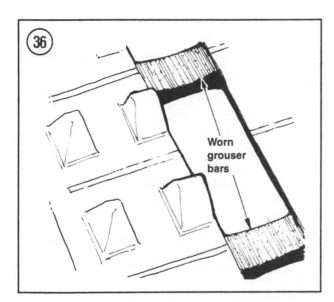

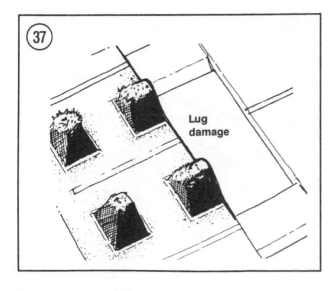

TRACK

Removal/Installation

1. Remove the rear suspension as described in this chapter.

2. Remove the driveshaft (A, **Figure 32**) as described in Chapter Twelve.

3. Remove the track (B, **Figure 32**).

4. Installation is the reverse of these steps. Note the following.

5. When installing the track, orient the track lugs to run in the direction shown in **Figure 33**.

Inspection

1. Check for missing or damaged track cleats (**Figure 34**). Replace cleats as described in this chapter.

2. Visually inspect the track for the following conditions:

   a. *Obstruction damage:* Cuts, slashes and gouges in the track surface are caused by hitting obstructions. These could include broken glass, sharp rocks or buried steel. See **Figure 35**.

   b. *Worn grouser bars:* Excessively worn grouser bars are caused by snowmobile operation over rough and non-snow covered terrain such as gravel roads and highway roadsides. See **Figure 36**.

   c. *Lug damage:* The lug damage shown in **Figure 37** is caused by lack of snow lubrication.

   d. *Ratcheting damage:* Insufficient track tension is a major cause of ratcheting damage to the top of the lugs (**Figure 38**). Ratcheting can also be caused by too great a load and constant "jack-rabbit" starts.

   e. *Over-tension damage:* Excessive track tension can cause too much friction on the wear bars. This friction causes the wear bars to melt and adhere to the track grouser bars. See **Figure 39**. An indication of this condition is a "sticky" track that has a tendency to "lock up."

14

f. *Loose track damage:* A track adjusted too loosely can cause the outer edge to flex excessively. This results in the type of damage shown in **Figure 40**. Excessive weight can also contribute to the damage.

g. *Impact damage:* Impact damage as shown in **Figure 41** causes the track rubber to open and expose the cord. This frequently happens in more than one place. Impact damage is usually caused by riding on rough or frozen ground or ice. Also, insufficient track tension can allow the track to pound against the track stabilizers inside the tunnel.

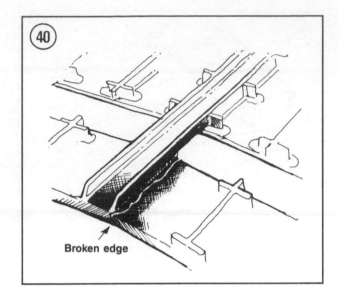

Broken edge

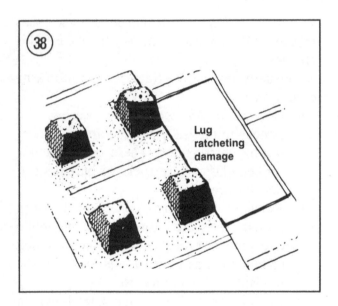

Lug ratcheting damage

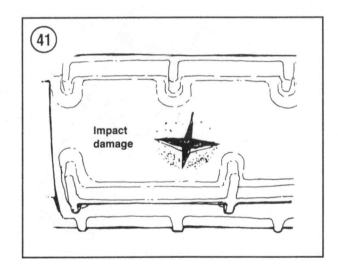

Impact damage

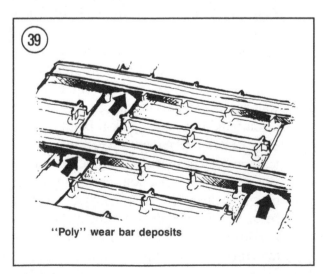

"Poly" wear bar deposits

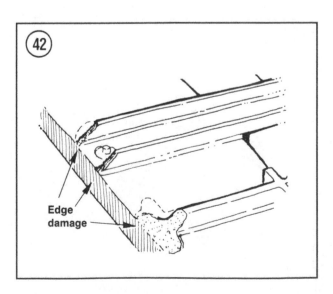

Edge damage

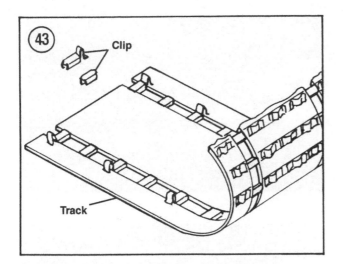

h. *Edge damage:* Edge damage as shown in **Figure 42** is usually caused by tipping the snowmobile on its side to clear the track and allowing the track edge to contact an abrasive surface.

Cleat Replacement

Before removing the cleats, note that 2 types are used alternately along the track (**Figure 43**). Follow the same pattern when installing new cleats. Method 1 describes the use of Ski-Doo tools to remove and install cleats. Method 2 describes cleat removal using universal tools.

Method 1

The Ski-Doo track cleat remover (part No. 529 0082 00) (**Figure 44**), small track cleat installer (part No. 529 0085 00) and the large track cleat installer (part No. 529 0077 00) (**Figure 45**) will be required to remove and install cleats.

1. Remove the track as described in this chapter.
2. Remove cleats as follows:
   a. Set the track cleat remover for small or large cleats. See **Figure 46**.
   b. Align tool with cleat and twist tool to open and remove cleat (**Figure 47**).

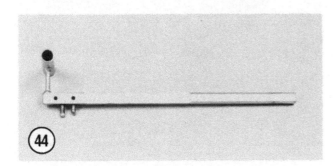

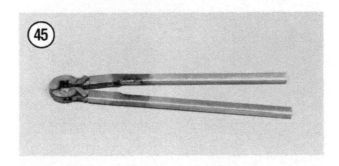

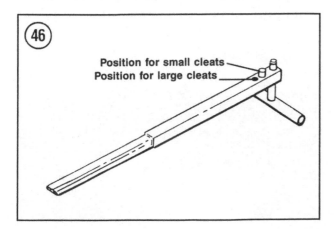

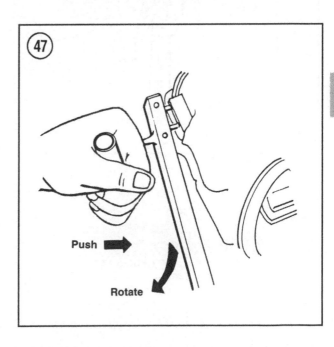

14

c. Align new cleat with track. Install suitable size track cleat installer and bend cleat (**Figure 48**) to secure cleat.

Method 2

A hand grinder, safety glasses and a universal track clip installer will be required to remove and install new cleats. See **Figure 49**.

> *WARNING*
> *Safety glasses must be worn when using a hand grinder to remove cleats.*

1. Using a hand grinder (**Figure 50**), grind a slit in the corner of the cleat. See **Figure 51**.
2. Pry the cleat off of the track (**Figure 52**).
3. Align the new cleat onto the track. Install the cleat with the cleat installation tool (**Figure 53**). Check the cleat to make sure it is tight.

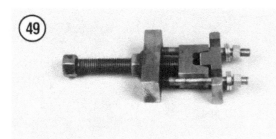

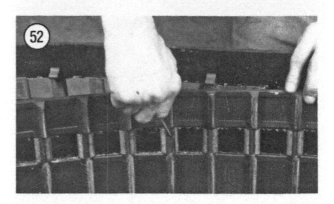

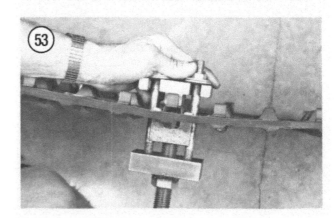

Table 1 TRACK SPECIFICATIONS

	mm	in.
Width		
1985-1986		
MX and MX (H/A)	38.1	15.0
Plus	41.9	16.5
1987-1988		
MX	38.1	15.0
MX LT	41.9	16.5
Plus	41.9	16.5
1989		
MX	38.1	15.0
MX LT	41.9	16.5
Plus	38.1	15.0
Plus LT	41.9	16.5
Mach I	41.0	16.1
Length		
1985		
All models	290	114
1986		
MX	290	114
MX (H/A)	315	124
Plus	290	114
1987-1988		
MX	290	114
MX LT	315	124
Plus	290	114
1989		
MX	290	114
MX LT	315	124
Plus	290	114
Plus LT	315	124
Mach I	307	121

Table 2 REAR SUSPENSION TIGHTENING TORQUES (MX, MX [H/A], MX LT, PLUS AND PLUS LT)

	N·m	ft.-lb.
Rear suspension		
Retainer plate bolts (all)	25	18
Shock pivot	48	35
Rear arm	48	35
Rear arm	48	35
Rear idler wheel	48	35
Shock bolts	48	35

Table 3 REAR SUSPENSION TIGHTENING TORQUES (MACH I)

	N·m	ft.-lb.
Rear suspension		
Retainer plate bolts (all)	15	11
Shock pivot	25	18
Rear arm	48	35
Shock bolts	48	35
Rear idler wheel	48	35

14

Table 4 SPRING SPECIFICATIONS

Center shock	
1985-1987	
Free length*	
Standard and optional	**241.3 mm (9.50 in.)**
Color code	
Standard	**Blue/blue**
Optional	**Orange/orange**
1988	
Free length	
Standard and optional	**241.3 mm (9.50 in.)**
Color code	
Standard	**Blue/blue**
Optional	**Orange/orange**
1989 (MX, MX LT, Plus and Plus LT)	
Free length*	
Standard and optional	**241.3 mm (9.50 in.)**
Color code	
Standard	**Blue/blue**
Optional	**Orange/orange**
1989 (Mach I)	
Free length*	**254 mm (10.0 in.)**
Color code	**Yellow/orange**
Rear shock	
1985-1987	
Free length*	
Standard	**241.3 mm (9.50 in.)**
Optional	**247.6 mm (9.75 in.)**
Color code	
Standard	**Green/green**
Optional	**Yellow/yellow**
1988	
Free length	
MX and MX LT	**247.6 mm (9.75 in.)**
Plus	**241.3 mm (9.50 in.)**
Color code	
MX and MX LT	**Yellow/yellow**
Plus	**Green/green**
1989 (MX, MX LT, Plus and Plus LT)	
Free length*	
Standard	**247.6 mm (9.75 in.)**
Optional	**241.3 mm (9.50 in.)**
Color code	
Standard	**Yellow/yellow**
Optional	**Green/green**
1989 (Mach I)	
Free length*	
Standard	**294 mm (11.58 in.)**
Optional	**302 mm (11.87 in.)**
Color code	
Standard	**Yellow/green**
Optional	**Black/white**

* **±3 mm (±0.12 in.)**

Chapter Fifteen

Off-Season Storage

One of the most critical aspects of snowmobile maintenance is off-season storage. Proper storage will prevent engine and suspension damage and fuel system contamination. Improper storage will cause various degrees of deterioration and damage.

Preparation for Storage

Careful preparation will minimize deterioration and make it easier to restore the snowmobile to service later. When performing the following procedure, make a list of replacement or damaged parts so that they can be ordered and installed before next season.

1. Remove the seat and clean the area underneath the seat thoroughly. Wipe the seat off with a damp cloth and wipe a preservative over the seat to keep it from drying out. If you are concerned about the seat during storage, store it away from the snowmobile in a safe place.

2. Flush the cooling system as described in Chapter Three. Before refilling the cooling system, check all of the hoses for cracks or deterioration. Replace hoses as described in Chapter Nine. Make sure all hose clamps are tight. Replace questionable hose clamps as required.

3. Change the chaincase oil as described in Chapter Three.

CAUTION
Do not allow water to enter the engine when performing Step 4.

4. Clean the snowmobile from front to back. Remove all dirt and other debris from the pan and tunnel. Clean out debris caught in the track.

5. Check the frame, skis and other metal parts for cracks or other damage. Apply paint to all bare metal surfaces.

6. Check all fasteners for looseness and tighten as required. Replace loose or damaged rivets.

NOTE
Refer to the appropriate chapter for the specified tightening torque as required.

7. Lubricate all pivot points with a low temperature grease as described in Chapter Three.

8. Unplug all electrical connectors and clean both connector halves with electrical contact cleaner. Check the electrical contact pins for damage or looseness. Repair connectors as required. After the contact cleaner evaporates, apply a dielectric grease to one connector half and reconnect the connectors.

CAUTION
Dielectric grease is formulated for electrical use. Do not use a regular type grease on electrical connectors.

9. To protect the engine from rust buildup during storage, the engine must be fogged in. Perform the following:

WARNING
This procedure must be done in an area with plenty of ventilation. The exhaust gases are poisonous. Do not run the engine in a closed area.

a. Jack up the snowmobile so that the track clears the ground.
b. Start the engine and allow it to warm-up to normal operating temperature. Then turn engine off.

NOTE
A cloud of smoke will develop in sub-step c. This is normal.

c. Press the primer button (to prevent gasoline leakage) and disconnect the outlet primer hose at the primer valve. See **Figure 1**. Connect a can of Ski-Doo Storage Oil (part No. 496 0141 00) into the primer outlet hose previously disconnected. Start the engine and allow to idle. Spray storage oil through the primer outlet hose until 1/2 of the can has been injected into the engine or until the engine stalls.
d. Turn all switches off.

CAUTION
Before removing spark plugs, blow away any dirt around the spark plug base. The

dirt could fall into the cylinder when the spark plug is removed, causing serious engine damage.

e. Remove the spark plugs (**Figure 2**). Reconnect the spark plugs at their plug caps to ground them.
f. Spray approximately 150 cc (5 oz.) of Ski-Doo Storage Oil (part No. 496 0141 00) into each cylinder. Pull the recoil starter handle to distribute the oil through the cylinder walls.
g. Wipe a film of oil on the spark plug threads and reinstall the spark plugs. Reconnect the high tension leads at the plugs.
h. Reconnect outlet primer hose.

CAUTION
During the storage period, do not run engine.

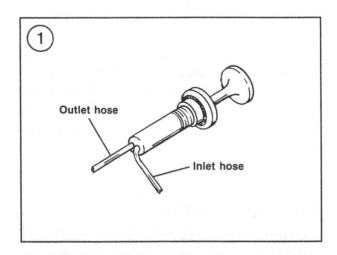

10. Plug the end of the muffler with a rag to prevent moisture from entering. Tag the machine with a note to remind you to remove the rag before restarting the engine next season.

11. Remove the drive belt and store it on a flat surface.

12. Remove and service the primary and secondary sheaves as described in Chapter Eleven. Replace worn or damaged parts as required.

13. Clean the jackshaft thoroughly. Then, apply a light coat of low-temperature grease to the jackshaft (**Figure 3**). Wipe off excess grease before installing the secondary sheave.

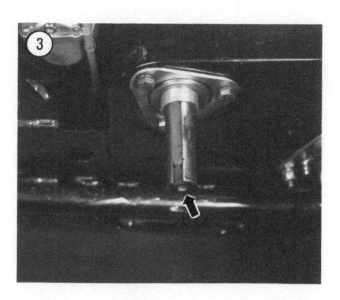

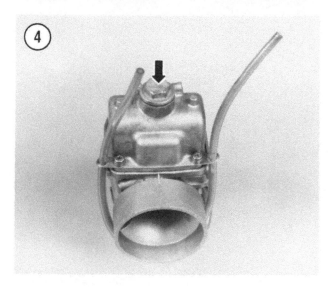

14. Reinstall the primary and secondary sheaves. Tighten the bolts to the torque specification listed in Chapter Eleven.

15. Close and secure the belt shield.

WARNING
Some fuel may spill in the following procedure. Work in a well-ventilated area at least 50 feet from any sparks or flames, including gas appliance pilot lights. Do not smoke in the area. Keep a BC rated fire extinguisher handy.

16. Using a suitable siphon tool, siphon fuel out of the fuel tank and into a gasoline storage tank.

17. When the fuel tank is empty, remove the drain plug (**Figure 4**) on each carburetor and drain the carburetors. Wipe up spilled gasoline immediately.

18. Protect all glossy surfaces on the chassis, hood and dash with an automotive type wax.

19. Raise the track off of the ground with wood blocks. Make sure the snowmobile is secure.

20. Cover the snowmobile with a heavy cover that will provide adequate protection from dust and damage. Do not cover the snowmobile with plastic as moisture can collect and cause rusting. If the snowmobile has to be stored outside, block the entire vehicle off the ground.

Removal From Storage

Preparing the snowmobile for use after storage should be relatively easy if proper storage procedures were followed.

1. Remove the rag from the end of the muffler.

2. Adjust track tension as described in Chapter Three.

3. Inspect the drive belt for cracks or other abnormal conditions. Reinstall the drive belt as described in Chapter Eleven.

4. Check the chaincase oil level. Refill as described in Chapter Three. If the oil was not changed before storage, change the oil as described in Chapter Three.

5. Check and adjust drive belt tension.

15

6. Check the coolant level and refill if necessary.

7. Check all of the control cables for proper operation. Adjust as described in Chapter Three.

8. Fill the oil injection tank. If the tank was dry or if a hose was disconnected, bleed the oil pump as described in Chapter Eight.

9. Perform an engine tune-up as described in Chapter Three.

10. Check the fuel system. Refill the tank.

11. Make a thorough check of the snowmobile for loose or missing nuts, bolts or screws.

WARNING
Be sure you are in a well ventilated area for Step 12. The exhaust gases are poisonous. Do not run the engine in a closed area.

12. Start the engine and check for fuel or exhaust leaks. Make sure the lights and all switches work properly. Turn the engine off.

13. After the engine has been initially run for a period of time, install new spark plugs as described in Chapter Three.

Index

16

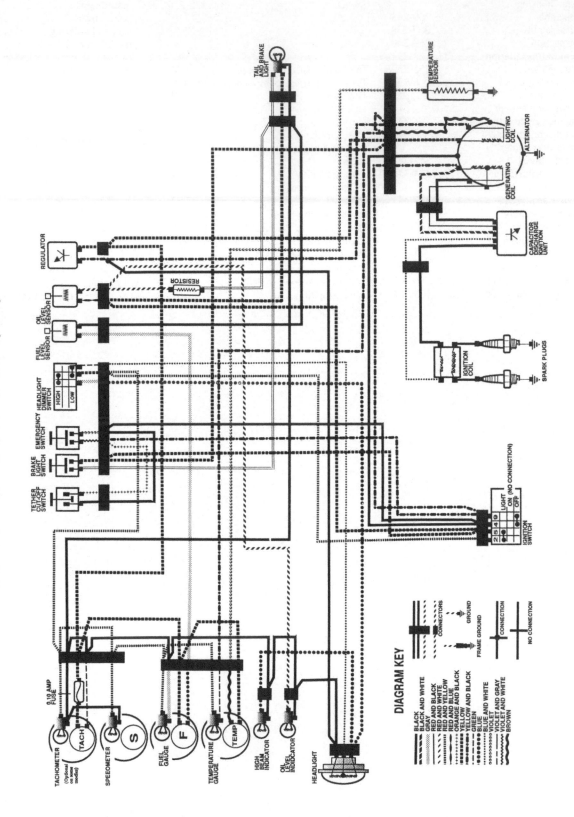

1985-1986 FORMULA MX/MX (H/A)/PLUS

TAIL AND BRAKE LIGHT

TEMPERATURE SENSOR

ALTERNATOR

LIGHTING COIL

GENERATING COIL

CAPACITOR DISCHARGE IGNITION UNIT

SPARK PLUGS

IGNITION COIL

REGULATOR

RESISTOR

OIL LEVEL SENSOR

FUEL LEVEL SENSOR

HEADLIGHT DIMMER SWITCH

HIGH   LOW

EMERGENCY SWITCH

BRAKE LIGHT SWITCH

TETHER CUTOFF SWITCH

LIGHT   (NO CONNECTION)
ON
OFF

IGNITION SWITCH

2 5 4 9

0.10 AMP FUSE

TACHOMETER (Optional on some models)

TACH

SPEEDOMETER

S

FUEL GAUGE

F

TEMPERATURE GAUGE

TEMP

HIGH BEAM INDICATOR

OIL LEVEL INDICATOR

HEADLIGHT

DIAGRAM KEY

BLACK
BLACK AND WHITE
GRAY
RED AND BLACK
RED AND WHITE
RED AND YELLOW
RED AND BLUE
ORANGE AND BLACK
YELLOW
YELLOW AND BLACK
GREEN
BLUE
BLUE AND WHITE
VIOLET
VIOLET AND GRAY
VIOLET AND WHITE
BROWN

CONNECTORS

GROUND

FRAME GROUND

CONNECTION

NO CONNECTION

1987 FORMULA MX/MX LT/PLUS

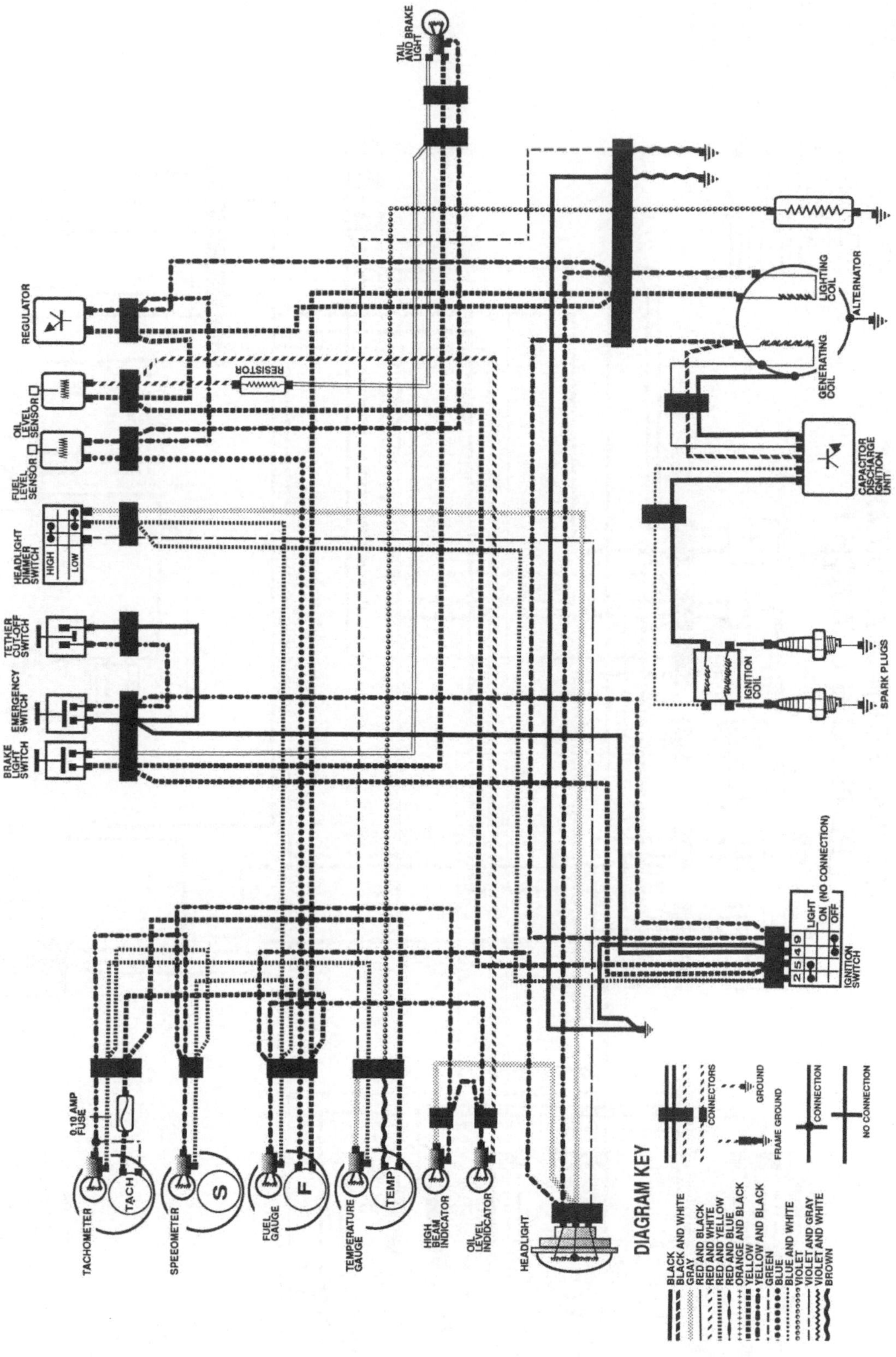

REGULATOR

OIL LEVEL SENSOR

FUEL SENSOR

RESISTOR

HEADLIGHT DIMMER SWITCH

HIGH

LOW

TETHER CUT-OFF SWITCH

EMERGENCY SWITCH

BRAKE LIGHT SWITCH

TAIL AND BRAKE LIGHT

LIGHTING COIL

ALTERNATOR

GENERATING COIL

CAPACITOR DISCHARGE IGNITION UNIT

IGNITION COIL

SPARK PLUGS

LIGHT (NO CONNECTION)

ON

OFF

IGNITION SWITCH

0.10 AMP FUSE

TACHOMETER

TACH

SPEEDOMETER

S

FUEL GAUGE

F

TEMPERATURE GAUGE

TEMP

HIGH BEAM INDICATOR

OIL LEVEL INDICATOR

HEADLIGHT

DIAGRAM KEY

CONNECTORS

GROUND

FRAME GROUND

CONNECTION

NO CONNECTION

BLACK
BLACK AND WHITE
GRAY
RED AND BLACK
RED AND WHITE
RED AND YELLOW
RED AND BLUE
ORANGE AND BLACK
YELLOW AND BLACK
GREEN
BLUE
BLUE AND WHITE
VIOLET
VIOLET AND GRAY
VIOLET AND WHITE
BROWN

17

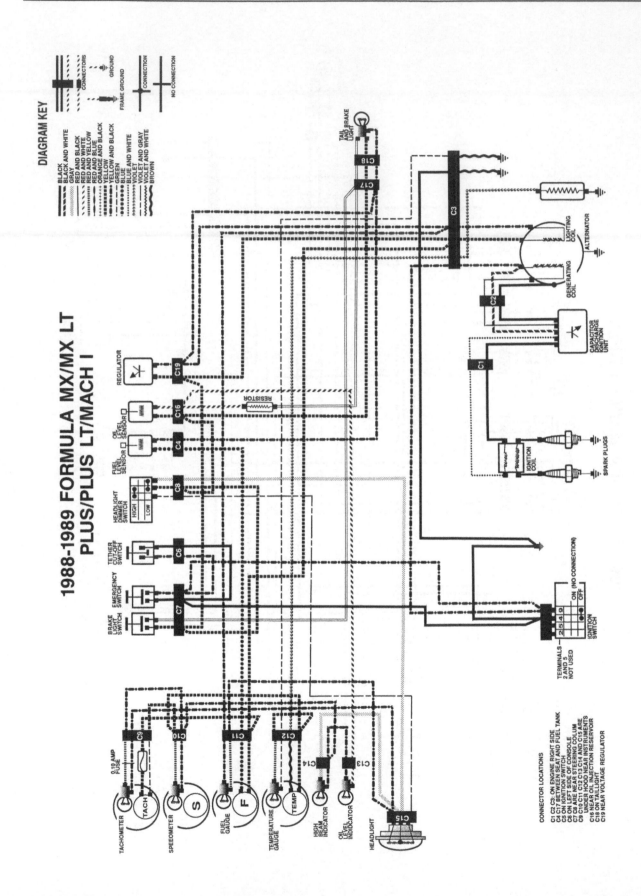

DIAGRAM KEY

BLACK
BLACK AND WHITE
GRAY
RED AND BLACK
RED AND WHITE
RED AND YELLOW
RED AND BLUE
ORANGE AND BLACK
YELLOW
YELLOW AND BLACK
GREEN
BLUE
BLUE AND WHITE
VIOLET
VIOLET AND GRAY
VIOLET AND WHITE
BROWN

CONNECTORS
GROUND
FRAME GROUND
CONNECTION
NO CONNECTION

1988-1989 FORMULA MX/MX LT PLUS/PLUS LT/MACH I

REGULATOR
OIL LEVEL SENSOR
FUEL LEVEL SENSOR
HEADLIGHT DIMMER SWITCH
HIGH   LOW
TETHER CUT-OFF SWITCH
EMERGENCY SWITCH
BRAKE LIGHT SWITCH

RESISTOR

C19
C16
C4
C8
C6
C7

TAIL AND BRAKE LIGHT
C18
C17
C3

LIGHTING COIL
ALTERNATOR
GENERATING COIL
C2
C1
CAPACITOR DISCHARGE IGNITION UNIT
IGNITION COIL
SPARK PLUGS

IGNITION SWITCH
ON   OFF
2 5 4 9
TERMINALS 2 AND 5 NOT USED

0.10 AMP FUSE
C9
C10
C11
C12
C14
C13
C15

TACHOMETER
TACH
SPEEDOMETER
S
FUEL GAUGE
F
TEMPERATURE GAUGE
TEMP
HIGH BEAM INDICATOR
OIL LEVEL INDICATOR
HEADLIGHT

CONNECTOR LOCATIONS
C1 C2 C3 ON ENGINE RIGHT SIDE
C4 C17 BETWEEN SEAT AND FUEL TANK
C5 ON IGNITION SWITCH
C6 ON LEFT SIDE OF CONSOLE
C7 C8 ARE NEAR STEERING COLUM
C9 C10 C11 C12 C13 C14 AND C15 ARE UNDER HOOD NEAR INSTRUMENTS
C16 NEAR OIL INJECTION RESERVOIR
C18 ON TAILLIGHT
C19 NEAR VOLTAGE REGULATOR

NOTES

NOTES

NOTES

NOTES